WAVE INSTABILITIES IN SPACE PLASMAS

ASTROPHYSICS AND SPACE SCIENCE LIBRARY

A SERIES OF BOOKS ON THE RECENT DEVELOPMENTS
OF SPACE SCIENCE AND OF GENERAL GEOPHYSICS AND ASTROPHYSICS
PUBLISHED IN CONNECTION WITH THE JOURNAL
SPACE SCIENCE REVIEWS

Editorial Board

J. E. BLAMONT, *Laboratoire d'Aeronomie, Verrières, France*

R. L. F. BOYD, *University College, London, England*

L. GOLDBERG, *Kitt Peak National Observatory, Tucson, Ariz., U.S.A.*

C. DE JAGER, *University of Utrecht, The Netherlands*

Z. KOPAL, *University of Manchester, England*

G. H. LUDWIG, *NOAA, National Environmental Satellite Service, Suitland, Md., U.S.A.*

R. LÜST, *President Max-Planck-Gesellschaft zur Förderung der Wissenschaften, München, F.R.G.*

B. M. McCORMAC, *Lockheed Palo Alto Research Laboratory, Palo Alto, Calif., U.S.A.*

H. E. NEWELL, *Alexandria, Va., U.S.A.*

L. I. SEDOV, *Academy of Sciences of the U.S.S.R., Moscow, U.S.S.R.*

Z. ŠVESTKA, *University of Utrecht, The Netherlands*

VOLUME 74
PROCEEDINGS

WAVE INSTABILITIES IN SPACE PLASMAS

PROCEEDINGS OF A SYMPOSIUM ORGANIZED WITHIN
THE XIXth URSI GENERAL ASSEMBLY HELD IN
HELSINKI, FINLAND, JULY 31 – AUGUST 8, 1978

edited by

PETER J. PALMADESSO and KONSTANTINOS PAPADOPOULOS

Naval Research Laboratory, Washington, D.C., U.S.A.

Sponsored by

COMMISSION H OF URSI

D. REIDEL PUBLISHING COMPANY

DORDRECHT:HOLLAND / BOSTON:U.S.A.
LONDON:ENGLAND

Library of Congress Cataloging in Publication Data

Main entry title:

    Wave instabilities in space plasmas.
    (Astrophysics and space science library ; v. 74)
    Bibliography: p.
    Includes index.
    1. Space Plasmas–Congresses. 2. Atmospheric Electricity–Congresses.
3. Atmosphere, Upper–Congresses. I. Palmadesso, Peter J. 1940–
II. Papadopoulos, Konstantinos, 1938–   III. International Union of Radio
Science. IV. Series.
QC809.P5W28   538'.76   79-20918
ISBN 90-277-1028-7

Published by D. Reidel Publishing Company,
P. O. Box 17, Dordrecht, Holland

Sold and distributed in the U.S.A., Canada and Mexico
by D. Reidel Publishing Company, Inc.
Lincoln Building, 160 Old Derby Street, Hingham,
Mass. 02043, U.S.A.

All Rights Reserved
Copyright © 1979 by D. Reidel Publishing Company, Dordrecht, Holland
No part of the material protected by this copyright notice may be reproduced or
utilized in any form or by any means, electronic or mechanical
including photocopying, recording or by any informational storage and
retrieval system, without written permission from the copyright owner

Printed in The Netherlands

TABLE OF CONTENTS

PREFACE vii

PART I: NATURAL NOISE IN SPACE

A. I. LIKHTER / ELF and VLF Noise Intensity and Spectra in the Magnetosphere — 3
H. KIKUCHI / ELF and VLF Activity Associated with High Latitude Hole — 21
R. A. HELLIWELL / Effects of Power Line Radiation into the Magnetosphere — 27
K. BULLOUGH and T. R. KAISER / Ariel 3 and 4 Studies of Power Line Harmonic Radiation — 37
B. T. TSURUTANI, E. J. SMITH, S. R. CHURCH, R. M. THORNE and R. E. HOLZER / Does ELF Chorus Show Evidence of Power Line Stimulation? — 51
B. T. TSURUTANI, E. J. SMITH, H. I. WEST, JR. and R. M. BUCK / Chorus, Energetic Electrons and Magnetospheric Substorms — 55

PART II: PLASMA TURBULENCE

M. ASHOUR-ABDALLA, C. F. KENNEL, and D. D. SENTMAN / Magnetospheric Multiharmonic Instabilities — 65
P. CARLQVIST / Some Theoretical Aspects of Electrostatic Double Layers — 83
S. TORVEN / Formation of Double Layers in Laboratory Plasmas — 109
F. W. CRAWFORD, J. S. LEVINE, and D. B. ILIC / Laboratory Simulation of Ionospheric Double-Layers — 129
C. T. DUM and R. CHODURA / Anomalous Transition from Buneman to Ion Sound Instability — 135
C. HANUISE and M. CROCHET / Marginal Plasma Waves in the Equatorial Electrojet Observed by HF Coherent Radar Techniques — 149

PART III: NONLINEAR EFFECTS

H. MATSUMOTO / Nonlinear Whistler-Mode Interactions and Triggered Emissions in the Magnetosphere: A Review — 163
R. A. HELLIWELL / Siple Station Experiments on Wave-Particle Interactions in the Magnetosphere — 191
D. NUNN / Non Linear Wave Particle Interaction Theory Applied to Siple Triggered Emissions — 205
N. CORNILLEAU-WEHRLIN and R. GENDRIN / Quenching of Natural Cyclotron Instability by Large Amplitude Monochromatic Waves Propagating in an Inhomogeneous Medium — 217

D. F. SMITH and D. R. NICHOLSON / Nonlinear Effects Involved in the Generation of Type III Solar Radio Bursts ... 225

M. L. GOLDSTEIN, K. PAPADOPOULOS and R. A. SMITH / A Theory of Solar Type III Radio Bursts ... 245

PART IV: IONOSPHERIC F-REGION IRREGULARITIES

S. L. OSSAKOW / A Review of Recent Results on Spread F Theory ... 265

M. C. KELLEY / Equatorial Spread F: A Review of Recent Experimental Results ... 291

INDEX OF SUBJECTS ... 307

PREFACE

This book contains lectures presented in the symposium on "Wave Instabilities in Space Plasmas" organized within the program of the XIX URSI General Assembly held in Helsinki, Finland, during the period of July 31 through August 1978.

The individual papers were deliberately limited in length; the author's cooperation in conforming with the guidelines is greatly appreciated. The contents of the book as well as of the symposium are organized along subject areas, although aspects of chapters II and III are overlapping. The contents of Chapter I were part of the session organized by Dr. Ja. I. Likhter for commissions E and H and chaired by Dr. F. Horner, dealing with natural noise in space. The main part of this chapter deals with evidence and effects of power line radiation in the magnetosphere as well as the morphology of ELF, VLF and kilometric radiation spectra in the magnetosphere. The contents of Chapter II were part of the two sessions organized by Dr. P. Palmadesso and chaired by Drs. S. L. Ossakow and M. L. Goldstein dealing with electrostatic turbulence. This chapter covers topics related to auroral acceleration processes such as anomalous resistivity, double layers, and ion sound and cyclotron turbulence. Experimental and theoretical studies of non-linear spectra of the electrojet instability, and whistler and cyclotron wave spectra are also discussed. The contents of Chapter III were part of the session organized and chaired by Dr. D. Gurnett. This chapter covers non-linear theories and reviews pertaining to kilometric radiation, type III solar radiobursts, and VLF, cyclotron and whistler waves. Chapter IV was part of the session organized by Dr. P. Dyson (Commission H) and Dr. J. W. King (Commission G), and chaired by Dr. S. Bowhill. This chapter contains studies of F-region irregularities. The symposium was convened under the aegis of Commission H of URSI, under the Chairmanship of Dr. R. Gendrin, and was directed by the joint URSI-IAGA working group on "Waves and Instabilities in Space Plasmas" co-chaired by Drs. P. Dyson, V. Karpman, K. Papadopoulos, and F. Scarf. The overall convenor was Dr. K. Papadopoulos. Many thanks are due to all of the above and to Dr. F. Crawford, then Vice-Chairman and now Chairman of Commission H.

Washington, D. C., February 1979

Peter Palmadesso
Konstantinos Papadopoulos

PART I

NATURAL NOISE IN SPACE

ELF AND VLF NOISE INTENSITY AND SPECTRA IN THE MAGNETOSPHERE

Ja. I. Likhter
Institute of Terrestrial Magnetism, Ionosphere and Radio
Wave Propagation (IZMIRAN)
p/o Akademgorodok
Moscow 142092 (USSR)

1. INTRODUCTION

A very large number of experimental and theoretical investigations has been devoted to study wave phenomena in the magnetosphere. Many reviews have been written regarding this subject (1-7).

In this paper, we will mainly discuss the experimental results obtained recently. As a basis for discussion, we will use exclusively the data obtained by satellite and rocket borne experiments, because the in situ measurements are the most adequate.

One of the first questions raised to experimentalists is the following : where are the appropriate waves excited in the plasma ? Solving this question can allow us to tackle another problem, namely, how can waves observed in given conditions and having the measured properties be excited ? This is why we first summarize the information about the noise source locations within the magnetosphere and how the wave measurement data are used to locate the source region.

The wave intensity is of first order significance for many processes taking place in the Earth magnetosphere (5), particularly, the wave intensity determines the energetic particle diffusion rate into the loss cone, and thus influences the state of radiation belts. Besides when the wave intensity becomes large enough, nonlinear effects start in the plasma, affecting the wave spectrum and again changing drastically the state of energetic particles. At present, controlled experiments with wave emission by rocket-and satellite borne transmitters are being carried out. In the near future, such kind of active experiments will develop. Under these circumstances, the natural ELF and VLF emissions will play a double role : they will act as interferences when receiving the transmitted signals, and the transmitted wave may cause changes in the natural emissions as was shown by Stenzel (8). For possible analysis of such situations, it is important to have a comprehensive notion about the intensity which the natural ELV and VLF noise usually have in various conditions. Such informations will be presented in the second part of this review.

2. THE REGION IN THE MAGNETOSPHERE WHERE THE NOISE IS EXCITED

Let us start with two figures on which the most probable wave source regions are shown. The first figure gives an overall picture of the day-night section of the magnetosphere. We can see there the source region for ULF waves (micropulsations) situated in the outer magnetosphere between the plasmapause and the magnetopause (1 - 9).

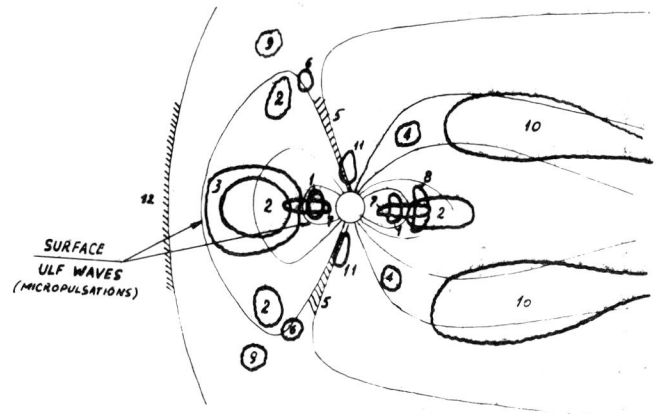

Figure 1. Sources of noise in the magnetosphere

1. ELF (VLF) plasmaspheric hiss. 2. Chorus. 3. Nonthermal continuum $f > f_{oe}$ (partly escaping); resonance line spectrum nf_{Be}; $(n + 1/2)f_{Be}$; n_{fg}; f_{oe}. 4. Auroral kilometric radiation. 5. Cusp ELF emissions $f < f_{Be}$. 6. Cusp-magnetosheath boundary micropulsations. 7. Equatorial ULF-ELF electromagnetic noise (line-spectra) $f < 200$ Hz. 8. Micropulsations. 9. Lions roar. 10. Broadband electrostatic noise; auroral field line turbulence. 11. ELF and VLF auroral hiss; 12. Upstream waves, plasma wave turbulence.

This wave source is not very localized because of specific modes of ULF waves (field lines oscillations and others). On the boundaries of this very vast region at the plasmapause and magnetopause, there are excited surface waves, as was shown by Kovner (10). The chorus sources evidences by Tsurutani and Smith, and Burtis and Helliwell (11-14) and the regions where the high frequency continuum radiation (15) and the plasma resonance line spectrum (16 - 18) are excited are also marked. At the plasmapause, there is a change of the type of noise excited and under the arch of the plasmapause, there exists a source region for the well known ELF (and VLF) plasmaspheric hiss (19 - 23).

Some of these sources are also acting in the night side of the magnetosphere; however, there exist new sources in this region. These are the auroral kilometric radiation, the auroral field line turbulence, the oblique resonance VLF noise (saucers), and different kinds of ion cyclotron waves. In the polar cusp, there are Cerenkov excited ELF and

VLF noise, lion roars, etc...

The second figure shows sources of noise in the topside ionosphere. On the day time, it is worth mentioning the source of ELF hiss which is situated at altitudes of 300 - 500 km and at L-values < 2,5 - 3.0. This source region was recently established by a group of authors working with the Intercosmos series of satellite data (24-26). These results will be discussed in more details later on.

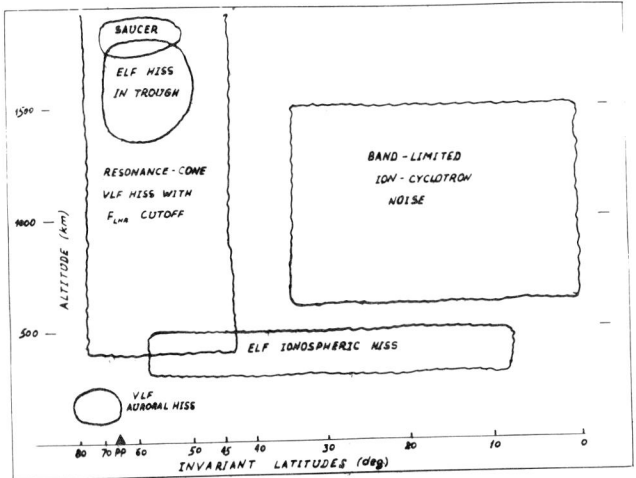

Figure 2. Sources of ELF and VLF noise in the topside ionosphere. Sources of ELF ionospheric hiss, band-limited ion-cyclotron noise and ELF hiss in trough are acting in the daytime. Other sources shown mainly act at night.

There is also a source of noise due to the main ionospheric and light ion trough evidenced by Tulunay and Hugues in the VLF band and by Likhter et al. at ELF frequencies (27-29).

At higher latitudes is the source of quasi-electrostatic noise band. The lower cut-off of this noise is very nearly equal to the local value of the lower hybrid resonance frequency. This is the low altitude part of the source of ELF and VLF electrostatic noise, streched along the auroral field lines, which was shown on the previous figure.

I would like to discuss now in more details the experimental evidences that these regions are sources for ELF and VLF noise. In other words, we discuss how the measured wave data may be used to locate the source. This is not a very easy task; nevertheless, there exists some ways of solving it more or less approximately. Naturally in all cases the interpretation is based on consideration of experimental data as well as on the comparison of the conclusions drawn with the consequences of the theories regarding wave instabilities, wave propagation and other wave-plasma processes.

It would be better if one could measure the field structure of the wave to determine the special properties of the field characteristics within the source. They may be analogous to the near field structure of the radiation aerial. However, for the spatial distributed source, the near field wave structure is unknown.

An approach to this method is the determination of the wave vector and Poynting vector orientations (30). This method was used in some experiments (for example on OGO satellites) where three magnetic aerials (31), or on FR-1 satellite where three magnetic and two electric aerials (32) were installed. On the OGO-5 satellite, Thorne et al. (20), analyzing the sense of polarization of the magnetic component of the wave, have established that the wave vector for ELF plasmaspheric hiss points away from equator, thus confirming the widespread opinion that the equatorial region of the magnetosphere is the source region, especially for this kind of ELF hiss. However, the number of wave experiments using aerials is still limited and the amount of publications regarding their results is also scarce.

2.1. Relationships between the wave frequency and the local plasma parameters at the source.

Onboard the GEOS-1 satellite, an electrostatic noise in the outer magnetosphere has been discovered between the plasmapause and the magnetopause. The frequency spectra of this noise contained usually many frequency lines. The sophisticated technique which combines the passive and active measurements on GEOS-1 gives the possibility to identify these lines as local plasma resonances (18). These are harmonics and half-harmonics of electron gyrofrequency, plasma frequency, the plasma frequency and the ω_q resonances.

At low altitudes in the auroral zone, the LHR noise is very often seen excited by interactions with low energy particles moving across the field lines (E ≃ 100 eV) (19). The spatial changes of the lower frequency cut-off of this noise is the same as the spatial changes of the LHR frequency, thus testifying the local generation of this noise. This effect is very well seen in the ionospheric trough region where the density of ion constituents of the plasma are changing quickly, therefore leading to a rapid change of the LHR frequency (33). At low and middle latitudes, LHR noise of more impulsive character is observed. This emission is excited by whistlers propagating in non-ducted mode and having Θ values in close vicinity of 90° (34, 35).

For some wave-particle interaction processes (e.g. cyclotron resonance, Cerenkov) the characteristic frequencies of the excited wave spectra (f_{max}; f_{up}) are proportional to the electron or ion gyrofrequency (36-39). Since the magnetic field strength changes monotonically in the radial direction in the magnetosphere, this sign is a good one for identification of the source. This property was previously used by Burtis and Helliwell to attribute chorus to sources located in the equatorial region (14).

Although in the course of propagation the frequency spectrum of the noise usually changes widely, the proportionality mentioned may be used successfully also in a remote region from the source. We have made a detailed statistical analysis of noise spectra measured at the INTERCOSMOS-5 satellite (26). The spectra has two maxima, one in a ELF and another in a VLF frequency band.

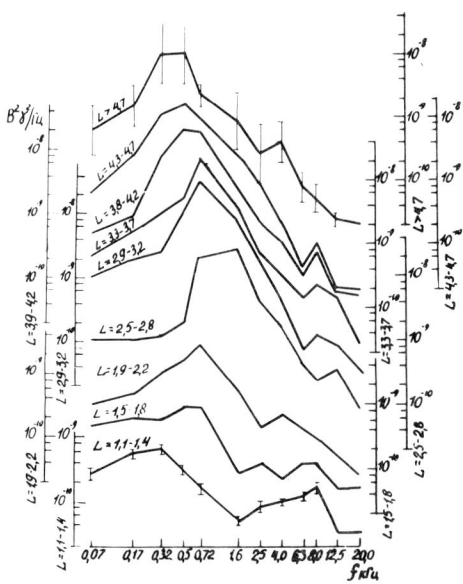

Figure 3 : Averaged ELF-VLF noise spectra in the topside ionosphere for quiet conditions, various L-values, and daytime (near noon). Note the regular variations of the frequencies of the maxima with L (INTERCOSMOS-5 data).

Frequencies of these maxima are changing with L in a regular way, following three branches. Along the uppermost one, the frequency is changing as $(0.25 - 0.50)\omega_{Be}$. This is similar to Burtis and Helliwell's finding (14). The second branch contains frequencies close to the equatorial value of the LHR frequency. Note that Etcheto et al., making self-consistent calculations for plasmaspheric hiss excitation in the equatorial region, have obtained the frequency maxima of ELF emission spectra in the neighbourhood of f_{LHR} (38).

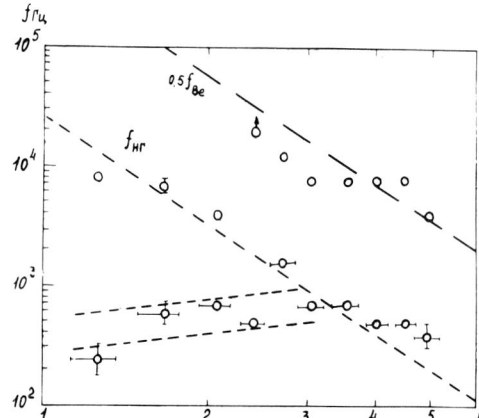

Figure 4. The frequency of the spectra maxima versus L parameter for the INTERCOSMOS-5 satellite position in the topside ionosphere. Two lines having a positive slope correspond to possible local values of the proton gyrofrequency for various positions of the INTERCOSMOS-5 satellite orbit above the Earth (magnetic anomaly effect).

Thus the waves belonging to these parts of the noise spectra may be regarded as generated near the equator and coming to low altitude by ducted propagations. The latter conclusion is also confirmed by ray-tracing calculations.

The most interesting is the third group of ELF maxima frequencies. This branch only exists for L lower than 2.5 and the frequencies along it are changing as the local value of proton gyrofrequency. This is one of the most important argument among many others, leading us to the conclusion that the topside ionosphere (300-500 km, L < 2.5 - 3.0) is also a source region for ELF hiss (24-26).

In some cases, the excited noise spectra may be limited by characteristic plasma frequencies. Thus on INTERCOSMOS-4 satellite, we have discovered a band limited noise, the border frequencies of which are the local values of the proton gyrofrequency and the cut-off frequency for left-hand polarized waves in a multicomponent plasma (n → 0), see figure 2 in (40). Again, the spatial variation of the border frequencies coinciding with the change of appropriate characteristic plasma frequencies, together with considerations regarding the propagation properties, allow us to conclude that the observed noise is the locally generated ion-cyclotron one.

2.1. The enhanced probability of noise occurrence or the enhanced amplitude of the noise in particular magnetospheric regions.

Many authors have used the enhanced occurrence rate of noise to ascertain the place where the noise intensity is the largest and to deduce in which regions the noise was excited. This is a good method in the near vicinity of the source. Thus, in a recent analysis of chorus in the upper magnetosphere, Tsurutani and Smith have used this method to distinguish between equatorial and high-latitude chorus sources (12).

However, having such kind of information in regions remote from the source, one must be very cautious because the wave propagation conditions in the magnetospheric plasma are very complicated. Appealing to them, in many papers where the data obtained at low altitude were discussed, the conclusion was drawn that the source is located somewhere above the satellite altitude. This means that for plasmaspheric hiss seen at $L \simeq 4$, and for chorus seen at auroral latitudes, the sources are situated at or near the equator at great altitudes. For VLF and ELF hiss at auroral and polar cap latitudes, the source is believed to be at altitudes between say 1000 km to $2-4$ R_E, depending on frequency (41-43).

Nevertheless, in some cases, when analyzing the data obtained at low latitude onboard orbiting satellites or rockets, the authors conclude that the source region is locally situated. Previously, we have mentioned the Lower Hybrid Resonance noise as an example of such case. We have also mentioned that, thoroughly analyzing the ELF hiss measured at low altitudes and at $L < 2.5 - 3$, we conclude that this noise is generated in the topside ionosphere at height of 300 - 500 km. The ELF and VLF noise generated at ionospheric altitudes ($\simeq$ 130 km) was also discovered by Holtet, Oguti, Ungstrup and Olsen et al. on the rockets launched in aurorae (3, 44, 45).

An additional example of ELF (and possible VLF) noise excitation in the ionosphere is believed to be in the main ionospheric trough or light ion trough (28, 29). On the INTERCOSMOS-14 satellite, we have observed enhancements up to 40 dB of wave amplitude at a frequency of 0.72 kHz, in the magnetic as well as in electric field component, when crossing the trough in the day-time at altitudes of about 1500 km. The ray tracing calculations in a non-homogeneous medium could not account for such a great growth of the emission amplitude. It is believed that the complicated processes leading to trough formation and especially the DC electric field existing in this region are responsible for ion acceleration and thus for wave generation. Dr. Kikuchi will set forth in his paper another case for ELF and VLF noise activity in the high latitude hole discovered onboard OGO-5 (this issue).

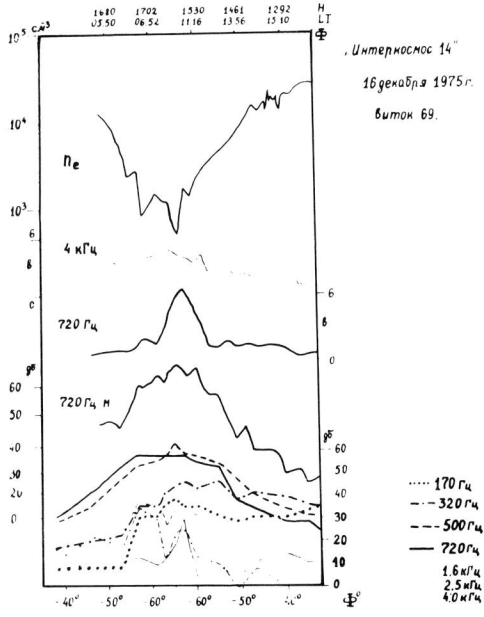

Figure 5. The ELF and VLF magnetic and electric field amplitude grow simultaneously when the satellite INTERCOSMOS-14 moves in the trough. Day-time, altitude ∿ 1500 km.

2.3. Relationship between the wave characteristics and other phenomena in magnetospheric plasma.

As a consequence of wave-particle interactions, the energetic particle distribution is changed. May be the more pronounced effect of this is the diffusion of particles into the loss cone and precipitation into the atmosphere. In the early experiment with the INJUN-3 satellite, Gurnett and O'Brien have established the close connection between the wave intensity and the streams of precipitating electrons having energies greater than 40 keV (46). Afterwards the connection between the state of waves and that of particles were investigated in more detail in a number of papers (see e.g. 47-54). It is certainly the best when the characteristics of waves and particles are measured in detail. This was the case on EXPLORER-45 and GEOS-1 satellites for example. Such kind of measurements not only give an information on source location but also a matter for thinking regarding wave excitation mechanism.

When waves and particles are measured at low altitudes, the spatial location of the stream of precipitated particles is more closely connected with the source region than the wave intensity maxima, because of the wave ray-path deviation from the corresponding magnetic field line (53).

There is another kind of wave-particle interaction when a wave of sufficiently large intensity modulates the precipitated electron flux. This is one of the non-linear effects connected with the influence of waves of large intensity on plasma distribution function. Parks and Winckler, as early as in 1969, have reported on modulation of particle flux with the micropulsation periods (55). Zakharov et al., analyzing the wave and particle measurements onboard INTERCOSMOS-5, reported that modulation is simultaneously observed at the same frequency in the wave amplitude and in the electron flux, with some phase difference (56, 57).

Thus we can conclude that this is a common phenomenon and confirm the theories of Coronity and Kennel and Trakhtengerts (58, 59) regarding the modulation of instability growth rate by magnetic field pulsations. At low altitudes, the modulation of ELF hiss and precipitating electrons is usually observed in the vicinity and inward the plasmapause in disturbed conditions, but chorus modulation was also observed in the upper magnetosphere as well as in auroral latitudes. These observations also confirm that the wave excitation processes are occurring in near equatorial regions.

The modification of the state of trapped particles leading to their precipitation is the one of the best known and more easily observable effects of wave-particle interaction. The consequences of wave-particle interactions on magnetospheric processes is more vital when there exists electrostatic waves. Incidentally the plasma becomes thermalized, the anomalous resistivity originates and the enhancement of DC electric fields takes place on auroral field lines, etc.. Some other papers in this volume will discuss this problem.

3. THE SURVEY OF THE NOISE INTENSITY DISTRIBUTION IN THE MAGNETOSPHERE

Before discussing the available information on the noise intensity distribution in the magnetosphere, I would like to make some comments about the methods which are used for measuring and evaluating the absolute field strength of the noise.

The ways for registering waves in space are described in a number of papers (see e.g. reviews by Russell et al., or in Vakhulov et al. (2, 50). It is well know that, to determine the real amplitude of the magnetic or electric field component, it is necessary to use three aerials of appropriate type. And for determining the wave intensity (energy density, or energy density flux), one has to use three magnetic and three electric aerials simultaneously (30, 60).

However, the main amount of existing data regarding the waves in the magnetosphere are obtained by using only one magnetic or electric antenna and in some cases two such antennas were used simultaneously as on INJUN-5 satellite (see table 1 in (2) and table 4 in (61)). Only the data measured at ELF frequencies with the triaxial search coils (31) is a pleasant exception to this rule. In some recent wave experiments, this limitation is eliminated.

When discussing the data obtained by using only one aerial, one has to take into account that the field pattern of the receiving aerial embedded in the magnetoactive plasma is more complicated than the pattern it has in free space.

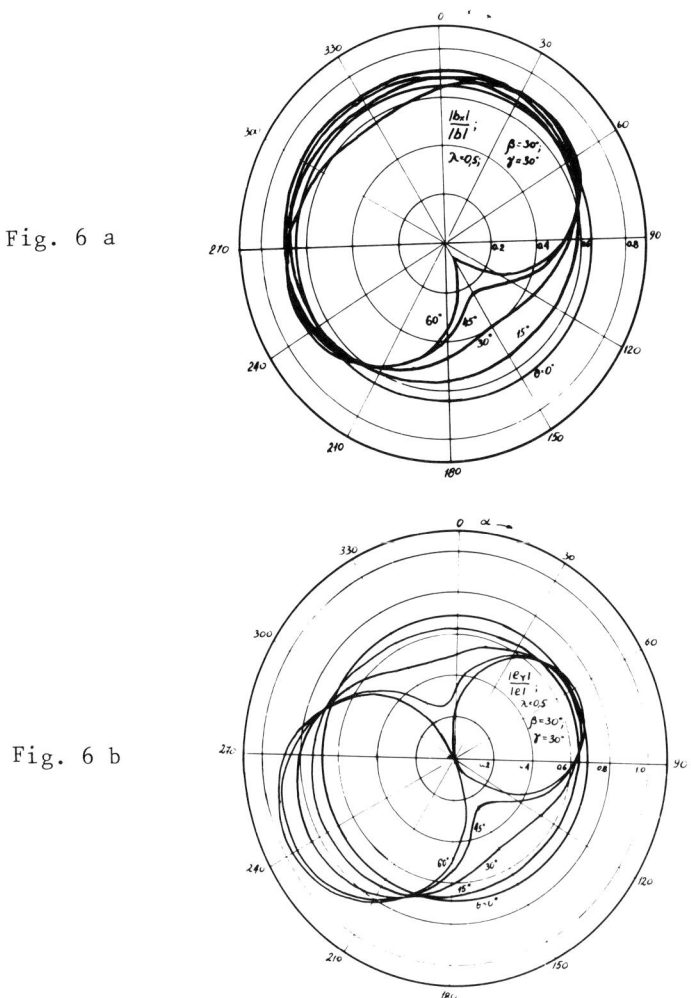

Fig. 6 a

Fig. 6 b

Figure 6. The receiving pattern of magnetic (fig. 6a) and

electric (fig. 6b) aerials in a magnetoactive plasma. It depends on the aerials α, β and γ angles, as well as on the wave normal (θ) orientation with respect to the magnetic field vector.

Figure 6 shows an example of receiving patterns of the magnetic and electric aerials for various orientations regarding the wave and for various values of the angle between the wave normal and the geomagnetic field (62). Bearing in mind the anisotropy of the wave patterns of receiving aerials, we conclude that the measured values are less than the absolute value of the corresponding field vector. The ratio of this two quantities depend on the wave normal orientation, as well as on the orientation of the aerial with respect to the magnetic field. Only for some special orientation of the aerial, when the frequency of the wave just coincides with the oblique or hybrid resonances, the real field strength is measured.

The analysis of the data measured with only one magnetic aerial onboard the not specially oriented satellite INTERCOSMOS-3 has shown that the ratio of measured quantity to the full field strength may be appreciated as a result of statistical evaluation of the orientation of the wave and the satellite. The evaluation of the measured data becomes more reliable when the three axes attitude of the satellite is known. Besides, the other difficulties which complicate the absolute measurements of wave field strength of the above mentioned one, originating from the anisotropy of wave patterns of the receiving antenna, must be taken into account when appreciating the wave intensity data available at present. Nevertheless, many papers contain useful information about the noise intensity in the magnetosphere. We bring here some of those data.

The typical data values of the ELF and VLF noise intensity in the topside ionosphere are seen on figure 3 (26). The data measured at the ARIEL 3 and 4 satellites at frequencies 0.75, 3.2, 9.6 kHz, and at altitudes of 550 km are of the same order of magnitude and may be found in (68-70). Under the arch of the plasmapause, near the equator, the amplitude of the ELF plasmaspheric hiss is of 5 - 50 mγ(0.1 < f < 1 kHz) (20). At VLF frequencies, the following values are measured : 0.22mγ.Hz$^{-1/2}$ (f = 8 kHz), and lower than 1.1 mγ.Hz$^{-1/2}$ for magnetic, and 2.8 mV.m^{-1}.Hz$^{-1/2}$ for electric field strengths (23). Characteristic change of ELF hiss amplitude in the main ionospheric trough is seen on figure 5 (28, 29).

VLF emission at the auroral field lines (h < 2000 km, Λ = 76°, 9-10 MLT) has rather small amplitudes of the order of 0.01 - 0.03 mV.m^{-1}.Hz$^{-1/2}$ in the frequency range 7.35 - 105 kHz. It is even weaker in the evening (0.005 mV.m^{-1}.Hz$^{-1/2}$) (72).

The chorus amplitude in the source regions in the outer magnetosphere varies between 1 mγ and 32 mγ (12), and at an altitude of 1000km and $4 < L < 5$ chorus amplitude is $1 - 10$ mγ (48, 66).

The weakest is the electrostatic radiation in the outer magnetosphere. The typical amplitude values are : 2.10^{-3} mV.m^{-1} for nonthermal continuous radiation ($f > f_{pe}$) (15, 17) and the resonance line radiation is of the order of some $10^{-5} - 10^{-2}$ mV.m^{-1}.Hz$^{-1/2}$ (18).

The most intense is the ion-cyclotron noise in the cusp, the amplitude of which is of the order of $1 - 10γ$ in the frequency band (0.67 - 0.87)f_{Bi} for $B_o = 340 - 550$ γ (71).

We tried to construct charts of time and spatial noise amplitude distributions. An example of such a chart for an altitude of 500 km and a frequency of 500 Hz is shown on figure 7. It is constructed by the use of invariant latitude and local time variations of the noise amplitude. We have used a number of published results obtained by many authors when constructing this chart.

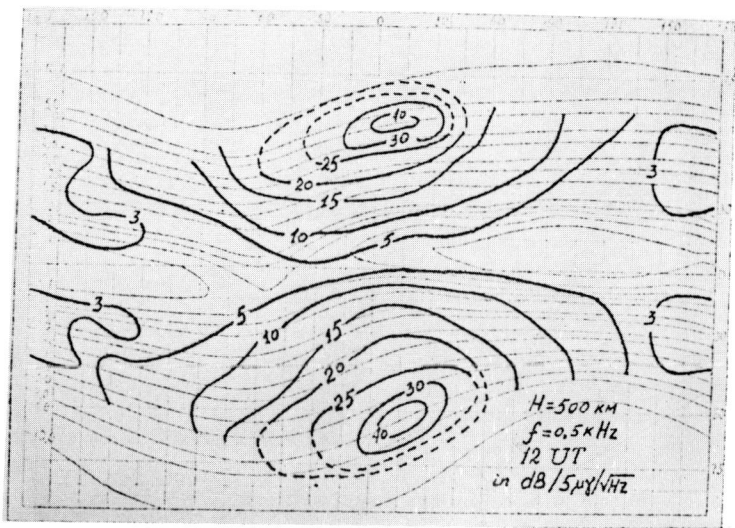

Figure 7. Spatial and local time distribution of ELF noise amplitudes (magnetic component in dB relative to 5.10^{-6}γ.Hz$^{-1/2}$). The frequency is 500 Hz and the altitude 500 km.

In all cases, the wave amplitude of the noise substantially grows during magnetospheric disturbances.

On figure 8, the growing amplitude of noise at three frequencies in dependence of Dst variation is shown. The data are averaged over five storms (63).

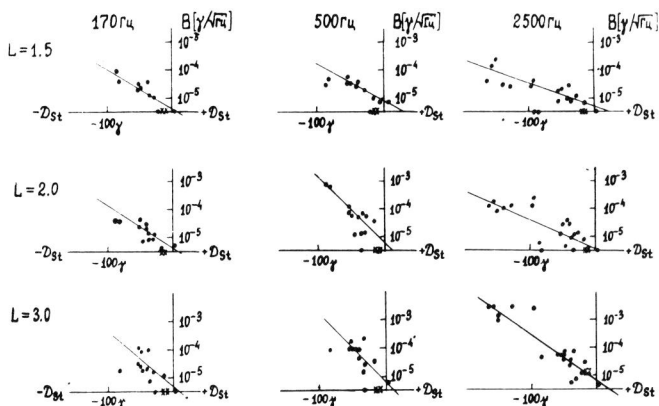

Figure 8. The noise amplitude in dB grows linearly when Dst variation rises during the magnetic storm. The alti altitude of the INTERCOSMOS-3 satellite was lower than 1500 km and the inclination of the orbit was 48.4°.

The storm time variation of noise intensity reflects the rise of energetic particle population in the magnetosphere, the variation in thermal plasma density and other processes taking place during the storm.

Particularly the spectra of ELF noise excited at the topside ionosphere extends to more higher frequencies and the source region extends to higher altitude during disturbances. This is not only a result of growing particle fluxes but also a consequence of a rising plasma density.

4. CONCLUSION

The available measurements of ELF and VLF noise intensity gathered in many space experiments allows to present a general picture of the location of the most important sources in the magnetosphere. The propagation characteristics of ELF and VLF waves are well known, thus allowing to calculate the contribution of various sources to the net amplitude of the wave at any point remote from the source.

It is felt that there is a need for more detailed measurements of the wave field structure at the source and far from it. This is important for detailed discussion of the generation mechanism and also for getting more reliable values of the wave field intensity. From the physical as well as from the radio interference point of views, it is important to measure the short term statistics of the field fluctuations (32).

Some of the experiments presently performed have enough elaborate technique to obtain appropriate data. As an example, we may refer to GEOS-1 (65).

In many experiments, waves, particles and the plasma characteristics are measured simultaneously. However, the number of papers devoted to comprehensive analysis of all the information is still not very large. In the near future, we may expect to read about the results of such analysis based on measureements on f.i. EXPLORER-45, HAWKEYE-1, GEOS-1 and other satellites. What is really needed are experiments on satellites crossing the magnetosphere at different altitudes and latitudes simultaneously. When reasonably equipped, such groups of satellites will bring valuable new results.

ACKNOWLEDGEMENTS

I acknowledge the receipt of many contributions sent by many scientists, who helped me preparing this review. Unfortunately, I could not use some of the received data, especially the information on ground measurements.

REFERENCES

1. McPherron, R.L., Russell, C.T., and Coleman, P.J. : 1972, Space Sci. Rev., 13, pp. 411.
2. Russell, C.T., McPherron, R.L, and Coleman, P.J. : 1972, Space Sci. Rev., 12, pp. 810.
3. Holtet, J.A. : 1973, Geophys. Norw., 30, pp. 88.
4. Gendrin, R. : 1975, Space Sci. Rev., 18, pp. 145.
5. Fredricks, R.W. : 1975, Space Sci. Rev., 17, pp. 449.
6. Fredricks, R.W. : 1975, in "Magnetospheres of the Earth and Jupiter" ed. by V. Formisano, D. Reidel Pub. Cy, pp. 113.
7. Scarf, F.L. : 1975, in "Physics of the Hot Plasma in the Magnetosphere, ed. by B. Hultqvist and L. Stenflo, Plenum Press, New-York, pp. 271.
8. Stenzel, R.L. : 1977, J. Geophys. Res., 82, pp. 4805.
9. Pudovkin, M.I., Raspopov, O.M., Kleimenova, N.G. : 1976, in "Geomanetic disturbances of the Earth magnetic field. 2. Micropulsations of geomagnetic field", ed. by M.S. Kovner, Leningrad, pp. 270 (in Russian).
10. Kovner, M.S., Kusnetsova, V.A., Lebedev, V.V., Plyasova-Bakounina, T.A., and Troitskaya, V.A. : 1976, Ann. Géophys., 32, pp. 189.
11. Tsurutani, B.T., and Smith, E.J. : 1974, J. Geophys. Res., 79, pp. 118.

12. Tsurutani, B.T., and Smith, E.J. : 1977, J. Geophys. Res., 82, pp. 5112.
13. Burtis, W.J., and Helliwell, R.A. : 1975, J. Geophys. Res., 80, pp. 3265.
14. Burtis, W.J., and Helliwell, R.A. : 1976, Planet. Space Sci., 24, pp. 1007.
15. Shaw, R.R., and Gurnett, D.A. : 1975, J. Geophys. Res., 80, pp.4259.
16. Kennel, C.F., Scarf, F.L., Fredricks, R.W., McGehee, J.H., and Coroniti, F.V. : 1970, 75, pp. 6136.
17. Gurnett, D.A.,and Shaw, R.R. : 1973, 78, pp. 8136.
18. Christiansen, P.J., Gough, M.P., Martelli, G., Bloch, J.J., Cornilleau, N., Etcheto, J., Gendrin, R., Jones, D., Beghin, C., and Décreau, P. : 1978, Nature, 272, pp. 682.
19. Trakhtengerts, V.Ju., and Shapaev, V.I. : 1977, Radiophysics (Russ.), 20, pp. 1004.
20. Thorne, R.M., Smith, E.J., Burton, R.K., and Holzer, R.E. : 1973, J. Geophys. Res., 78, pp. 1581.
21. Thorne, R.M., and Barfield, J.N. : 1976, Geophys. Res. Lett., 3, pp. 29.
22. Likhter, Ja. I., and Mogilevski, M.M. : 1974, in "Low frequency waves and signals in the topside ionosphere", ed. by N.G. Kleimenova, Nauka Pub., Apatity, pp. 117 (in Russian)
23. Koons, H.C., and McPherson, D.A. : 1972, J. Geophys. Res., 77, pp. 3475.
24. Vakulov, P.V., Dobrowolska, A.V., Zakharov, A.V., Kovner, M.S., Kuznetsov, S.N., Kuznetsova, V.A., Larkina, V.I., and Likhter, Ya. I. : 1975, J.E.T.P. Lett., 22, pp. 213.
25. Kovner, M.S., et al. : 1977, in "Geomagn. Issledovania", ed. by Nauka Publ., Moscow, 20, pp. 57.
26. Zakharov, A.V., Likhter, Ja. I. and Kusnetzov, S.N. : 1979, in "Low frequency waves and signals in the magnetosphere and the ionosphere of the Earth", ed. by Nauka Publ., Moscow (in press).
27. Tulunay, Y., and Hugues, A.R.W. : 1973, J. Atmosph.Terrestr. Phys., 35, pp. 153.
28. Likhter, Ya. I., et al. : 1979, Proceedings of the Cospar XXIth Plenary Meeting, Innsbrück, 1978 (in press).
29. Gdalevitch, G.L., Likhter, Ja. I., Larkina, V.I., and Mikhailov, Ja. M. : 1979, in "Low frequency waves and signals in the magnetosphere and ionosphere of the Earth", ed. by Nauka Publ., Moscow (in press).
30. Shawhan, S.D. : 1970, Space Sci. Rev., 10, pp. 689.
31. Frandsen, A.M.A., Holzer, R.E., and Smith, J.E. : 1969, IEEE Trans. Geosci. Electron., GE-7, pp. 61.
32. Storey, L.R.O. : 1967, Cospar Space Research VII, pp. 588.
33. Bettac, D., Gdalevich, G.L., Gubsky, V.F., Debabov, A.S., Jiricek, F., Kaputsin, I.N., Klimov, S.I., Lemann, H., Likhter, Ja. I., Mikhailov, Yu. M., Savin, S.P., Titova, E.E. and Triska, P. : 1976, Cospar Space Research XVI, pp. 575.
34. Vernova, L.V., et al. : 1972, Kosm. Issledovania, 10, pp. 82 (russian).

35. Jiricek, F., and Triska, P. : 1976, Cospar Space Research, 16, pp. 567.
36. Trakhtengertz, V. Ju : 1963, Geomagnetism i Aeronomia, III, pp.442.
37. Kennel, C.F., and Petschek, H.E. : 1966, J. Geophys. Res., 71, p.1.
38. Etcheto, J., Gendrin, R., Solomon, J. and A. Roux : 1973, J. Geophys. Res., 78, pp. 8150.
39. Schulz, M. : 1972, Phys. Fluids, 15, pp. 2448.
40. Likhter, Ja. I., Sobolev, Ja. P., and Vernova, L.V. : 1978, J. Atmos. Terrestr. Phys., 40, pp. 1047.
41. Siwft, D.W., and Kan, J.R. : 1975, J. Geophys. Res., 80, pp. 985.
42. Palmadesso, P., Coffey, T.P., Ossakov, S.L., and Papadopoulos, K.: 1976, 81, pp. 1762.
43. Gurnett, D.A., and Frank, L.A. : 1978, J. Geophys. Res., 83, pp.1447.
44. Oguti, T. : 1975, J. Atmosph. and Terrestr. Phys., 37, pp. 761.
45. Olesen, J.K., Primdahl, F., Spangslev, F., Ungstrup, E., Bahnsen,A., Fahleson, U., Fälthammar, C.G., and Pedersen, A. : 1976, Geophys. Res. Lett., 3, pp. 399.
46. Gurnett, D.A., and O'Brien, B.J. : 1964, J. Geophys. Res., 69, pp. 65.
47. Kivelson, M.G., Farley, T.A., and Aubry, M.P. : 1973, J. Geophys. Res., 78, pp. 3079.
48. Holzer, R.E., Farley, T.A., Burton, R.K., and Chapman, M.C. : 1974, J. Geophys. Res., 79, pp. 1007.
49. Gurnett, D.A., and Frank, L.A. : 1974, J. Geophys. Res., 79, pp. 2355.
50. Vakulov, P.V., et al. : 1974, Kosm. Issledovania, 12, pp. 707.
51. Tsurutani, B.T., Smith, E.J. and Thorne, R.M. : 1975, J. Geophys. Res., 80, pp. 600.
52. Ungstrup, E., Bahnsen, A., Olesen, J.K., Primdahl, F., Spangslev, F., Heikkila, W.J., Klumpar, D.M., Winningham, J.D., Fahleson, U., Fälthammar, C.G., and Pedersen, A. : 1975, Geophys. Res. Lett., 2, pp. 345.
53. Laaspere, T., and Hoffmann, R.A. : 1976, J. Geophys. Res., 81, pp. 524.
54. Zakharov, A.V., and Kusnetzov, S.N. : 1978, Geomagnetism i Aeronomia, 18, pp. 352.
55. Parks, G.K., and Winckler, J.R. : 1969, J. Geophys. Res., 74, pp. 4003.
56. Zakharov, A.V., Kusnetzova, V.A., Larkina, V.I., and Likhter, Ya.I.: 1974, in "Low frequency waves and signals in the topside ionosphere" ed. by N.G. Kleimenova, Nauka Pub., Apatity, pp. 110.
57. Vakulov, P.I., et al. : 1975, in "International Symposium on Solar-Terrestrial Physics", Sao-Paulo, Brazil, 2, pp. 274.
58. Coroniti, I.V., and C.F. Kennel : 1970, J. Geophys. Res., 75, pp. 1279.
59. Trakhtengertz, V. Ju. : 1975, in "Ionospheric investigations", N-22, Moscow, pp. 8 (russian).
60. Storey, L.R.O. and F. Lefeuvre : 1974, Cospar Space Research, 14, pp. 381.

61. Likhter, Ja. I. : 1975, in "Geomagnetism i Visokie Sloi Atmospheri", 2, ed. by Viniti, Moscow, pp. 169.
62. Likhter, Ja. I. : 1979, in "Low frequency waves and signals in the magnetosphere and ionosphere of the Earth", ed. by Nauka Pub., Moscow (in press).
63. Larkina, V.I., and Likhter, Ja. I. : 1976, in "Symposium KAPG on Solar-Terrestrial Physics", Tbilissi, USSR, 3, pp. 38.
64. Smith, E.J., Frandsen, M.A., Tsurutani, B.T., Thorne, R.M., and Chan, K.W. : 1974, J. Geophys. Res., 79, pp. 2507.
65. GEOS S-300 Experimenters : 1979, Planet. Space Sci., 26, to be published.
66. Thorne, R.M., Smith, E.J., Fiske, K.J., and Church, S.R. : 1974, Geophys. Res. Lett., 1, pp. 193.
67. Parady, B.K., Eberlein, D.D., Marvin, J.A., Taylor, W.L., and Cahill, Jr., L.J. : 1975, J. Geophys. Res., 80, pp. 2183.
68. Bullough, K., Hugues, A.R.W., and Kaiser, T.R. : 1974, in "Magnetospheric Physics", ed. by B.M. McCormac, D. Reidel, pp. 231.
69. Kaiser, T.R., and Bullough, K. : 1971, Cospar Space Research, 11, pp. 1323.
70. Hayakawa, M., Bullough, K., and Kaiser, T.R. : 1977, Planet. Space Sci., 25, pp. 353.
71. Fredricks, R.W., and Russell, C.T. : 1973, J. Geophys. Res., 78, pp. 2917.
72. Gurnett, D.A., and Frank, L.A. : 1972, J. Geophys. Res., 77, pp. 172.

ELF AND VLF ACTIVITY ASSOCIATED WITH HIGH LATITUDE HOLE

H. Kikuchi
Max-Planck-Institut für Aeronomie, Katlenburg-Lindau
W. Germany*

ABSTRACT

   Based upon observations from a series of polar orbiting OGO satellites, the properties and structure of a "high latitude trough or hole" (HLT) or (HLH) in the topside ionosphere are summarized and correlated with the following terms, specifically in relation to ELF and VLF activity: (1) distinction from the midlatitude light ion trough (LIT) in H^+, associated with the plasmapause; (2) a narrow, abrupt depletion in the concentrations of atomic ions H^+, He^+, O^+, and N^+; (3) molecular ion enhancements in NO^+, O_2^+, and N_2^+; (4) the > 30 keV electron trapping boundary; (5) the > 30 keV electron precipitation; (6) the electric field reversal; (7) the convective flow reversal; (8) the polar cap boundary. These observations are often well correlated with ELF and VLF activity, exhibiting enhancements in broadband auroral hiss, ELF noise and/or auroral kilometric radiation in the HLH region.

   <u>The September 29, 1969, OGO-6 event</u>[1]. Fig. 1 shows a typical example of simultaneous observations for a dawn-dusk pass of OGO-6 over the northern hemisphere on 29 September, 1969 during a rather intense storm. The top and the second panel show ELF and VLF data, the middle panel concentrations of atomic ions, and the lower two panels low energy electron data.

   In the middle panel, high latitude holes are identified by a narrow, abrupt depletion in concentrations of atomic ions H^+, N^+ and O^+ at nearly 68° at local dawn and again H^+, He^+, O^+ and N^+ at nearly 65° in dipole latitude at local dusk. In the HLH region at local dawn, significant intense fluctuations are observed from the lower two panels for both trapped and precipitated electrons in a very similar fashion. Though this makes it difficult to define the trapping

---

*Permanent address: Nihon University, College of Science and Technology, Tokyo, Japan

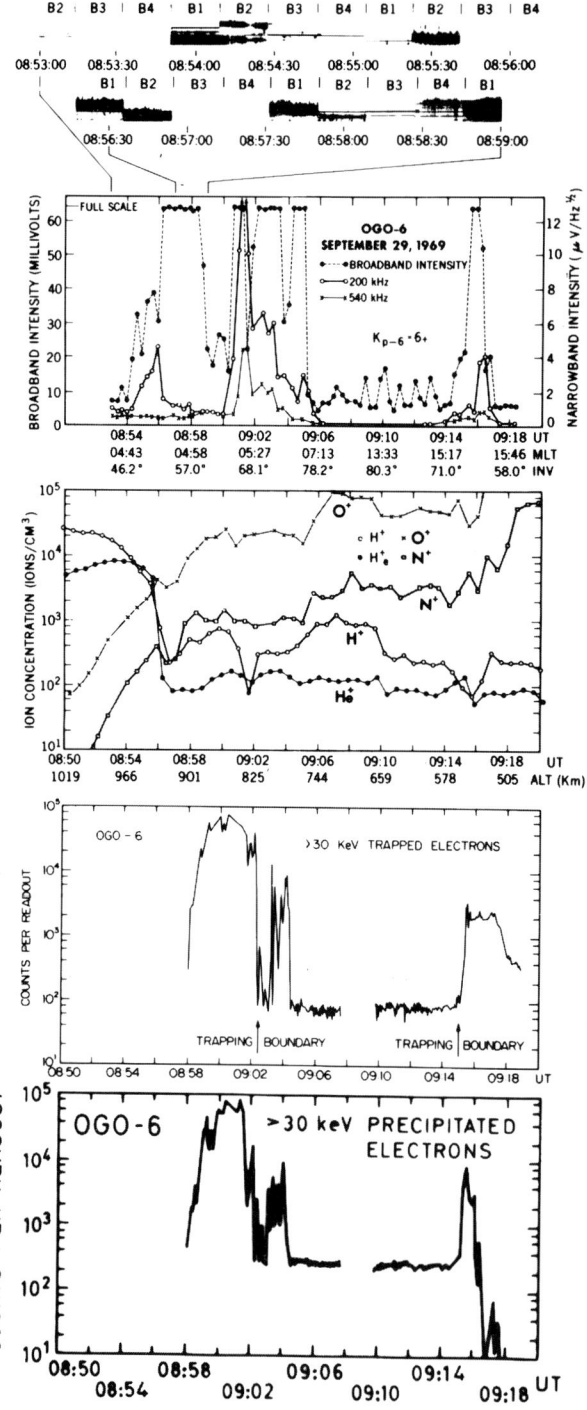

Fig. 1. Simultaneous measurements of ELF·VLF, thermal ion and low energy electron data from OGO-6.

boundary, the first equatorward boundary certainly falls into the HLH. In contrast, the electron flux at local evening shows no significant fine structure irregularities and the trapping boundary is well defined into the hole region. A very narrow and sharp localization of the > 30 keV electron precipation is also observed in the dusk sector just at the center of the HLH and at the electron trapping boundary.

With regard to simultaneous ELF and VLF records, the second panel shows digital data from two narrow-band receivers at 200 kHz and 500 kHz and a broadband intensity detector covering a frequency range from about 20 Hz to more than 1 MHz. The 200 kHz signal level correlates better than the 500 kHz level and the broad-band intensity best correlates with the HLH and trapped and precipitated low energy electrons at both local dawn and dusk, indicating a greater contribution of ELF and/or a lower range of VLF. Particularly, broad-band intensity at local dawn exhibits separated activity into three regions which exactly coincide with three separations of fine structure irregularities in both trapped and precipitated electrons. Concurrent sonagrams on the top panel also shows intense activity in a lower range of frequency, Band 1 (20 Hz - 15 kHz) and 2 (15 - 30 kHz) rather than in a higher range of frequency Band 3 (92.5 - 107.5 kHz) and 4 (280 - 295 kHz). Enhancements in ELF and VLF activity are observed at the middle latitude LIT or the plasmapause and emissions are a continuous, band-limited noise, most likely identified as the lower hybrid resonance[2]. In contrast, ELF and VLF emissions at the HLH are rather broad-band in a type of auroral hiss and/or irregular chorus. In the polar cap region, low energy electron intensities are very low and essentially absent and ELF-VLF activity is also calm.

The June 16, 1969, OGO-6 event. Fig. 2 shows another example of simultaneous ELF-VLF and ion data, particularly representing molecular ion concentrations from OGO-6 over the northern hemisphere on 16 June, 1969 during rather quiet magnetic activity ($Kp_6 = 3_-$). Large enhancements in molecular ions are observed in the HLH region[3]. The first peak in broad-band intensity at approximately 64° is supposed to occur at the LIT or the plasmapause. However, the 2nd peak at nearly 72° is not clear whether this emission comes from the HLH or not, since fine structures in ion data are not available in this region. Next peak in broad-band intensity and 200 kHz level at 76° comes from the HLH where enhancements in molecular ions, NO^+, O_2^+ and N_2^+ are significant, as well as depletions in atomic ions H^+, He^+, N^+ and O^+. The last peak in broad-band intensity at 60° appears to occur at the midlatitude LIT or the plasmapause where the decrease in the atomic ions H^+, He^+, N^+ and O^+ are rather mild under quiet conditions.

Summary. Although the D. C. electric field data are not available for the September 29, 1969, event, it has been recognized that the D.C. electric field reversal may occur at the low energy electron trapping boundary, based upon other OGO events[4]. Incorporating these results, the locations of the > 30 keV electron trapping boundary and precipitation, the electric field reversal, consequently the convective flow reversal and the polar cap boundary, all fall into the HLH region which is also

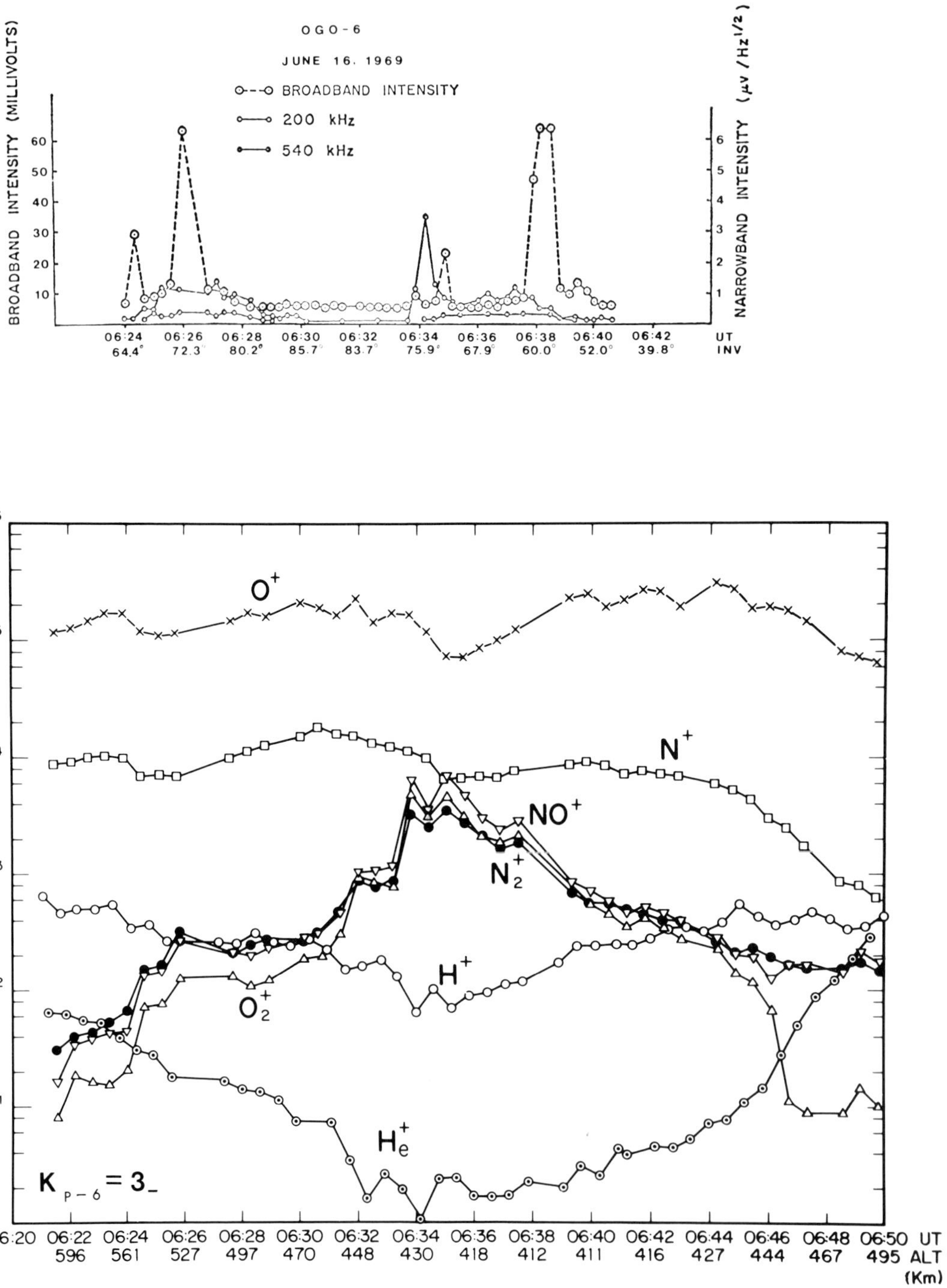

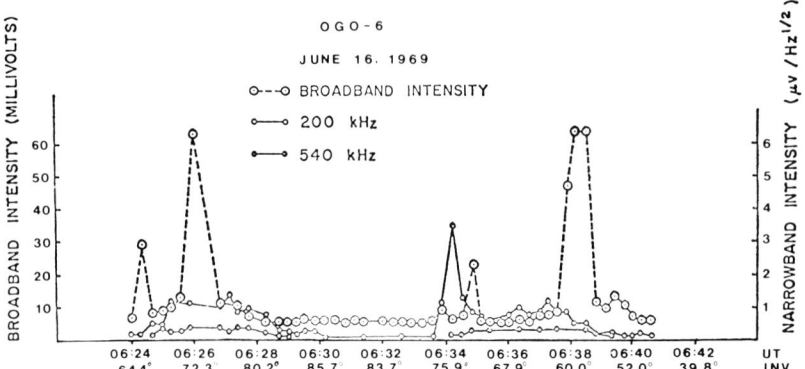

Fig. 2. Simultaneous ELF·VLF and ion data from OGO-6, particularly representing molecular ion concentrations.

well correlated with ELF and VLF activity, exhibiting enhancements in broad-band emissions. This is in contrast to continuous band-limited noise at the LIT or the plasmapause.

In view of the close interrelationships of the HLH, low energy electron trapping boundary, precipitation, electric field and convective flow reversals and polar cap boundary, these factors are all supposed to be equally responsible for the pronounced ELF and VLF activity at the HLH. Simultaneously, intense ELF and VLF activity would make re-action back to the hole formation as a result of strong wave-particle interactions. These results indicate that a variety of instabilities may be involved in ELF and VLF generation together with the HLH formation, thus making the generation processes fairly complicated. According to observations that a frequency range for ELF and VLF emissions appears intensively to be lower than the local electron plasma frequency, it can be stated that new mechanisms might be essential for the HLH formation and intense ELF and VLF generation. Namely, when the electrostatic noise generated by the low energy precipitated electrons exceeds a certain level, nonlinear parametric processes might develop so as to form a "caviton" or "solitary hole", identified as a "high latitude hole".

REFERENCES

1. H. Kikuchi et al., Trans. AGU 52, 307 (1971).
2. T. Laaspere et al., J. Geophys. Res. 76, 4477 (1971).
3. H. A. Taylor, Jr. et al., J. Atmos. Terr. Phys. 37, 613 (1975).
4. J. M. Grebowsky et al., J. Geophys. Res. 81, 690 (1976).

EFFECTS OF POWER LINE RADIATION INTO THE MAGNETOSPHERE

R. A. Helliwell
Radioscience Laboratory
Stanford University
Stanford, CA 94305

ABSTRACT

   VLF power line radiation (PLR) can cause triggering, suppression and entrainment of whistler-mode chorus emissions in the magnetosphere. High-altitude OGO-3 spectral data show evidence of enhanced chorus activity over centers of population and industry, indicating that PLR may play a significant role in the excitation of VLF whistler-mode noise and the associated precipitation of electrons. Sampled OGO-3 data showed chorus elements starting in multiples of 50 Hz in the European-Siberia sector, of 60 Hz in the United States-Canada sector, of either 50 or 60 Hz in the Alaska (60-Hz system)-New Zealand (50-Hz system) sector. Low-altitude Ariel III data on the intensity of all types of VLF activity at 3.2 kHz showed enhanced noise over populated areas and their conjugates, but the relative contributions of lightning and PLR were not assessed. VLF data from the low-altitude OGO-4 scanning receiver (150-1500 Hz; 200 orbits; 1966-67) detected PLR from both 50-Hz and 60-Hz systems with average intensities near 1.0 mγ; PLR intensities greater than 10 mγ were detected less than 10% of the time. Intensities were low in the midnight and dawn sector.
   Antarctic data from Eights (L = 3.9; 1963-65) and Siple Station (L = 4.2; 1973-75) show strong control of chorus frequencies by PLR. Activity peaks at 3 kHz and at 12 LT. 2-4 kHz noise (mostly chorus) shows a distinct minimum on Sunday, as does average power consumption in the conjugate area.
   To explain the observed geophysical variations in VLF wave activity, it is suggested that PLR acts to lower the threshold for wave growth, just as in the Siple Station controlled triggering experiments. Geographical localization of VLF waves is expected to cause corresponding localization of particle precipitation. However, global average precipitation would not necessarily be affected.

I.   INTRODUCTION

   Much of the electromagnetic energy in the magnetosphere appears as whistler-mode waves at very low frequency (VLF). Whistler-mode wave sources include natural lightning, very-low-frequency transmitters, and energetic particles that generate emissions, such as plasmaspheric hiss and chorus.

Recently it was found that power lines also excite whistler-mode waves [Helliwell et al., 1975]. Power line radiation (PLR) has been detected both on the ground and in satellites. This paper reviews observations of PLR made on the high-altitude OGO-3 satellite, the low-altitude Ariel III satellite, the low-altitude OGO-4 satellite, and at ground stations in the Antarctic. An experiment to simulate power line radiation, using the Siple VLF transmitter located at $L = 4.2$, is described.

The propagation of PLR in the magnetosphere involves both ducted and non-ducted paths, as depicted in Figure 1. The 'duct' is thought to consist of a field-aligned enhancement of ionization that traps VLF waves the way light is trapped in an optical fiber. Many ducts may be present at one time. The 'non-ducted' path is determined by the overall distribution of the refractive index. Here the waves reflect in the vicinity of the lower hybrid resonance and may bounce back and forth between the hemispheres many times before dropping below the noise level. Power line radiation excites both kinds of paths.

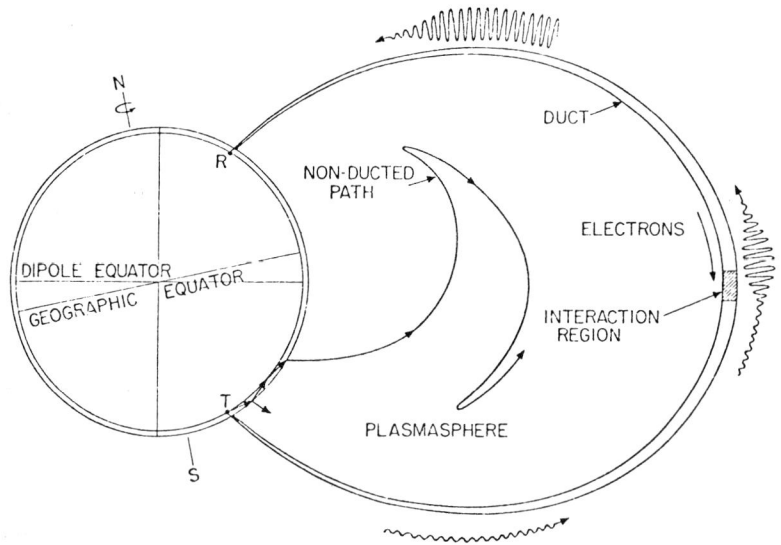

Figure 1. Sketch showing ducted and non-ducted whistler-mode paths in the magnetosphere. Three 0.5 s long ducted wave trains are shown before, during and after amplification on the equator. A group velocity of C/20 is assumed [After Helliwell, 1977].

The power line radiation effect was unexpectedly found during experiments on emission triggering using the Siple Station VLF wave-injection facility. Some triggered emissions were repeatedly reversed or inflected at harmonics of the 60-Hz Canadian system that supplies power to Siple's conjugate station at Roberval, Quebec. Subsequent study of simultaneous recordings from Siple and Roberval showed situations in which the same lines were present at both stations at the same time. Sometimes these lines were intensity-modulated in anti-phase at the 2-hop whistler period [Helliwell et al., 1975], demonstrating their magnetospheric origin. It was found that power line-associated effects included triggering, suppression and entrainment.

One of the first questions raised by these findings was whether power line effects extended to the more distant parts of the magnetosphere not covered in the Siple-Roberval experiments. To help answer this question, an available set of OGO-3 VLF chorus data was reexamined, as described in Section II.

II. OBSERVATIONS OF PLR PHENOMENA

A. OGO-3 Observations

VLF chorus was observed on OGO-3 throughout the region of closed field lines beyond the plasmapause in the frequency range 300 to 12,500 Hz. Virtually all passes of OGO-3 for the period 15 June 1966 to 14 June 1967 were studied [Burtis and Helliwell, 1976]. The original tapes from this data set were re-analyzed to provide higher frequency resolution than was needed in the earlier study. An example of the chorus elements observed in this study is shown in Figure 2. In only about 15% of the available passes was it possible to accurately measure the starting frequencies of the chorus elements. For these cases the starting frequencies clustered tightly about harmonics of the power systems located at the two feet of the field line passing through the satellite. Over the U.S. and Canada the starting frequencies were at multiples of 60 Hz; over Europe and Siberia at 50 Hz; over the Alaska-New Zealand region at both 50 and 60 Hz. In a few cases the actual

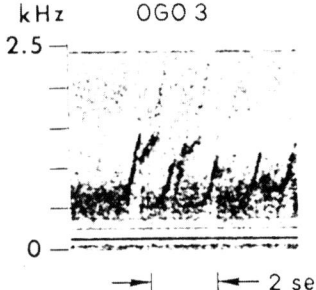

Figure 2. Spectrum of chorus elements observed at 2337:30 UT on 12 Aug 66 (Alaska-New Zealand meridian) on OGO-3 at L = 6, R = $4R_E$, Inv. Lat. = 47°, Mag. Long. = 280°.

harmonic lines were observed on the OGO-3 sweeping and broadband receivers with the electric antenna. Their intensities were near 10 μV/m.

The chorus occurrence data were organized according to the geomagnetic latitude and longitude of the northern hemisphere foot of the field line passing through OGO-3. The southern hemisphere measurements were included with those from the northern hemisphere. In each latitude-longitude bin (10° x 10°), the number of passes in which any chorus was observed was counted. The data are plotted in invariant dipole coordinates in Figure 3, and show peaks in activity over the industrial regions of the earth. The percent occurrence of chorus in each 10° geomagnetic longitude interval is shown beneath the map. The four main peaks appearing over the eastern U.S., Europe, Siberia, and Alaska-New Zealand suggest that PLR influences the distribution of chorus in the magnetosphere. Conclusive statistical tests could not be performed because the data set is small, 4668 samples, and it was difficult to determine the extent to which the individual samples were statistically independent of one another.

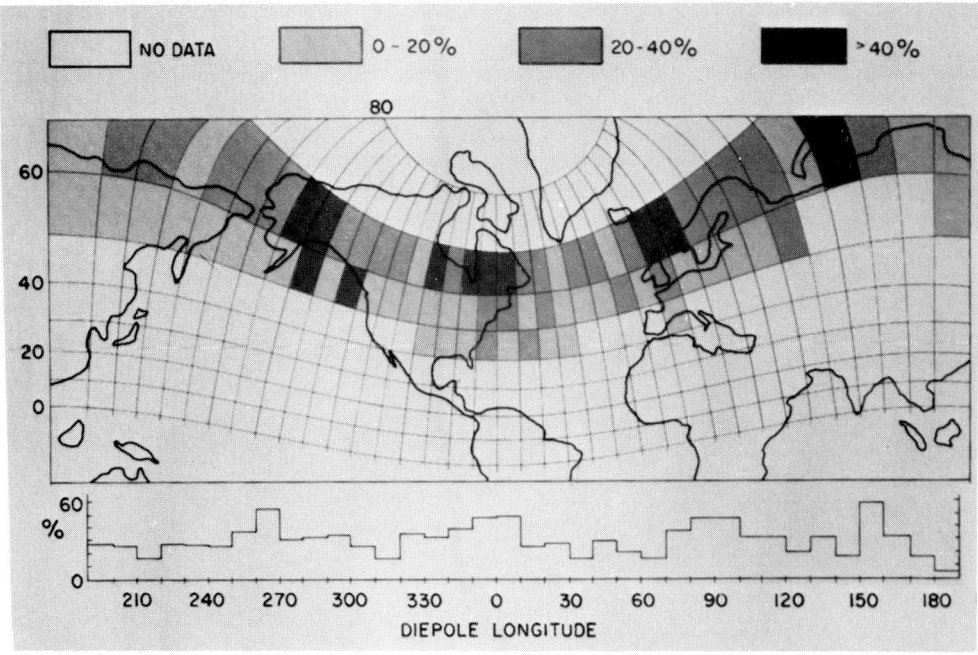

Figure 3. Map of chorus occurrence frequency in invariant dipole coordinates for period from June 15, 1966, to June 14, 1967. The upper histogram shows longitudinal variations in percent occurrence averaged over invariant latitude. The lower histrogram shows the number of samples in each 10° longitudinal interval.

B. Aeriel III Observations

A geographical distribution of VLF noise intensity, at 3.2 kHz, was obtained by the Ariel III satellite [Bullough et al., 1976], as shown in Figure 4. A permanent zone of enhanced VLF activity is seen over North America and has been attributed to a combination of whistlers and PLR. Lesser peaks appear over Europe and Siberia. Added to the map are the peaks of the OGO-3 observations shown in Figure 3. A fairly close association in longitude is observed between the peaks in occurrence from these two studies. However, the Ariel III data include all types of noise (hiss, whistlers, chorus, PLR), whereas the OGO-3 data are limited to identifiable chorus. The relative importance of power line radiation, whistlers and spontaneous emissions in the Ariel III data is not known.

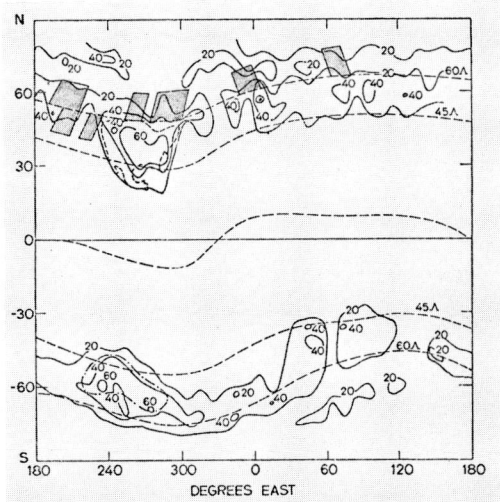

Figure 4. Percent occurrence of emissions on Ariel III with intensity $> 4.8 \times 10^{-15}$ wm^{-2} Hz^{-1} at 3.2 Hz; May-July, 1967. Shaded regions correspond to peak chorus activity detected by OGO-3.

C. OGO-4 Observations

Further information on the noise spectrum at low altitude is provided by the OGO-4 polar-orbiting satellite. The data covered the year 1966-67 and included 200 orbits from which spectral data were extracted over the 150-1500 Hz frequency range. An example of the power line radiation spectrum found in these data is given in Figure 5, taken over South America. The spectral lines are multiples of 50, indicating the presence of 50-Hz power systems on the ground. The average intensity of the power line harmonics measured in this way is about 1 mγ, with 10 mγ fields occurring less than 10% of the time

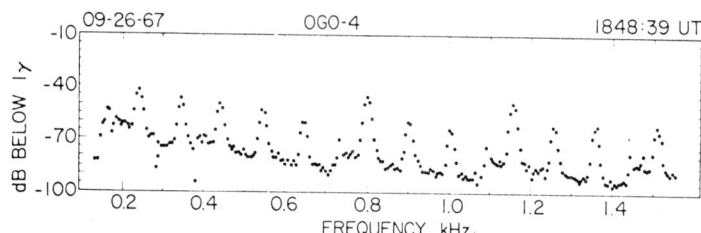

Figure 5. Spectrum of 50 Hz PLR measured by OGO-4 sweeping receiver. Lines 1-5 and 9-11 are odd harmonics, while lines 6-8 are even harmonics. Satellite's location: altitude 500 km, dipole latitude 4°, dipole longitude 338°.

[Luette, private communication]. The intensities tend to be low in the midnight-to-dawn sector, in qualitative agreement with other PLR information.

Unfortunately, the OGO-4 VLF sweeping receiver was often saturated by strong signals that made the identification of PLR impossible. A study of the geographical distribution of this saturation effect showed that it occurred whenever the receiver was near or over industrial areas. Observations just prior to the onset of saturation often showed clear evidence of PLR lines. It was tentatively concluded, therefore, that the saturating signals were mainly PLR.

III. GROUND OBSERVATIONS

A different kind of survey is possible from the ground because virtually all ground observations depend upon ducting (see Figure 1). PLR effects on ducted paths were surveyed using broadband data recorded on tape at Eights Station, Antarctica, in 1965 and from Siple Station, Antarctica, and Roberval, Quebec, from 1973 through 1976 [Park and Helliwell, 1978]. If PLR effects were present in a one-minute signal interval, taken every 15 mins, it was counted as an event and the frequencies at which PLR activity occurred were recorded. Significant PLR activity occurred in the range 1-8 kHz, peaking at 3-3.5 kHz. Whistler analysis showed equatorial crossing radii for the PLR paths of 3.4-5.5 R_E, with a most probable value between 4.5 and 5.5 R_E. The corresponding equatorial resonant electron energy range was 0.5-10 keV, representing the minimum energies to be expected in PLR-induced precipitation flux.

PLR activity usually occurred during quiet times ($0 < k_p \leq 2$) following moderate substorms ($1 < k_p \leq 4$), these being periods of good VLF propagation conditions coupled with the enhanced particle fluxes needed for wave-particle interaction. Especially pronounced was the diurnal fluctuation of PLR events, their occurrence probability rising sharply at dawn and decaying steadily in the afternoon, covering a

6:1 range in occurrence probability. The rise was attributed to the increase in power consumption near dawn, while the fall was explained by the restricted access of energetic electrons to the late-afternoon side of the magnetosphere.

An interesting new result from the ground data is the appearance of a weekend effect in PLR occurrence [Park and Miller, 1979]. A two-year (1974 and 1975) study of 2-4 kHz chorus data from Siple Station showed that the occurrence of power line radiation effects on Sunday was significantly less than on the average of the other days of the week, as shown in Figure 6. By comparing these data with the power consumption records from eastern Canada, it was found that the variation in chorus activity has the same phase as that of the power consumption in eastern Canada, each having a sharp minimum on Sunday. The Sunday minimum was pronounced for $k_p < 2$ and absent for $k_p \leq 2$, indicating that both substantial electron fluxes and PLR are required for chorus generation.

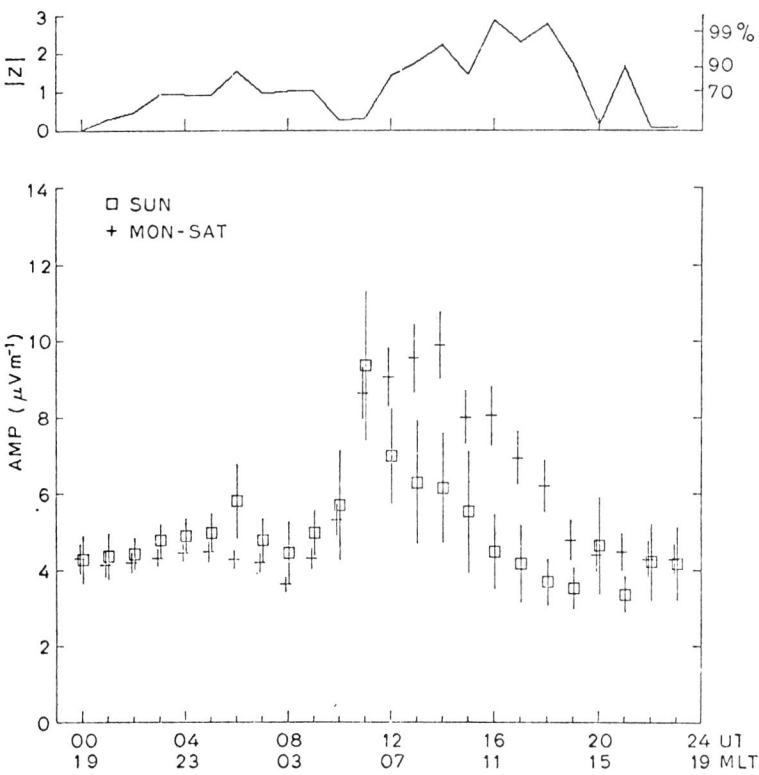

Figure 6. Daily variations of 2-4 kHz wave amplitude showing differences between Sunday and Monday-Saturday. Vertical bars indicate ±1 standard error of the mean. Top curve shows 'z' score, a measure of statistical significance [After Park and Miller, 1979].

IV. CONTROLLED EXPERIMENTS

A simulation of PLR effects was attempted, using the Siple Station transmitter [Park and Chang, 1979]. Spectral components having frequency spacings of 50 and 100 Hz were produced by frequency modulation techniques, as shown in Figure 7. The observed amplification and emis-

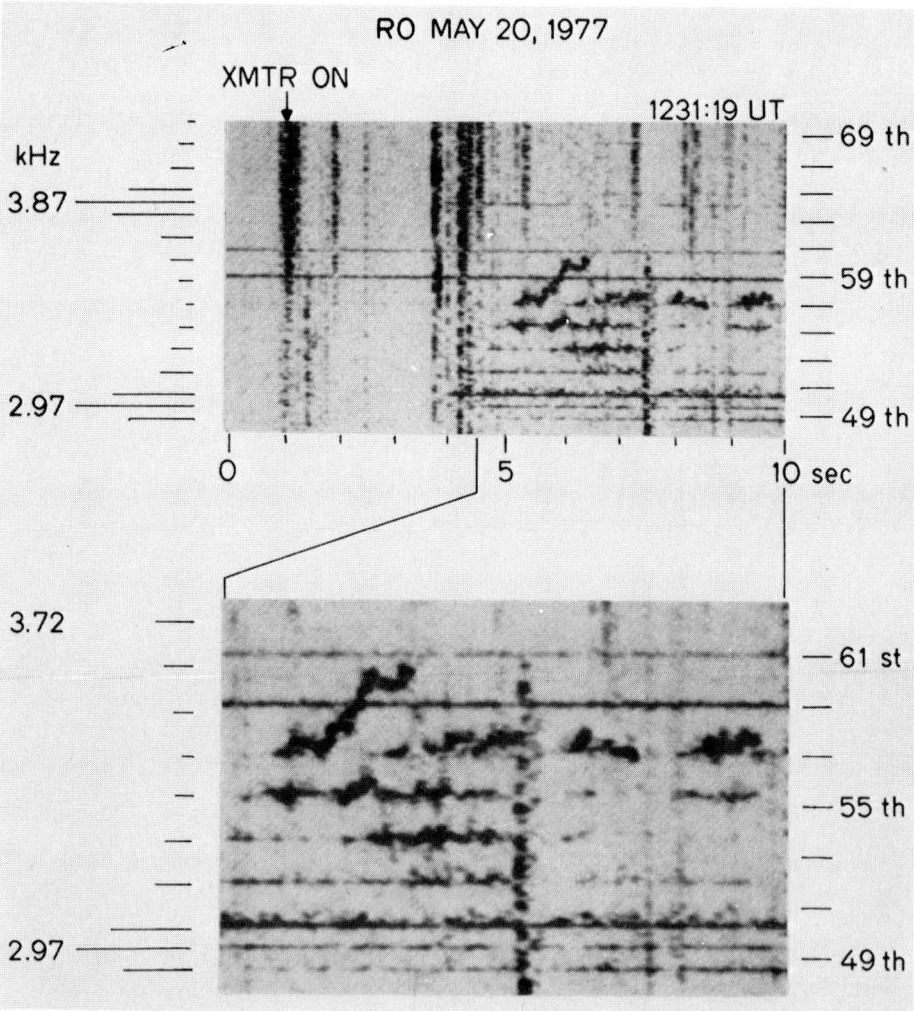

Figure 7. Frequency-time spectrogram showing emissions triggered by sidebands transmitted by Siple Station. Transmitted signals begin at 1 s and reach Roberval at about 4 s. Horizontal bars at left give relative amplitude of transmitted sidebands. Right-hand scale gives odd harmonics of 60 Hz [After Park and Chang, 1979].

sion structure was found to be similar to that from power lines. An interesting feature of these experiments is the fact that the frequency of strongest effect was not necessarily the frequency of maximum radiation. Also observed in these experiments was the creation of double sidebands (not shown) at frequencies of 0.5 to 30 Hz removed from the carrier. This result may help to explain PLR-induced line emissions from frequency separations less than the fundamental power frequency [Helliwell et al., 1975].

V. CONCLUSIONS AND RECOMMENDATIONS

Although PLR is clearly evident in satellite and ground data, the full extent of its occurrence is not known. Satellite receiver sensitivities have not been optimized for PLR detection and anlysis techniques have not been sufficiently detailed to identify PLR without ambiguity. Thus some new experiments are needed in which both the unamplified input PLR is measured along with the amplified output and associated triggered emissions. Such a study could usefully cover the entire magnetosphere, but might concentrate on low-altitude, high-inclination orbits for the input signals and high-altitude, near-equatorial orbits for the output signals. A frequency resolution of about 5 Hz would be desirable.

Since PLR-induced emissions are often comparable in strength to other types of whistler-mode emissions, we can expect PLR to have some effect on the precipitation of energetic electrons. Then we should see enhancements of precipitation (at least at energies less than 10 keV) over industrial areas of the earth and also a measurable difference in precipitation between Sunday and the remaining days of the week. It would be interesting to examine available energetic particle data for evidence of this effect.

Finally, it should be noted that even if global precipitation patterns are affected by PLR, it may still be true that the average global precipitation rate is unaffected by PLR. Thus, if we assume that the average precipitation rate is related to the average injection rate of energetic particles, and that the former is not under power line control, the global average precipitation rate should not depend on power line radiation. However, more experimental and theoretical work is required to answer this question.

Acknowledgements: I thank J. P. Luette and T. R. Miller for discussions and aid with the figures. This research was supported in part by the National Aeronautics and Space Administration under grant NGL-05-020-008, and in part by the Atmospheric Sciences Section of the National Science Foundation under grants ATM74-20084, ATM75-07707 and ATM78-05746.

REFERENCES

Bullough, K., Tatnall, A. R. L., and Denby, M.: 1976, Nature, 260, p. 401.
Burtis, W. J., and Helliwell, R. A.: 1976, Planet. Space Sci., 24, p. 1007.
Helliwell, R. A., Katsufrakis, J. P., Bell, T. F., and Raghuram, R.: 1975, J. Geophys. Res., 80(31), p. 4249.
Park, C. G., and Chang, D. C. D.: 1979, Radioscience Lab., Stanford Univ., Stanford, CA 94305, in preparation.
Park, C. G., and Helliwell, R. A.: 1978, Science, 200(4343), p. 727.
Park, C. G., and Miller, T. R.: 1979, Radioscience Lab., Stanford Univ., Stanford, CA 94305, in preparation.

ARIEL 3 AND 4 STUDIES OF POWER LINE HARMONIC RADIATION

K. Bullough and T. R. Kaiser
Department of Physics,
University of Sheffield.

ABSTRACT

Continuous on-board tape-recording of ELF/VLF signals on the Ariel 3 and 4 satellites has made it possible to study the world-wide distribution of PLHR and associated emissions above the ionosphere and their frequency of occurrence as a function of location, intensity, storm-time, season and solar cycle.

A particularly prominent zone of emission is obtained at VLF (3.2 kHz) over N. America where frequent PLHR induced magnetospheric wave amplification up to 50 dB and typically 10 to 20 dB, is obtained at invariant latitudes 45 to 55° (2 < L < 3) centred on the electron slot. It appears that PLHR and associated emissions are responsible for pitch angle diffusion of energetic electrons ($E \geq 100$ keV) at large pitch angles by first order gyroresonance and thereby contribute to the formation of the electron slot.

A marked lack of PLHR induced emission in winter is attributed to the inability of the waves to become entrapped in ducts. The dependence of signal intensity on storm-time is similarly attributed to duct formation and wave-trapping therein. One and two (or multi) hop emissions occur with about equal probability at 3.2 kHz; at 9.6 kHz one hop are predominant.

There is evidence of a possible association between PLHR emissions and thunderstorm activity at American longitudes.

INTRODUCTION

This paper presents an outline of studies reported in greater detail elsewhere (Tatnall, Matthews, Bullough and Kaiser, 1978).

The satellites Ariel 3 and 4 had low altitude (500 – 600 km) orbits of high inclination (80.2, 83° respectively) and were designed, by means of on-board tape-recording, for morphological studies of various ionospheric, wave and particle phenomena (Bullough et al, 1975). Their launch dates were 5 May, 1967 and 11 December, 1971 and we obtained good data recovery for 6 and 8 months respectively. Their high orbital inclination and associated precession gave us comprehensive coverage in geomagnetic latitude, longitude and local time in each 3-monthly period.

Similar elf/vlf narrow-band receivers were flown on both satellites and measurements were made of the peak, mean and minimum signal intensities (magnetic component) in several narrow band channels each successive 28s period ($\equiv$ 200 km) around the orbit. The time constants of the peak and minimum reading circuits (f $\leq$ 9.6 kHz) were 0.01 s and 0.1 s respectively, each sampling circuit being reset immediately after read-out at the end of each 28 s period. The mean circuit had a running mean of 30 s. On Ariel 3 the centre frequencies and pass-bands were 3.2 (1.0) kHz, 9.6 (1.0) kHz and 16 (1.0, 0.1)) kHz and on Ariel 4: 0.75 (0.5), 1.25 (0.5), 3.2 (1.0), 9.6 (1.0), 16 (1.0, 0.1) and 17.8 (0.17) kHz. In addition Ariel 4 had an impulse counter on the 3.2, 9.6 kHz channels for more detailed study of thunderstorm activity. Our analysis is restricted to frequencies $\leq$ 9.6 kHz. Measurements of the local plasma density allowed us to determine the local refractive index and hence, on the assumption of near longitudinal whistler-mode propagation, the equivalent wave power flux (w m^{-2} Hz^{-1}).

Thus,
$$1 \gamma^2 \text{ Hz}^{-1} \text{ (free space)} \equiv n \gamma^2 \text{ Hz}^{-1} \text{(satellite) where n = local refractive index.}$$
$$\equiv 4.8 \times 10^{-4} \text{ watts m}^{-2} \text{ Hz}^{-1}.$$

To obtain maps of emission activity a count was made of the number of observations (typically 35 to 55) in each 10° x 2° longitude-latitude sector and the percentage of such occasions, on which the signal intensity (mean reading) exceeded a preselected value, found. A latitudinal profile was obtained by fitting, with appropriate weighting, a 15 degree polynomial. Computer contouring was then performed on the "smoothed" values obtained for each 10 x 2° box.

THE OCCURRENCE AND DISTRIBUTION OF ELF/VLF EMISSIONS

The frequency of occurrence of 3.2 kHz emissions of intensity greater than 4.8×10^{-16} w m^{-2} Hz (Kp $\leq$ 2 +) is shown in fig. 1. This, our most sensitive level, is normally 5 to 10 dB above the receiver noise level. General features apparent on the map include, at low latitude, the thunderstorm regions with extensive minima over the oceans and a broad band, at 100°E, extending over the equator. At high latitudes there are emissions associated with the auroral regions. The most significant features however are the strong maxima over North America and its southern geomagnetic conjugate which we attribute to power line harmonic radiation (PLHR). There are also three maxima over Euro-Asia (N_1, N_2, N_3), a weaker, sharp maximum over the UK and also one to the magnetic north of Japan. The southern region S_1 is approximately conjugate to N_1. All the northern maxima

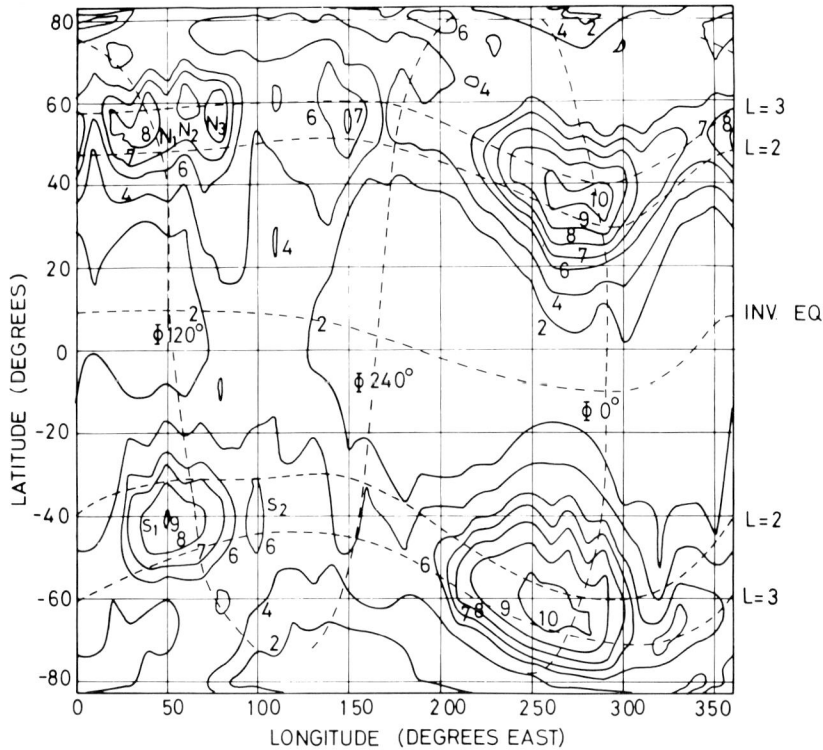

Figure 1   1972 summer (days 117 - 213)   The percentage occurrence of emissions (3.2 kHz) of intensity greater than 30 dB above 4.8×10^{-19} W m^{-2} Hz^{-1}; Kp ≤ 2 +.   Contour values 1, 2, 3, 4 ... 10 ≡ 10, 20, 30, 40, ... 100%.   - - - Contours of Invariant Longitude Φ and L values at 600 Km.

may be associated with industrialised regions of high generation/ consumption of electric power (fig. 2) and all the emission zones, north and south, peak in the range of L-shells, 2 < L < 3, associated with the electron slot between the Inner and Outer Radiation Belts.

At a signal level 10 dB higher (fig. 3) emissions associated with thunderstorms, the UK and Japan are no longer observed and the Euro-Asian maxima are displaced to significantly higher latitudes.   At still higher signal intensities (+ 20 dB, fig. 4) only the American zones remain clearly identifiable with that in the north displaced to 3 < L ≤ 4.   We attribute this poleward displacement of the more intense emissions to the occurrence of greater fluxes of energetic electrons on these higher L-shells.   Note the much smaller displacement to high latitudes in the south.   This suggests that we have predominantly 1-hop (rather than ≥ 3-hop) signals in the south some of which return on higher L-shells to give the amplified 2-hop signals in

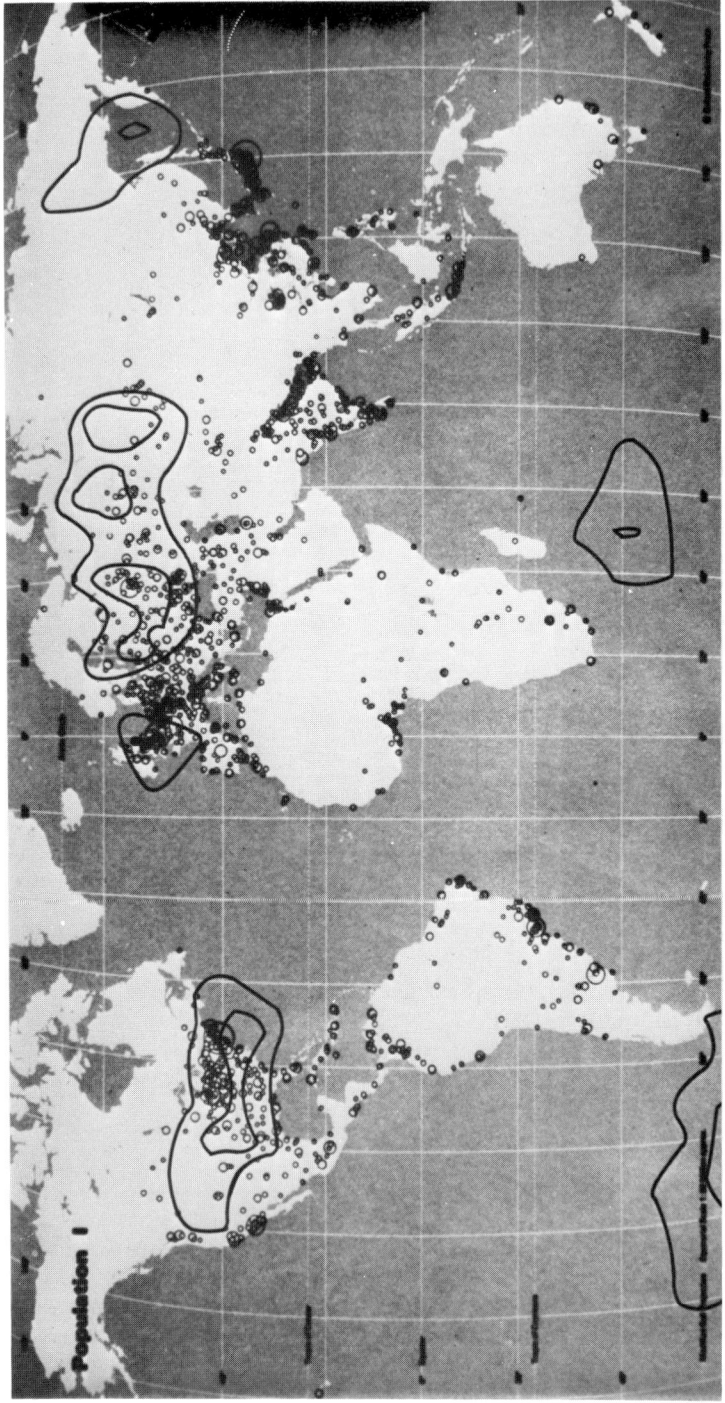

Figure 2 Zone maxima (from fig. 1) superposed on a map showing the distribution of towns of at least 100,000 population.

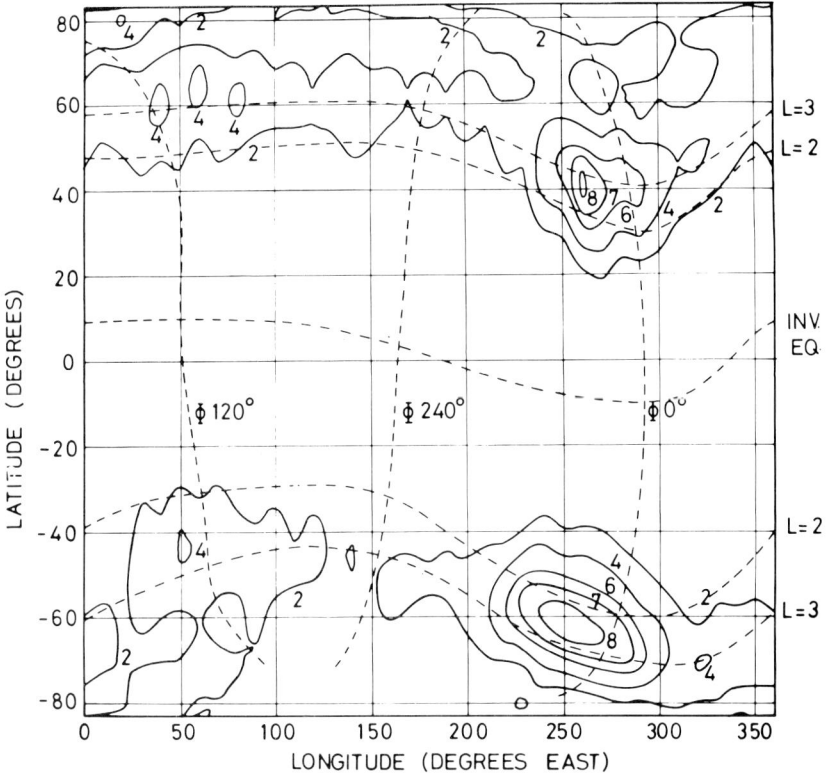

Figure 3 1972 summer
3.2 kHz; 40 dB above 4.8×10^{-19} W m^{-2} Hz^{-1}; Kp $\leq$ 2 +.

the north. Parks (1977) and our own (Matthews et al 1978) studies indicate that the PLHR signals observed in Antarctica (L ~ 4) tend to occur during multi-hop (indicated by whistlers) conditions.

At 0.75 kHz the maxima in occurrence centre around 2 < L < 3 and naturally generated/stimulated emissions ($I \geq 4.8 \times 10^{-15}$ W m^{-2} Hz^{-1}), probably mainly plasmaspheric hiss, dominate with greater than 70 to 80% occurrence almost everywhere in the electron slot (fig. 5). The American zone becomes clearly identifiable at 1.25 kHz and at 9.6 kHz the one-hop signal in the south is significantly stronger and more frequent than the 2-hop in the north. At all frequencies mean power contours exhibit little evidence of longitudinal localisation being dominated, even in so-called quiet periods (Kp $\leq$ 2 +), by the fewer but very strong events of natural origin.

The seasonal dependence in the PLHR phenomenon is particularly

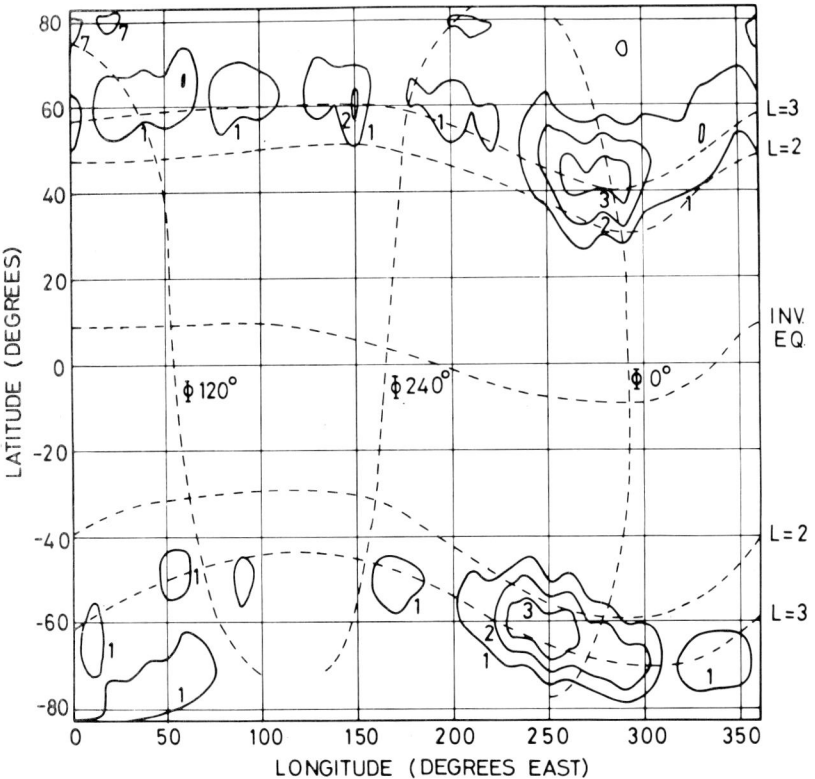

Figure 4   1972 summer
3.2 kHz; 50 dB above 4.8×10^{-19} W m^{-2} Hz^{-1}; $K_p \leqslant 2+$.

striking — it is virtually absent in the northern winter (fig. 6). Plots of signal intensity vs. local time (figs. 7(a), (b)) indicate that even the base-line level is reduced in the winter period. Measurements of power-line harmonics and their seasonal variation (attributed to summer insect activity (La Forest, 1968)) on the ground are in accord with the satellite base-line levels. The latter also show clearly the effect of day-time D-region absorption on the upgoing signal. The local time variations in signal intensity in the south conjugate zone differ for the 1967 (solar max.) and 1972 (solar min.) summer periods (figs. 7(c), (d)). In fig. 7(c) the base-line level reproduces the day-time (D-region) signal reduction seen in the north whereas in fig. 10 the level rises towards evening. An outstanding feature of all these plots (fig. 7) is, except for local midnight in the north (see later), the large spread in signal intensity which is greater than that which we could attribute to spatial variation in the power line distribution. This indicates that PLHR amplification/

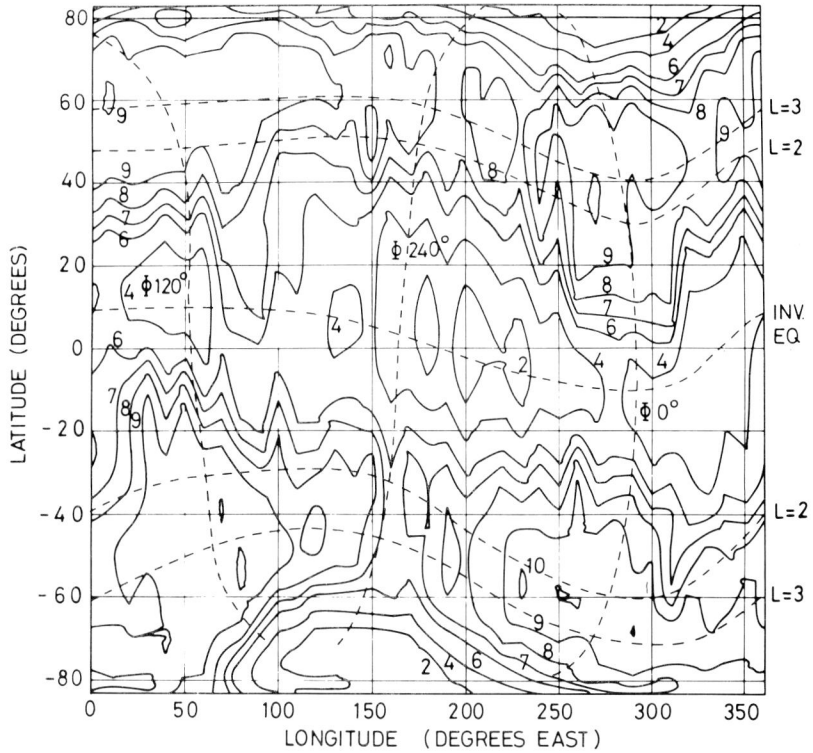

Figure 5   1972 summer
0.75 kHz;   40 dB above 4.8×10^{-19} W m^{-2} Hz^{-1};   Kp ≤ 2 +.

stimulation of emission is almost always present.

It was found that emission intensity correlated most strongly with the D_{ST} index with a lag of 0 – 5^h in the morning and ~10^h in the evening. This is shown in fig. 8 where the maximum correlation coefficient is plotted as a function of local time. The corresponding lag times are indicated on the figure. In both 1967 and 1972 the zero correlation coefficient at local midnight occurred at a time of high base-line level and minimum range in signal intensity. This would indicate a 2-hop amplification/stimulated emission signal smaller than and/or obscured by the upgoing signal in this local time period.

The storm-time delay (few hours) in onset of activity at 2 < L < 3 is short relative to that (few days) required for particles to diffuse to lower L-shells for typical substorm activity and we suggest that the waves are made to interact with energetic electrons, already present in the slot region and mirroring close to the equator, by the

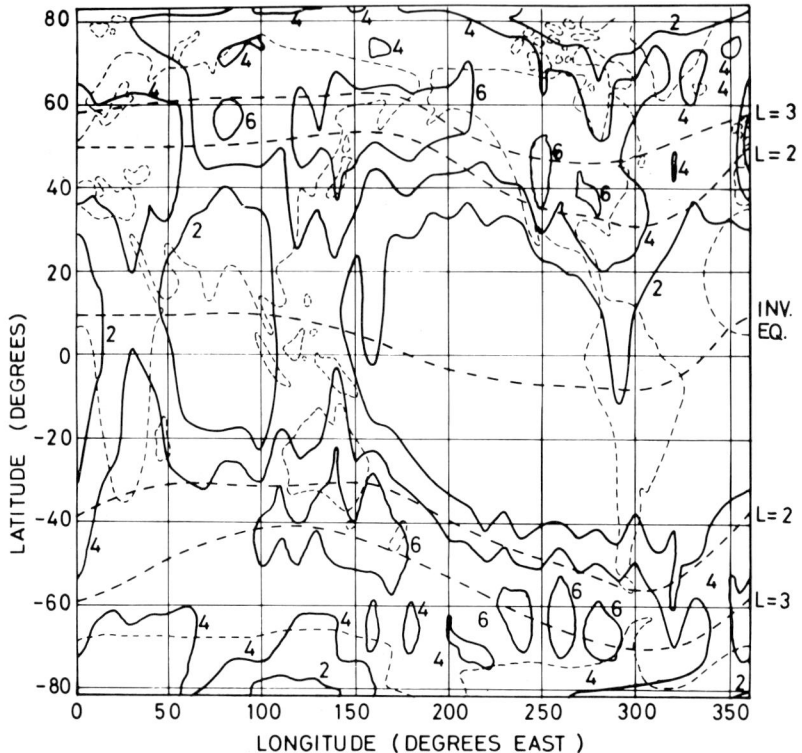

Figure 6   1971/2 winter (days 355 – 79)
3.2 kHz; 30 dB above 4.8×10^{-19} W m^{-2} Hz^{-1}, Kp ≤ 2 +.
The outline of landmasses are shown.

creation of field-aligned ducts by the flux-tube interchange mechanism. There is supporting evidence for the creation of such ducts, at even lower latitudes, from whistler observations (Kotaki et al, 1977). Ground-based studies in Antarctica also indicate that ducting of the coherent, or partially coherent PLHR gives rise to a strong wave-particle interaction where signal amplification takes place through phase-bunching of the energetic electrons.

Our assumption that ducting of the PLHR is important affords a possible explanation of the seasonal variation in activity. This is illustrated in fig. 9 where the wave-normal angle to the local magnetic field is shown as a function of altitude for the northern hemisphere upgoing ray in summer and winter. The upgoing wave-normal stays closely field-aligned and is more likely to be trapped in a duct over a much greater range of altitude in summer. Also, the interaction region is located to the summer hemisphere side of the equator.

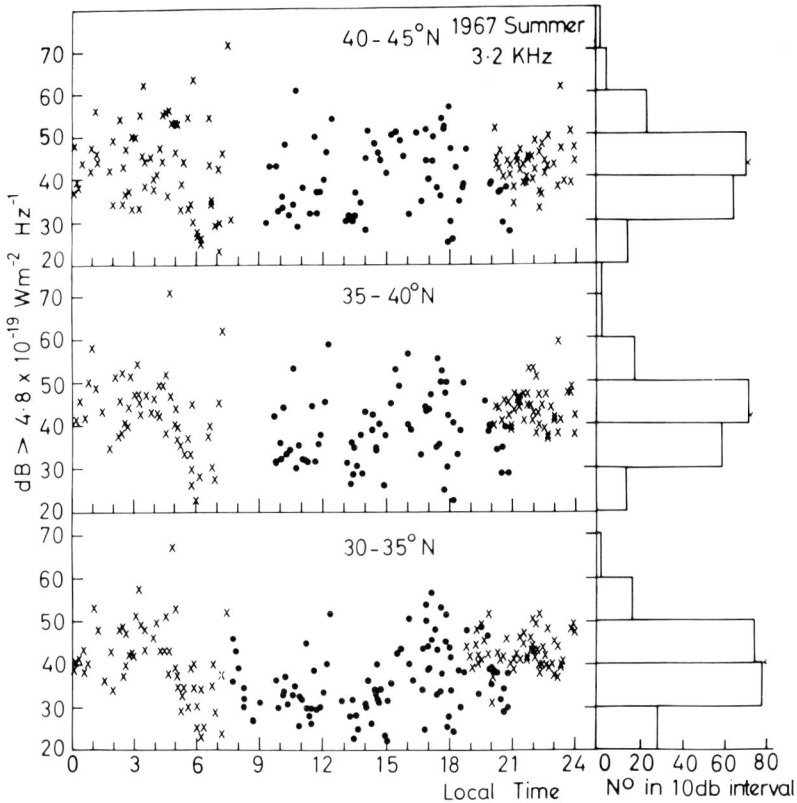

Figure 7   The local time variation in signal intensity (3.2 kHz) in the N. American PLHR zone (240 - 300°E; 30 - 45°N) and its conjugate (220 - 280°E, 45 - 60°Λ).   x N-S pass, ● S-N pass. ⊗, ⊙   signal below receiver noise level.   All Kp.
(a) 1967 summer 30 - 35°N, 35 - 40°N, 40 - 45°N.

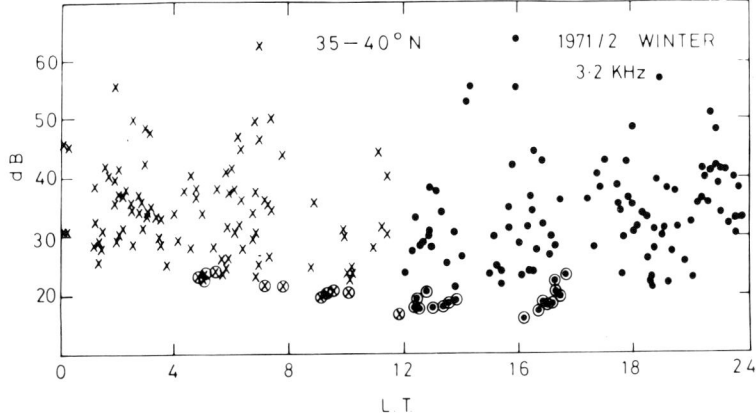

(b) 1971/2 Winter 35 - 40°N.

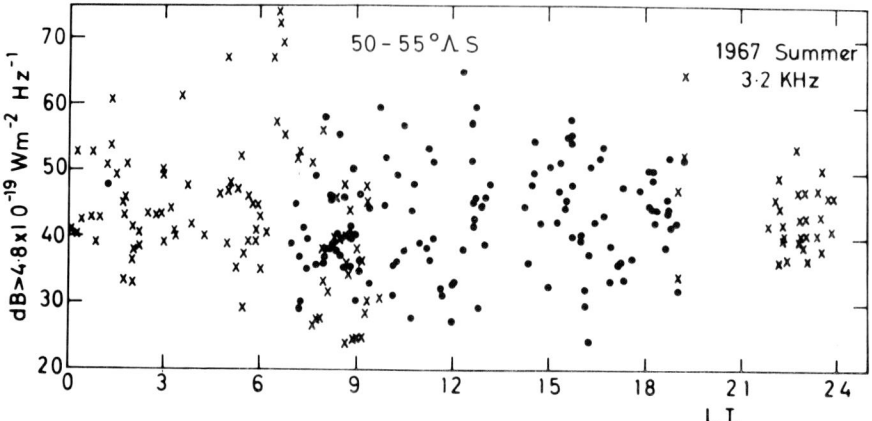

Figure 7(c)   1967 Summer 50 – 55° Λ S.

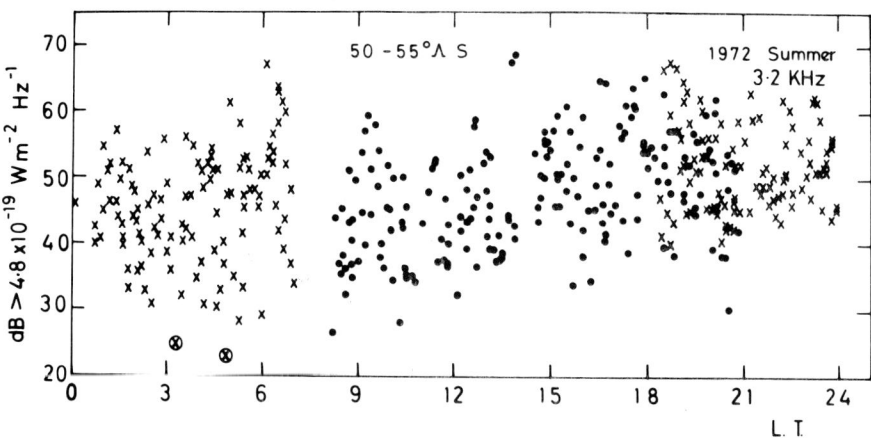

Figure 7(d)   1972 Summer 50 – 55° Λ S.

The importance of PLHR stimulated emission for electron pitch-angle diffusion in the electron slot is shown in fig. 10 where the bounce-averaged pitch-angle diffusion coefficients for ELF hiss and the PLHR stimulated emission are compared (first-order gyroresonance). It can be seen that PLHR emissions have an important contribution to make to diffusion for electrons of large equatorial pitch angles where the ELF hiss is ineffective. There is evidence of an excess of such electrons in the slot region (Lyons, Thorne and Kennel, 1972). The PLHR stimulated emission is relatively more important for lower electron energies, on lower L-shells and at lower electron densities. At $L < 2$ gyroresonance is no longer possible and the lifetimes of the Inner Radiation Belt electrons become very large (Roberts, 1969).

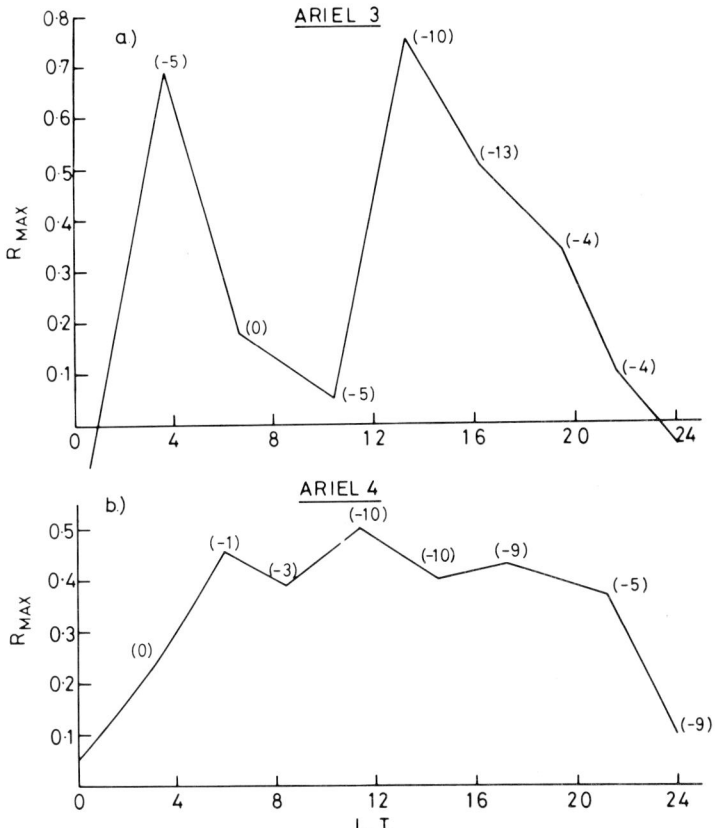

Figure 8 The maximum correlation coefficient, $R_{(max)}$, between D_{ST} and emission intensity (3.2 kHz) plotted as a function of local time. The corresponding time interval (hours): $\Delta t = UT(D_{ST}) - UT(I(3.2\ kHz))$ is shown in brackets ( ).
(a) 1967 Summer   (b) 1972 Summer

Since first preparing this short paper Lyons (1978; private communication) has drawn the authors' attention to the fact that strong thunderstorm activity over North America at latitudes above $40°N$ in the summer has a spatial distribution rather similar to that of our "PLHR zone" of VLF emission; the activity is also very low in winter (Guttman, 1971). The integrated mean signal intensity due to sferics, even in the world's most active thunderstorm area (East Indies, see fig. 1), is well (10 to $\sim$ 20 dB) below our baseline level (fig. 7) over N. America in the summer period. The latter is clearly

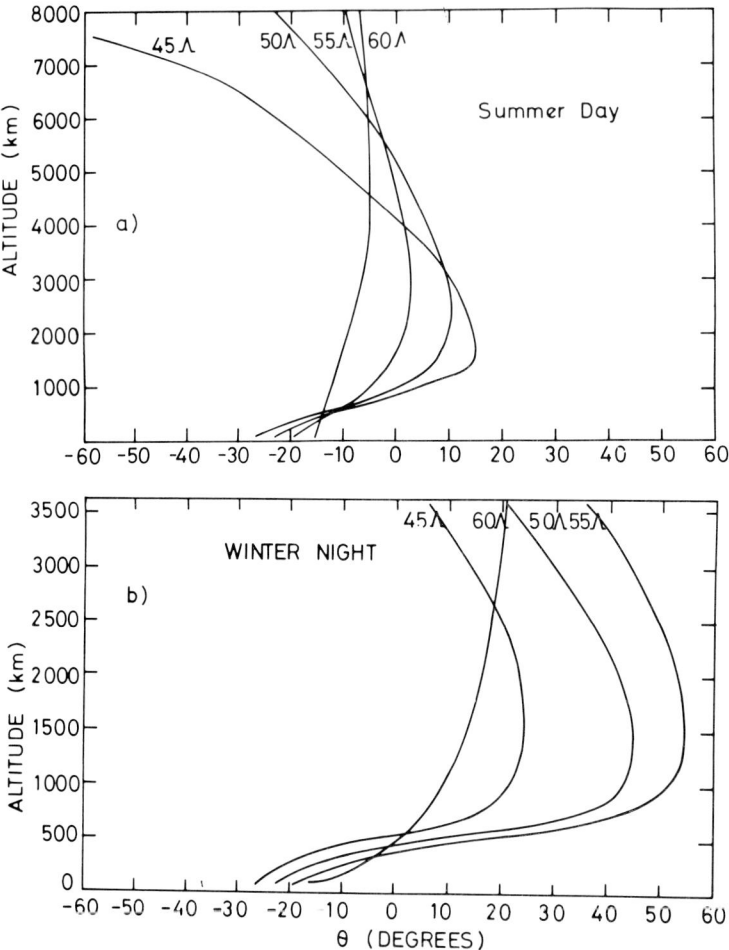

Figure 9 The angle, θ, between the wave-normal and the local magnetic field as a function of altitude for upgoing rays in the northern hemisphere starting at 45, 50, 55 and 60° Λ. The initial wave-normal is assumed vertical at 100 Km.

due to a sustained source of emission for which PLHR is the only possible candidate. However the thunderstorm activity may contribute indirectly to PLHR amplification/stimulated emission events by the creation of ducts by flux-tube interchange (Park and Helliwell, 1971). The extent to which thunderstorms may affect PLHR emission activity is currently being investigated including the inverse possibility that Brehmstrahlung radiation from the greater precipitation events on higher L-shells (L ≳ 3 - 4) may occasionally be sufficiently intense to modify the atmospheric conductivity and hence the atmospheric electric field and thunderstorm activity in these

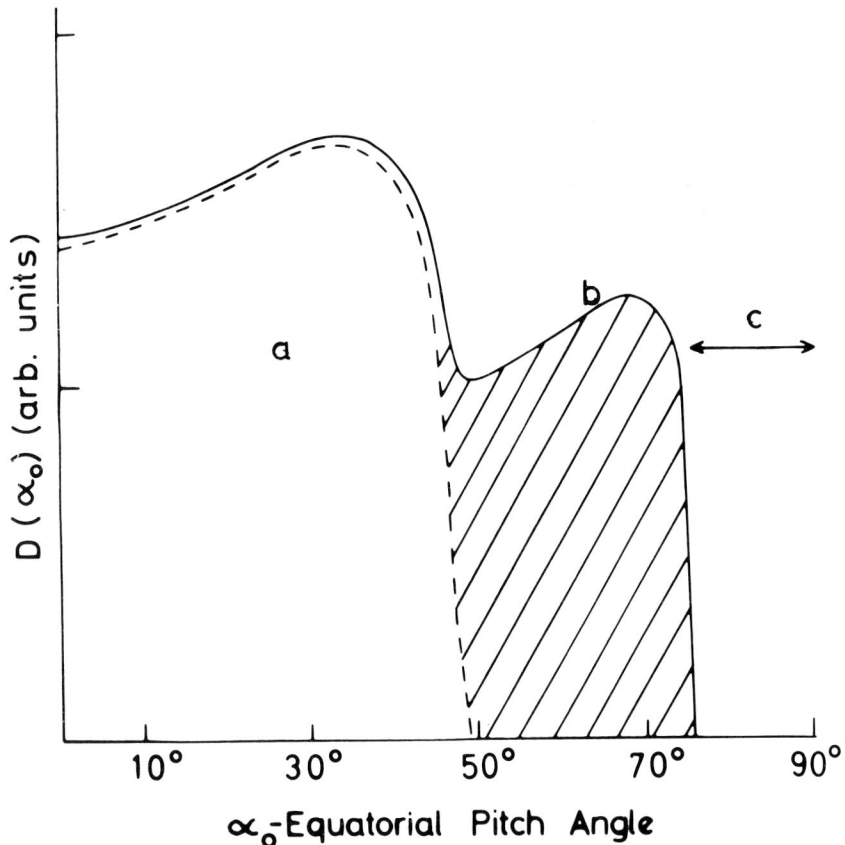

Figure 10   The bounce orbit averaged pitch-angle diffusion coefficient, $D(\alpha_o)$, at $L = 2.6$ of 200 keV electrons as a function of the equatorial pitch angle (α_o). $N_{eq} = 2.0 \times 10^9$ m^{-3}
(a) ELF hiss (first order resonance only; otherwise modelled after LTK (1972)).
(b) PLHR (3 to 5 kHz).
(c) Landau resonance.

regions (Markson, 1978). It is also interesting to note that the South Atlantic Anomaly region of electron precipitation ($L = 1.25$, 45°W, 33°S), in the pre-1925 thunderstorm data for southern winter (Brooks, 1925), is precisely coincident with the only intense (> 20% of days) thunderstorm centre south of 10°S.

REFERENCES

Brooks, C. E. P., 1925, Met. Office Geophys. Mem. London 24.
Bullough, K., Denby, M., Gibbons, W., Hughes, A. R. W., Kaiser, T. R., 1975, Proc. Roy. Soc. A., 343, pp.207 – 226.
Guttman, N. B., 1971, U.S. Naval Weather Service Comman, Navair 50 – 1C – 60.
La Forest, J. J., 1968, IEEE Trans. Power App. Syst. PAS7-87, pp.928 – 931.
Lyons, L. R., Thorne, R. M. and Kennel, C. F., 1972, J. Geophys. Res. 77, pp.3455 – 3474.
Markson, R., 1978, Nature 273, pp.103 – 109.
Matthews, J. P. Smith, A. J. and Smith, I. D., 1978, URSI Conference, Helsinki, 1978.
Park, C. G. and Helliwell, R. A., 1971, Radio Sci., 6, pp.299 – 304.
Roberts, C. S., 1969, Rev. Geophys. 7, pp.305 – 337.
Tatnall, A. R. L., Matthews, J. P., Bullough, K. and Kaiser, T. R., 1978, Scientific Report 1978 No. 1, Sheffield University Space Physics Group.

DOES ELF CHORUS SHOW EVIDENCE OF POWER LINE STIMULATION?

B. T. Tsurutani and E. J. Smith
Jet Propulsion Laboratory, California Institute of
Technology, Pasadena, CA 91103
S. R. Church, R. M. Thorne, R. E. Holzer
University of California, Los Angeles
Los Angeles, CA 90024

Abstract It has previously been reported (Luette et al, 1977) that electromagnetic chorus exhibits a longitudinal dependence, with enhanced occurrence over population centers (Alaska-New Zealand, Eastern U.S.-Canada, Western Europe and Western Siberia). This result has been cited as possible evidence of Power Line Harmonic Radiation (PLR) control of magnetospheric chorus. In this paper we report an analogous study using chorus data from OGO-5 to test this result. Chorus is found to exhibit maxima over the Eastern USSR, Greenland and Central Siberia and minima over central and Eastern Canada, a distribution significantly different than the OGO-3 result. This gross discrepancy is explained as an effect of data oversampling (persistence) in the method of analysis used in the previous study. The OGO-5 data are reanalyzed with the oversampling removed. It is found that none of the longitudinal maxima or minima are then statistically significant ($<2\sigma$). Thus, we find no statistically significant correlation between longitude and chorus occurrence which implies that there is little or no evidence of PLR effects on chorus triggering.

Luette et al. (1977)[1] have recently analyzed the geographic longitudinal distribution of chorus using a broad band (0.3-12.5 hz) magnetic loop on OGO-3. They reported enhanced chorus occurrence in "regions threaded by geomagnetic field lines that intersect industrialized areas". This was interpreted as evidence for chorus triggering by power line harmonic radiation (PLR) leaking from the Earth. As an independent check of this result, we have undertaken an analogous study using ELF (10-1500 hz) chorus detected on OGO-5.

Our analysis used one full year of analog data acquired by the OGO-5 search coil magnetometer[2]. Signals other than chorus, such as plasmaspheric hiss, lightning-generated whistlers and lion roars, were eliminated from the data base. The occurrence or lack of occurrence of chorus was determined for each 10-minute interval when OGO-5 was in the outer magnetosphere and the emission occurrence frequency was calculated

for 10° intervals in invariant dipole longitude. The result of our
analysis, shown in Figure 1, indicates a high degree of longitudinal
variability but there appears to be little correlation with industrialized areas.

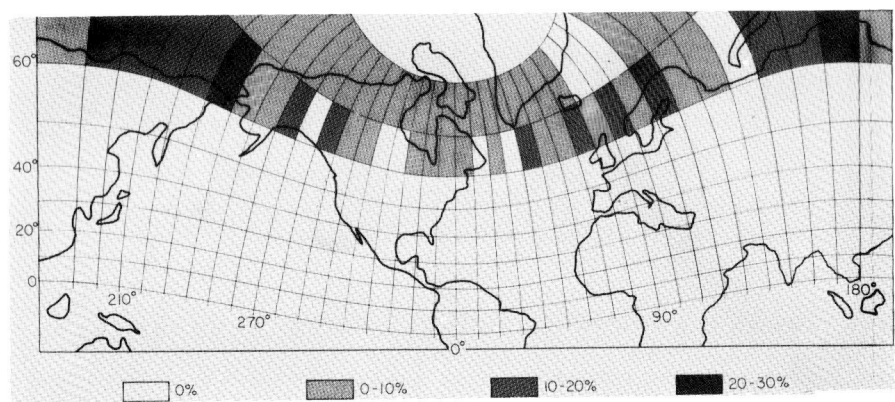

Figure 1 - Chorus occurrence frequency in invariant dipole coordinates.
Both hemispheres of the magnetosphere have been mapped into this northern hemisphere mercator plot. The shading indicates the percent of 10
minute intervals containing chorus. The corresponding legend is given
at the bottom of the figure. No outstanding geographical longitude
dependence of chorus is observed.

A direct comparison between the OGO-3 and OGO-5 analyses is shown
in Figure 2. The top panel is a reproduction of the OGO-3 study showing
the four regions of enhanced chorus occurrence reported by Luette et al
(1977). The OGO-5 chorus distribution is contained in the center panel.
If one assumes that each of our 10 minute samples represents statistically independent data, confidence levels can be assigned to the variability. On this basis three local maxima (mapping into the Eastern
USSR, Greenland and Central Siberia) and two local minima (mapping to
central and Eastern Canada) would meet the 2σ (95%) confidence level
for significant deviation from the mean.

The OGO-3 and OGO-5 distribution, however, show little or no correspondence. This gross discrepancy can best be reconciled if the
persistent nature of chorus is properly taken into account. It is well
documented that chorus is often continuously present for hours (especially during substorms). The occurrence of chorus in any 10 minute
(or 5 minute) interval is not independent of chorus occurrence in
adjacent time intervals, and should therefore not be treated as an
independent event. Methods of analysis which disregard persistence
suffer from oversampling. To overcome this difficulty, the OGO-5 data
were reanalysed taking only one event per 10° of longitude during each

satellite pass (inbound or outbound). If chorus was detected at any
time during the pass through the longitude bin, it was recorded as a
chorus event. The results of our analysis, taking persistence into
account, is shown in the bottom panel of Figure 2. While the overall
distribution remains similar to that shown by the oversampling data
(center panel) the longitudinal maxima are found to be less than 2σ
from the mean. Thus in our OGO-5 study, chorus occurrence is found to
be independent of geographic longitude, to first order. We believe
that Luette et al (1977) would have come to a similar conclusion had
they removed the effects of oversampling from their data.

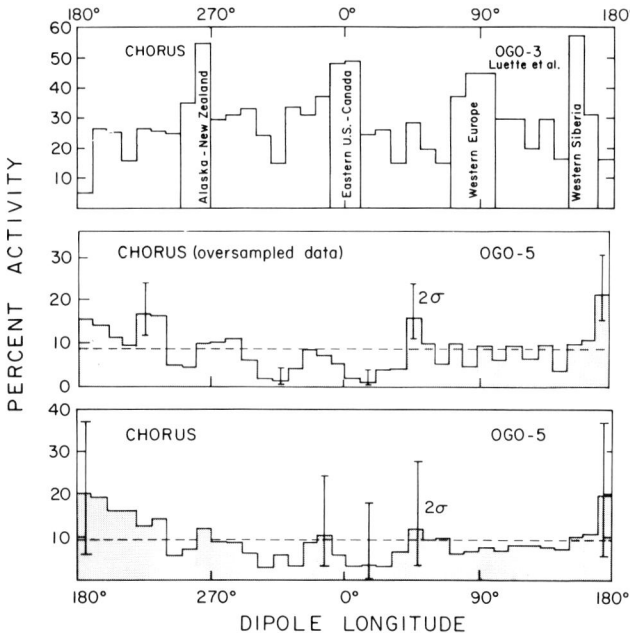

Figure 2 - A comparison of the present study's geographical longitude
dependence of chorus with the previous results of Luette et al (1977).
The percent chorus occurrence is plotted as the ordinate and the
invariant dipole longitude as the abscissa. The Luette et al OGO-3
results are shown at the top panel. The four longitude regions with
enhanced chorus activity are indicated. The center panel contains the
results of the present OGO-5 study, done in a manner identical to the
OGO-3 study. Three local maxima and two local minima are present in
the longitudinal distribution. However, these maxima and minima show
little correspondence with those of the OGO-3 distribution. The gross
discrepancy can be reconciled if the persistent nature of chorus is
taken into account. The chorus has been reanalyzed with data oversampling removed, and the results are displayed in the bottom panel.
None of the longitudinal maxima or minima now meet the two sigma test
for significance. The maxima and minima are then best ascribed to

random statistical fluctuations.

OGO-5 analog data were also examined to determine if chorus starting frequencies were coincident with harmonics of 50 to 60 hz. Care was taken to select events with low ($\leq$ 500 hz) starting frequencies, so that absolute errors in the starting frequency would be minimized. In addition, chorus tones with flat frequency-time profiles were selected for study. No preference was found for chorus elements to start at power line harmonics.

In summary, we were unable to find any evidence for a significant effect of power line radiation of the stimulation of chorus. Previously reported evidence favoring this hypothesis is considered invalid because persistence effects were ignored, leading to an overly optimistic assessment of the statistical significance of peaks in the occurrence frequency.

Acknowledgements

R. A. Helliwell and C. G. Park of Stanford University and J. C. Luhmann of the Aerospace Corporation encouraged us to undertake this study. J. P. Luette very kindly supplied the blank northern hemisphere mercator plot used in Figure 1. J. D. Mannan did the computer programming. This research was supported by NSF Grant ATM 75-01431 A02 and NSF Grant ATM 77-22346. This paper represents results on one aspect of research done at the Jet Propulsion Laboratory, California Institute of Technology, for NASA under contract NAS7-100.

References

(1)
Luette, J. P., Park, C. G., and Helliwell, R. A.: 1977, Geophys. Res. Lett., 4, pp. 275-278.
(2)
Frandsen, A. M. A., Holzer, R. E., and Smith, E. J.: 1969, IEE, Trans. Geosci. Electron., 7, pp. 61-74.

Tsurutani, B. T., and Smith, E. J.: 1977, J. Geophys. Res., 82, pp. 5112-5128.

CHORUS, ENERGETIC ELECTRONS AND MAGNETOSPHERIC SUBSTORMS

B. T. Tsurutani and E. J. Smith
Jet Propulsion Laboratory
California Institute of Technology
Pasadena, CA 91103

H. I. West, Jr. and R. M. Buck
Lawrence Livermore Laboratory, Livermore, CA 94550

ABSTRACT

The origin(s) of whistler mode chorus in the outer region of the terrestrial magnetosphere has been investigated using simultaneous measurements of chorus, energetic (79 ± 23 keV) electron fluxes and pitch angle distributions and ambient magnetic fields obtained with OGO 5. It is found that chorus occurring within 15° of the magnetic equator (equatorial chorus) was detected during magnetospheric substorms and is closely related to enhanced, anisotropic fluxes of energetic electrons. The observations are consistent with wave generation by a loss-cone instability associated with freshly injected 10-100 keV electrons. Chorus observed at higher latitudes appear to have several causes. 1) Nightside emissions are substorm related and are observed when the magnetic field changes from a tail-like to a more dipolar configuration. Possible explanations are an onset of chorus growth due to a change in the electron pitch angle distribution or a decrease in the Landau damping of the waves as they propagate to higher latitudes. 2) Dayside high latitude emissions, which are substorm related, are found to be correlated with high fluxes of energetic electrons. Local, high latitude generation in "minimum B pockets" or equatorial generation and subsequent propagation to higher latitudes by wave ducting can both occur. It is possible to determine the proper mechanism for each individual case if the measurements of the value f/f_c (f_c is the local electron gyrofrequency) and the ambient plasma densities can be made. 3) High latitude chorus also occurs during prolonged geomagnetic quiet but is not well understood. These emissions were not related to features in the electron flux or pitch angle distribution. Solar wind pressure fluctuations and magnetic flux cutting were ruled out as major causes of these emissions. Several possible generation mechanisms are proposed. One possibility is that lower energy electrons (E < 55 keV) are responsible. Another possible generation mechanism requires the existence of boundary layer plasma. This enhanced, thermal plasma could lower the local wave phase velocity, leading to enhanced wave-particle interactions and to chorus growth.

Further study is necessary to test the latter hypothesis.

INTRODUCTION

ELF (10-1500 Hz) whistler-mode chorus occurs principally in two regions, one near the equator, and the other at high latitudes, with a distinct minimum separating the two regions at $\simeq 15°$ magnetic latitude[1]. Equatorial chorus occurs primarily during magnetospheric substorms. High latitude chorus, on the other hand, is not strongly dependent on the AE index and often occurs during intervals of prolonged quiet (AE < 100γ for the previous 12 hours or more). A number of possible mechanisms leading to wave generation in the two regions have been previously suggested (1 and references therein).

It is the purpose of this paper to investigate possible chorus generation mechanisms using simultaneous OGO-5 measurements of chorus, energetic (79 $\pm$ 23 keV) electron fluxes and pitch angle distributions[2], and ambient magnetic fields. Additionally, we have examined solar-wind data during high-latitude dayside chorus events. The data used in this study were obtained during the first year of operation. The highly eccentric orbit had an apogee of 24 R_e and an inclination of $\simeq 30°$ to the geographic equator.

EQUATORIAL CHORUS

Many features of equatorial chorus suggest that a close correspondence with trapped, energetic electrons is to be expected. Figure 1 reveals that the whistler-mode emission is most intense and occurs most frequently in the post-midnight and dawn-to-noon hours at auroral L values. The sharp onset just postmidnight has been ascribed to a rapid eastward gradient and curvature drift of 10-100 keV electrons after substorm injection[1]. At times other than during magnetic storms, this azimuthal drift will dominate radially-inward drifts associated with convective electric fields. The chorus enhancement observed postdawn can be explained by enhanced wave-particle interactions as electrons drift into that local time sector. An increase in the ambient plasma density, caused by the heating of the sunlit ionosphere, leads to a lowering of the wave phase-velocity. This effect, in turn, lowers the energy of electrons which are cyclotron resonant with the chorus. Because the electron energy spectra are typically exponential in the outer magnetosphere[3], the overall effect is an increase in the wave-particle interaction in this local time sector[4]. The radial location of chorus occurrence increases from L = 5-8 postmidnight to L = 7-11 postdawn. This spatial pattern is consistent with generation by electrons which have undergone drift-shell splitting[5].

There are several additional features of chorus which suggest generation by energetic electrons via the loss-cone instability[6]. The onset of chorus is delayed from substorm onsets by an amount that is proportional to the satellite local time. Assuming electron injection near midnight, the delay times are consistent with an azimuthal gradient drift of ~ 25 keV electrons. The occurrence of chorus is a maximum at the equator, |magnetic latitude| < 5°, a region where cyclotron resonance is most efficient.

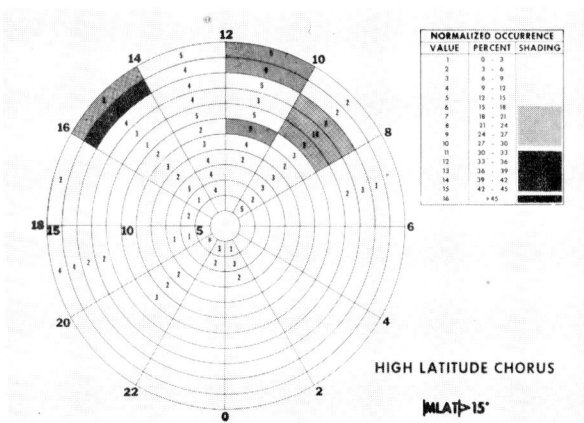

Figure 1 The normalized L-LT distribution for equatorial (|MLAT| ≤ 15°) chorus. The L value is represented by the radius, increasing from L = 4 to L = 15 outward from the center of the figure. Local time is represented by the azimuth angle, with midnight at the bottom and dawn at the right. The number in each area corresponds to the normalized percent occurrence of chorus. A conversion from the numbers to percent occurrence is given in the legend. A sharp onset is observed just postmidnight, from 5 ≤ L ≤ 8, with a maximum occurrence of 54%. A second sharp chorus onset is found postdawn from 7 ≤ L ≤ 11, with a maximum occurrence of 56%. The L value of the region of most frequent chorus occurrence is observed to increase steadily with local time from midnight to noon. No equatorial chorus is detected from 1600 to 2400 LT. The L-LT equatorial chorus distribution is similar to the distribution of precipitating energetic (10-100 keV) electrons.

EQUATORIAL CHORUS AND ENERGETIC ELECTRONS

From an examination of the 28 available events, it was found that equatorial chorus is, in fact, closely related to enhanced, anistropic fluxes of energetic electrons. On the average, chorus intensities are well correlated with the flux and the pitch angle anisotropies of electrons. The maximum chorus intensity is often reached at or near the time when the electron intensity and/or anisotropy reaches a maximum. Such correlations indicate that further quantitative studies[7] of loss cone generation of chorus are likely to be productive and should be pursued.

Especially clear correlations between the fluxes of energetic electrons and chorus occurrence have been observed in the midnight-to-dawn hours. The onset of intense chorus events are accompanied by

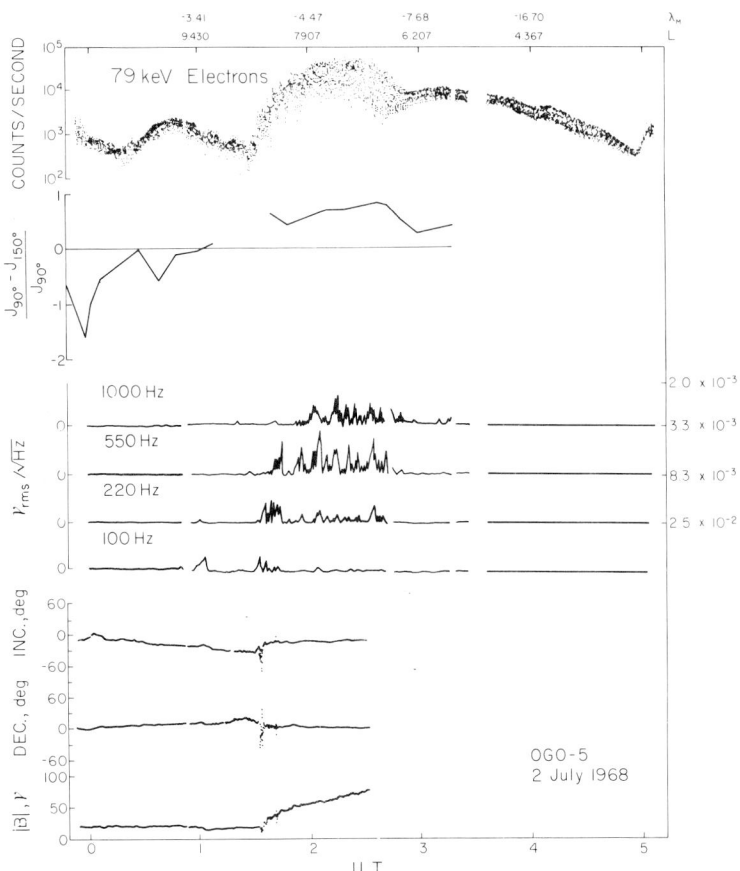

Figure 2 An example of the correlation between drifting clouds of ~ 79 keV electrons and chorus on 2 July 1968. The top two panels are the electron flux (geometric factor = 0.18 cm^2-keV-str) and the pitch angle anisotropy values. The latter quantity is expressed as the difference in flux at 90° from the flux at 150°, normalized to the flux at 90°. The center panels are the simultaneous spectrum analyzer data centered at frequencies of 1000, 550, 220 and 100 Hz. The bottom panels contain the OGO-5 magnetic field data. Ground based magnetograms indicate a substorm onset at 0050 UT. The substorm effects first become apparent at OGO 5 from the increase in electron flux and magnetic field changes at 0130 UT. The substorm electron event lasts from 0130 to ~ 0245 UT. The pitch angle distribution during the event is one of a loss-cone. Chorus is detected throughout this time period. The upward sweep of chorus frequency with time is caused by the inward motion of the satellite, from L = 8.7 to 6.7. Chorus is generated at relatively constant fraction of the local equatorial electron cyclotron frequency, f_{chorus} = ~ 0.25 - 0.30 f_{ge}.

drifting clouds of substorm electrons. The electrons typically exhibit a loss cone pitch angle distribution. Figure 2 contains an example of such a correlation. Ground based magnetograms recorded near midnight indicate a substorm onset at 0050 UT. Forty minutes later, at ~ 0130 UT, a cloud of drifting electrons was detected at OGO 5. The radially extended, azimuthally drifting electron cloud continuously enveloped the spacecraft from 0130 to 0245 UT as OGO 5 travelled radially inward from $L = 8.7$ to 6.7. Throughout this period, the electrons exhibited a loss-cone pitch angle anisotropy. Chorus was continuously observed throughout this period. The sporadic amplitude variability of the chorus is a characteristic feature of the emission. The electron flux profile is typically smooth and exhibits little or no variability.

HIGH LATITUDE CHORUS

The distribution of high-latitude chorus in L and local time is displayed in Figure 3. The emission occurs predominantly on the dayside at $0800 \leq LT \leq 1600$ and often at large L, within 1-2 R_e of the magnetopause. The emission is also detected less frequently at lower L values and essentially at all other locations. Some events occur during substorm periods and others are detected during extreme geomagnetic quiet. Because of the great variability in location and geomagnetic conditions of high latitude chorus, it is quite possible that several different generation mechanisms are involved.

HIGH LATITUDE CHORUS AND ENERGETIC ELECTRONS

We have examined 27 events for which there are simultaneous energetic particle and field data. It is found that by separating high-latitude chorus into different types based on local time and substorm dependence, the different relationships between chorus and particles and fields become more apparent. The three types which were empirically distinguished are: 1) high-latitude chorus that occurs on the nightside 2) dayside high-latitude chorus that occurs during intense AE (substorm periods) and 3) dayside high latitude chorus that occurs during intervals of prolonged geomagnetic quiet (AE < 100γ for 12 hrs or more). For category 3) events, hourly averages of the interplanetary solar wind velocity, density, magnetic field and velocity, density and field variances were also examined.

Although only a few nightside high-latitude chorus events were available for analysis, they all displayed similar features. The events occurred during magnetospheric substorms. Chorus was detected only after the magnetic field topology changed from being tail-like to dipolar. One possible explanation is that the dipolar field geometry allows the waves to propagate more parallel to the magnetic field lines as they travel away from the equatorial generation region. Chorus would thus experience less Landau damping and would be observable at higher latitudes. A second possibility is that the change in field geometry is accompanied by a change in the pitch angle distribution of the

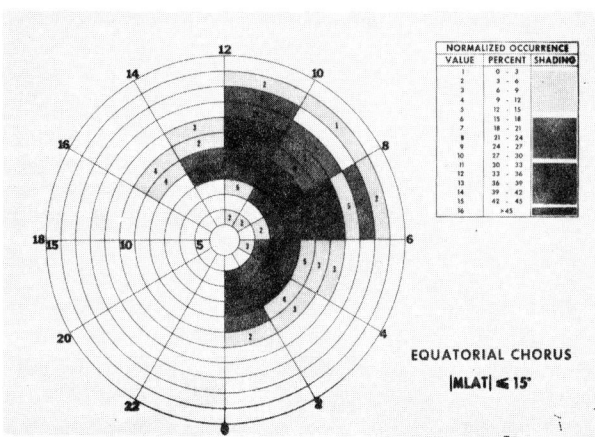

Figure 3 The normalized L-LT distribution for high-latitude ($|MLAT|\geq 15°$) chorus. The format is the same in Figure 1. High-latitude chorus is detected at all local times and all L values from 4 to 15. However, it occurs predominantly on the dayside with a distribution roughly symmetric about noon. There is a tendency for the emissions to occur close to the magnetopause within 1-2 R_e of the boundary.

electrons, which in turn causes chorus growth[8]. Unfortunately, the electron pitch angle information was not adequate to test this hypothesis for these events.

The dayside high latitude chorus which occurred during substorms were generally found to be associated with high fluxes of energetic electrons. One or both of two different physical processes could be occurring. The emission could be locally generated in a high-latitude weak field region, called a "minimum B pocket"[9], or it could be generated near the equator on dipole-like field lines with subsequent propagation to higher latitudes by wave ducting[10]. The proper mechanism for each individual case can be determined by calculation of the value f/f_c (the wave frequency relative to the local electron gyrofrequency) and a simultaneous examination of the ambient plasma densities[1].

For high-latitude chorus occurring during prolonged geomagnetic quiet, a number of possible causes have been excluded: (1) Solar wind pressure functions do not appear to be the cause. The pressure and pressure fluctuations are generally lower than normal during these chorus events. (2) Magnetic flux cutting[11] is also apparently ruled out. The interplanetary magnetic field B_z value was typically more positive during the chorus events than at other times. (3) The possibility that the chorus is generated by enhanced fluxes of substorm electrons can be excluded because substorm electron events were not detected during these periods. Thus, high latitude substorms, centered above the auroral zone, which are not detected by the AE index, are not influential in causing these chorus events. Furthermore, the electron pitch angle distributions did not consistently exhibit any feature that would be associated with wave generation by the loss cone instability.

Other possible quiet-time high-latitude chorus generation mechansims involve lower energy plasma. The electrons which are responsible for the generation of chorus may have energies below the OGO-5 electron detector energy threshold, 55 keV. Also, the existence of enhanced thermal plasma, such as boundary layer plasma[12] just inside the magnetopause, can affect a marginally stable plasma by increasing the wave-particle interaction via a mechanism briefly discussed previously. It is quite possible that either one of the above or both phenomena contribute to chorus generation during quiet periods.

ACKNOWLEDGEMENTS

R. E. Holzer is co-investigator of the OGO 5 search coil experiment. Discussions with him and A. M. A. Frandsen are greatly appreciated. C. T. Russell kindly provided the fluxgate magnetometer data. The work at Lawrence Livermore Laboratory was done under the auspices of the Department of Energy under contract W-7405-ENG-48. This paper presents results of one aspect of research done at the Jet Propulsion Laboratory, California Institute of Technology, for NASA under contract NAS7-100.

REFERENCES

1. Tsurutani, B. T. and Smith, E. J.: 1977, J. Geophys. Res., 82, pp. 5112-5128.

   Likhter, Y. A.: 1979, Astrophy. Spa. Phys. Lib., this issue.

2. West, H. I., Jr,, Buck, R. M., and Walton, J. R.: 1973, J. Geophys. Res., 78, pp. 1064-1081.

3. Frank, L. A., Saflekos, N. A. and Ackerson, K. L.: 1976, J. Geophys. Res., 81, pp 155-167.

4. Brice, N. and Lucas, C.: 1971, J. Geophys. Res., 76, pp. 900-908.

   Jentsch, V.: 1976, J. Geophys. Res., 81, pp. 135-146.

5. Stone, E. C.: 1963, J. Geophys. Res., 68, pp. 4156-4166.

6. Kennel, C. F. and Petschek, H. E.: 1966, J. Geophys. Res., 71, pp. 1-28.

   Curtis, S. A.: 1978, J. Geophys. Res., 83, pp. 3841-3848.

7. Anderson, R. R. and Maeda, K.: 1977, J. Geophys. Res., 82, pp. 135-146.

8. Scarf, F. L., Fredricks, R. W., Kennel, C. F. and Coroniti, F. V.: 1973, J. Geophys. Res., 78, pp. 3119-3130.

    West, H. I., Jr., Buck, R. M. and Walton, J. R.: 1973, J. Geophys. Res., 78, pp. 3093-3102.

9. Roederer, J. G.: 1970, "Dynamics of Geomagnetically Trapped Radiation", Springer, New York.

10. Helliwell, R. A.: 1965, "Whistlers and Related Ionospheric Phenomena", Stanford University Press, Stanford, Calif.

    Burton, R. K. and Holzer, R. E.: 1974, J. Geophys. Res., 79, pp. 1014-1023.

11. Dungey, J. W.: 1961, Phys. Rev. Lett., 6, pp. 47-48.

    Tsurutani, B. T. and Meng, C. I.: 1972, J. Geophys. Res., 77, pp. 2964-2970.

12. Haerendel, G., Paschmann, G., Sckopke, N., Rosenbauer, H. and Hedgecock, P. C.: 1978, J. Geophys. Res., 83, pp. 3195-3216.

PART II

PLASMA TURBULENCE

MAGNETOSPHERIC MULTIHARMONIC INSTABILITIES

M. Ashour-Abdalla[1,2], C. F. Kennel[1,2,3], and D. D. Sentman[1]
[1]Institute of Geophysics and Planetary Physics
[2]Department of Physics
[3]Center for Plasma Physics and Fusion Engineering
University of California, Los Angeles

In this review, we discuss the status of our calculations of linear convective growth rates of instabilities of electrostatic multiple electron cyclotron harmonic waves in a plasma consisting of a hot electron component with a loss-cone type of free energy source and a cold electron component of presumably ionospheric origin. When T_C/T_H, the ratio of cold to hot electron temperature is small, the cold upper hybrid frequency controls the harmonic bands that can be non-convectively unstable. The band containing the cold upper hybrid frequency and those below it can be non-convectively unstable, however convective instability is possible in bands whose frequencies exceed the cold upper hybrid frequency. When T_C/T_H increases above a few times 10^{-2} non-convective instability disappears more or less simultaneously for each harmonic band, when the density ratio N_C/N_H is less than unity. A consistent interpretation of the spatial localization and harmonic frequency bandwidths of the observed waves can be made assuming linear convective saturation, provided that the cold electrons have temperatures considerably in excess of those in the ionosphere. Nonlinear saturation and cold electron heating mechanisms are briefly discussed.

I. INTRODUCTION

OGO-5 had the first broadband electric field detector to operate successfully beyond the plasmapause. The OGO-5 electric field experiment detected a number of essentially longitudinal electrostatic plasma waves with frequencies ω exceeding the local electron cyclotron frequency (Kennel *et al.*, 1970). The dominant class had frequencies between multiples of the electron cyclotron frequency and have come to be known as odd-half harmonic emissions since ω was near $(n + \frac{1}{2})\Omega$, though nothing in the data or subsequent theories suggests that ω should be exactly $(n + \frac{1}{2})\Omega$. Of the odd half-harmonic emissions, the 3/2 waves were the dominant class, although examples of multiple harmonics were given. Due to a relatively poor threshold sensitivity, OGO-5 detected waves with integrated amplitudes typically exceeding 1mv/m and these predominantly near the geomagnetic equator on the earth's morning side.

Waves at higher magnetic latitudes were occasionally observed (Fredricks and Scarf, 1973). More recently, multiple half-harmonic emissions have been detected on IMP-6 (Shaw and Gurnett, 1973), on S^3 (Anderson and Maeda, 1977), GEOS (R. Gendrin, private communication, 1978) and ISEE (Gurnett et al., 1978). The threshold sensitivities on these more recent experiments were very signifantly improved, so that now, more multiple harmonics are more frequently observed over a wider range of magnetic latitudes than on OGO-5. A question still remains, however, whether their amplitudes tend to peak near the geomagnetic equator, as OGO-5 found.

In figure 1, we show a frequency-time spectrogram of results from the ISEE broadband electric field detector (Gurnett et al., 1978). Many modes other than half-harmonic emissions are detected, which enable us to interpret this display. Of particular interest is the diffuse electromagnetic noise called nonthermal continum radiation (Gurnett, 1975) which is trapped between the plasmapause (at 1230 UT) and the magnetosheath (at 1700 UT) where its frequency equals the local plasma or upper-hybrid frequency. The lower frequency limits of this noise enables us to follow the behavior of the local plasma or upper hybrid frequency. Just below the lower cutoff of the continum radiation, one sees a collection of relatively narrow band electric field emissions, which are approximately near $(n + \frac{1}{2})\Omega$. These are weak, or $\sim 1\mu V/m$, within the magnetosphere, and suddenly intensify closer to the magnetopause at about 1900 UT. Here their amplitudes are more nearly .1 mV/m (D. Gurnett and W. Kurth, private communication). This intensification of the electrostatic odd half-harmonic noise corresponds to an intensification of the fluxes of few keV electrons (L. A. Frank, 1978, private communication). Other waves, such as whistlers near $\frac{1}{2}\Omega$ are easily discernible on this plot. For our purposes here, it is sufficient to note that multiple odd-harmonic emissions are nearly always present between the plasmapause and magnetopause, when the detector has a low threshold sensitivity as on ISEE. Moreover, the amplitude rose to values where nonlinear effects could be significant near the magnetopause on this pass.

The facts that odd half-harmonic waves interact with the main part of the electron velocity distribution and not a high energy tail, that that they were often present near the equator on auroral field lines, and that they intensified during substorms led Kennel et al. (1970) to suggest that these waves were responsible for the diffuse auroral electron precipitation, a suggestion supported by Lyons' (1974) direct calculation of the electron precipitation rate from an observed plasma wave spectrum, and other circumstantial evidence (see Ashour-Abdalla and Kennel, 1978b). Thus there has long been substantial theoretical interest in the linear and nonlinear properties of electrostatic odd half-harmonic emissions.

Fredricks (1971) showed that a "loss cone" distribution of energetic electrons could be unstable to odd half-harmonic wave growth. In a land-mark paper, Young et al. (1973) showed that instability was

MULTIHARMONIC INSTABILITIES 67

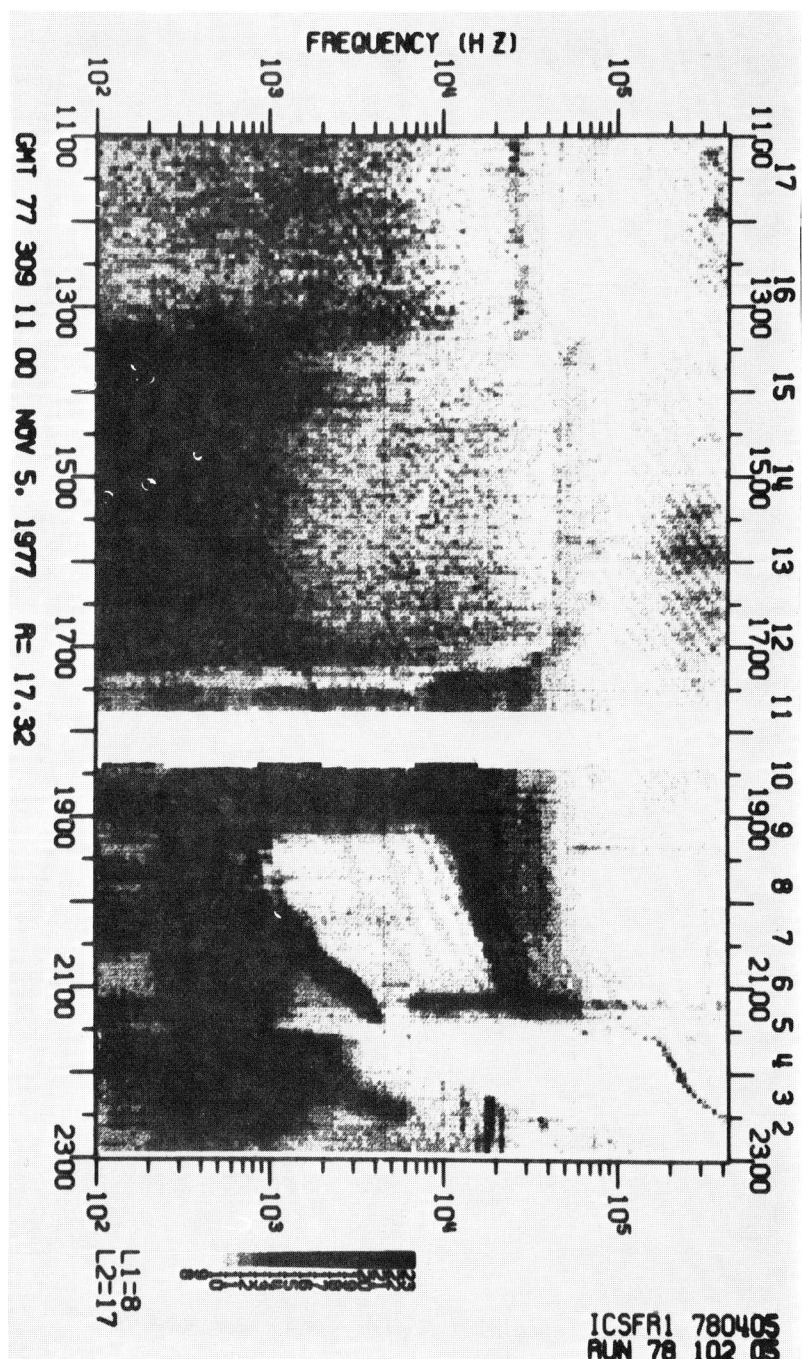

Figure 1. Frequency-time spectrogram from ISEE. Odd half-harmonic emissions are observed between the plasmapause (at 2130 UT) and the magnetopause at 1700 UT). (From Gurnett et al., 1978.)

possible for the gentle loss cone electron distributions expected in the magnetosphere at low harmonics only if cold electrons were present. They thus defined the basic model used by all subsequent investigators – a mix of a cold plasma, of density N_c and T_c, of presumable ionospheric-origin, and a hot plasma of density N_H, effective temperature T_H, and a loss cone source of free energy for wave growth. Using this model, Karpman et al. (1973, 1975) showed that the upper hybrid frequency based upon the cold electron density alone controls the frequency band of strong growth: the unstable waves lie below the cold-upper hybrid frequency. However, Hubbard and Birmingham (1978) have pointed out that this conclusion is true only for a lower frequency branch of the dispersion relation. In a study of the 3/2 band alone, Ashour-Abdalla and Kennel (1976, 1978a) showed that the cold electron temperature T_c controls the convective nature of the instability. If the ionospheric electron energy distribution is not modified, the 3/2 waves would be strongly non-convective and <u>must</u> saturate by nonlinear processes. If T_c/T_H exceeds $1-5 \times 10^{-2}$, the waves turn convective and could saturate by simple propagation out of a locally unstable region.

In this paper we report an extension of our linear instability calculations to the multiharmonic case, both because it provides a timely addition to discussions of the newest observational data in which multi-harmonic emissions are often found, and also because it provides necessary information on the possible models of saturation. In particular, we must know whether the instability is convective or non-convective when it is typically observed. In the final portions of this paper, we discuss in general terms the nonlinear saturation of the waves and cold electron heating.

II. LINEAR INSTABILITY CALCULATIONS

For all our calculations, we have chosen the distribution function model of Ashour-Abdalla and Kennel (1975, 1976, 1978a). In figure 2, we plot an example of such a distribution function, which is strongly unstable, in a three dimensional display. The parameters chosen refer to those in Ashour-Abdalla and Kennel (1978a). One sees a strong peak in the distribution due to cold electrons and a small second peak in the perpendicular velocity distribution of hot electrons which is the free energy source for wave growth. Figure 3 shows a plot of the real frequency normalized to Ω vs. perpendicular and parallel wavenumber, $K_\perp \rho_H$ and $K_\parallel \rho_H$ respectively, normalised to the hot electron characteristic Larmor radius (top inset). The bottom inset shows the spatial growth rate $k_i \rho_H$ normalized to the hot electron Larmor radius, as a function of $k_\perp \rho_H$ and $k_\parallel \rho_H$. We note that ω is a weak function of $k_\perp \rho_H$ and $k_\parallel \rho_H$ in this unstable range. Therefore the group velocity is everywhere small and spatial growth rates can be large even for weak loss cone distributions such as that in figure 2. This is a general property of both ion and electron cyclotron harmonic waves, whatever their source of free energy, and especially when cold electrons are present. In fact, there are two saddle points in the ω vs. $(k_\perp, k_\parallel)$ plot where the group

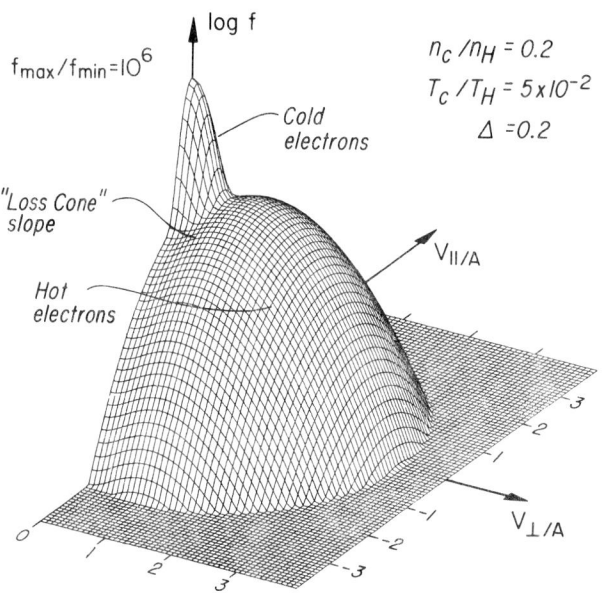

Figure 2. Model electron distribution function

Odd half-harmonic emissions are observed between the plasmapause (at 1230 UT) and the magnetopause (at 1700 UT). From Gurnett *et al.* (1978).

velocity goes to zero. Here one expects non-convective instability; the corresponding strong peaks in $k_i \rho_H$ exists in the lower inset.

In figures 2 and 3 we presented an example of one instability calculation. By itself, it is not very illuminating, because the hot electron distribution can certainly vary, because the cold electron density and tempurature are extremely difficult to measure in space and so our experimental knowledge of them is limited, and because electrons coming from the ionosphere can have their distribution function modified either as they move from ionosphere to the equator, or by the presence of the electrostatic waves themselves. In our opinion, meaningful instability calculations should conform to the following ground rules:
1) Spatial and not temporal growth rates should be computed for reasons cited above.
2) It should be determined whether the instability is convective or non-convective.
3) For a given set of plasma parameters, one should search over the entire unstable range of $k_\perp \rho_H$ and $k_\parallel \rho_H$ to find either the maximum convective growth rate or the point or points of non-convective instability.

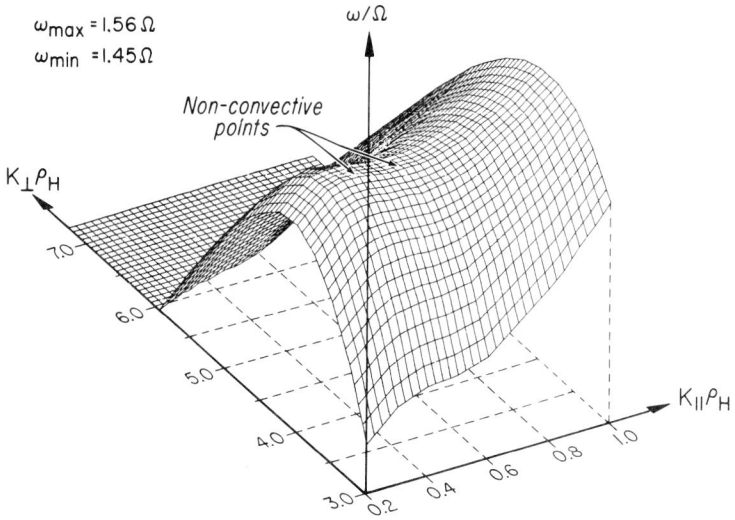

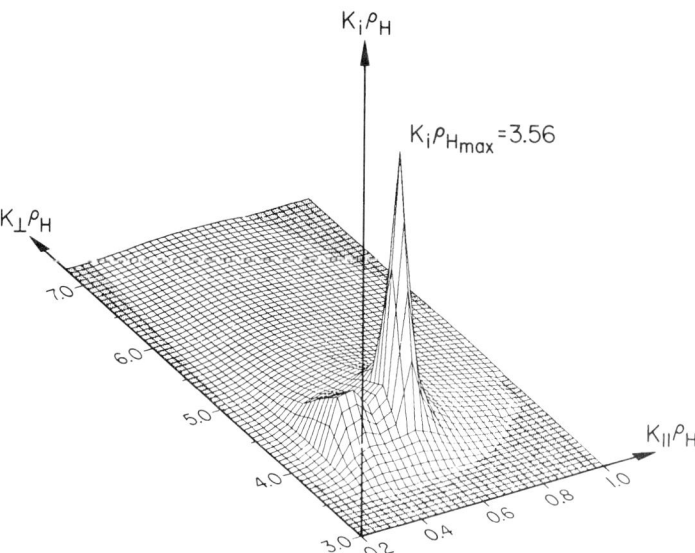

Figure 3. Dispersion relation and corresponding spatial growth rates.

The upper panel shows the dispersion relation ω vs. $k_\perp$, $k_\parallel$. The frequency is normalized to the electron gyrofrequency and $k_\perp$ and $k_\parallel$ are normalized to the hot electron Larmor radius. The frequency does not vary much over the unstable range ($1.45 < \omega/\Omega < 1.56$). There are two saddle points. In the lower inset we plot the spatial growth rates, and find two peaks in $k_i \rho_H$ corresponding to the two saddle points. One peak is dominant because there the temporal growth rate is much larger.

4) It is not sufficient to consider only one set of N_C/N_H and T_C/T_H. One must calculate the range of these parameters for which significant spatial growth is possible, using the maximum growth rates.

Thus, it is necessary to do many calculations such as that shown in figure 3 to gain a feeling for the behavior of odd half-harmonic waves in the magnetosphere.

If we are interested in explaining multiharmonic emissions, we must ensure that our calculations can answer the following questions:
a) Can several harmonics be simultaneously non-convective?
b) If they are simultaneously convective, under what conditions can they have comparable spatial growth rates?

If one harmonic is non-convective and its neighbors convective, or if one has a predominant growth rate, it would be the one most likely to be observed. Furthermore, since all harmonics interact resonantly with the same electrons, they will quasilinearly stabilize as an intercoupled system with either the last band to go convective or the one with the largest spatial growth pacing the stabilization of all the bands. Thus, if multiharmonic emissions are commonly observed, they should have similar growth rates in similar regions of parameter space.

In figure 4, we survey the dependence upon N_C/N_H and T_C/T_H of the instability properties of the first two harmonic bands. In figure 4a we show the results for the first harmonic and in figure 4b we show results for the second harmonic. For each pair in N_C/N_H and T_C/T_H space we first performed calculations for all wavenumbers for which there is an instability and then searched either for the nonconvective region or the wave with the largest spatial growth rates. The shaded region shows the region of non-convective instability. We show the lines of constant frequency by dotted lines, and the solid lines show lines if constant spatial growth rates. Although we call this the frequency of the instability we do not wish to imply that the instability necessarily will be narrowband. This is the frequency of peak growth rate. Looking at figure 4, we find a broad region of NCI extending below $T_C/T_H \sim 7 \times 10^{-2}$ for the first harmonic, and $T_C/T_H \sim 8.8 \times 10^{-2}$ for the second harmonic. At $T_C/T_H = 3 \times 10^{-2}$ NCI extends between N_C/N_H = 0.1 and 1.6 for the first harmonic, and between $N_C/N_H \sim 0.1$ and between 2.0 for the second harmonic. In other words, the range of N_C/N_H for which NCI occurs increases with decreasing T_C/T_H. This conforms generally with our previous expectations; that decreasing the cold electron temperature T_C makes the mode more non-convective (Ashour-Abdalla and Kennel, 1978a).

In figure 4 we see that the regions of NCI are virtually the same for both the first and second harmonic bands. This probably reflects the fact that the cold electrons control the parallel group velocity near the $\partial\omega/\partial k_\perp = 0$ line, and that the parallel group velocity depends mostly upon the cold electron thermal speed. In the region of strongest

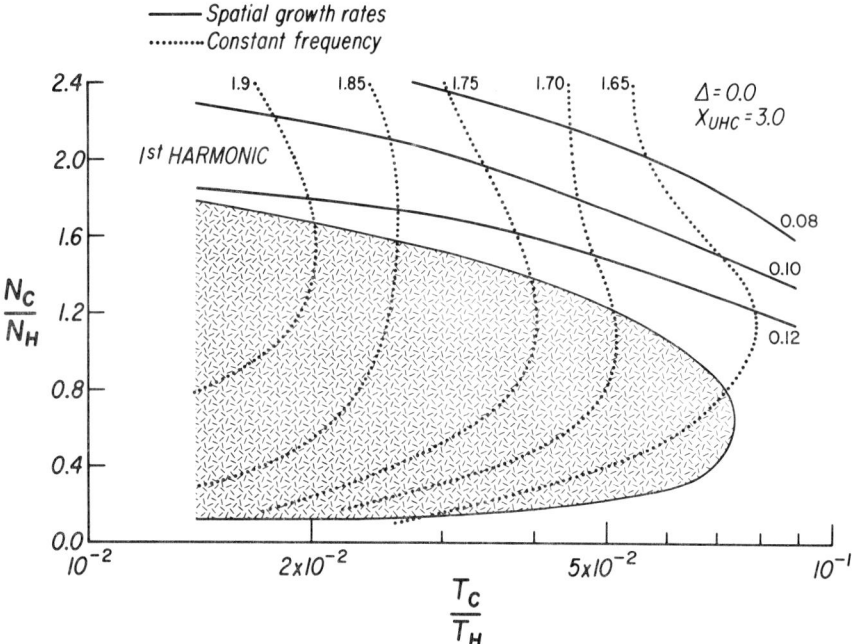

Figure 4a. Regions of non-convective and convective instabilities in T_C/T_H and N_C/N_H space (1st Harmonic).

In this figure we plot contours of maximum spatial growth rates for the first harmonic normalized to the hot electron Larmor radius (solid lines) or the non-convective region (shaded region) in the T_C/T_H, N_C/N_H parameter space, where T_C/T_H is the ratio of cold to hot electron temperature and N_C/N_H is the ratio of cold to hot electron density. The value of the cold upper hybrid frequency $X_{UHC} = 3.0$. Dotted lines denote contours of the frequency of the most rapidly growing - or non-convective - waves normalized to the electron cyclotron frequency.

NCI, where $N_C/N_H \sim 0.8$, the two modes become convective simultaneously. For $N_C/N_H > 0.8$, there is a difference in the T_C/T_H at the convective non-convective transition. In any case, by the time $T_C/T_H = 0.09$, all non-convective behavior has disappeared. We also note that as T_C/T_H increases, the frequencies of the instability which originally exceeded the half-harmonic frequencies do approach $(n + \frac{1}{2})\Omega$, where Ω is the electron cyclotron frequency. Thus, observations of waves near $\omega \sim (n + \frac{1}{2})\Omega$ may be prima facie evidence that T_C/T_H is not small, and the waves are turning convective.

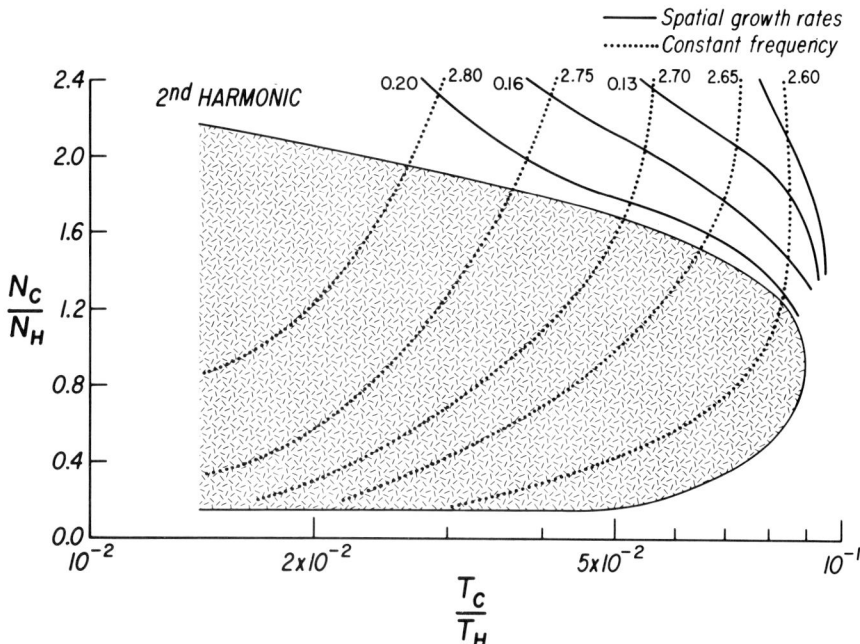

Figure 4b. Regions of non-convective and convective instabilties in T_c/T_H and N_c/N_H space (2nd Harmonic).

Same as for 4a but for the second harmonic.

In figure 5, we investigate the dependence of the first three harmonic instabilities upon the cold electron density parameterized by the cold upper hybrid frequency normalized to the electron cyclotron frequency, X_{UHC}, and upon N_c/N_H. We chose T_c/T_H to be 5×10^{-2} in all three cases, both in order to keep the volume of parameter space for which non-convective instability (NCI) occurs reasonably small, and because of our theoretical expectation that T_c/T_H may not be small when the waves are observed. From figure 5 we conclude that if $N_c/N_H \sim 0.5$, simultaneous NCI for all three bands is possible for a wide range of X_{UHC}, provided the cold upper hybrid frequency exceeds the wave frequency. When X_{UHC} is in a given harmonic band, that harmonic can be non-convective for a wide range of N_c/N_H. Otherwise, NCI is limited to $N_c/N_H < 1$. For $N_c/N_H > 1.2$ and $X_{UHC} \sim 3$, the second harmonic band can be separately non-convective. For $N_c/N_H > 1.6$ and $X_{UHC} \simeq 4$ the third harmonic band can be separately non-convective.

In general the contours of constant spatial growth rate in the convective regime follow the boundary of non-convective instability. Since the volume of parameter space for which the first harmonic is NCI, and the volume for which the first harmonic mode has significant

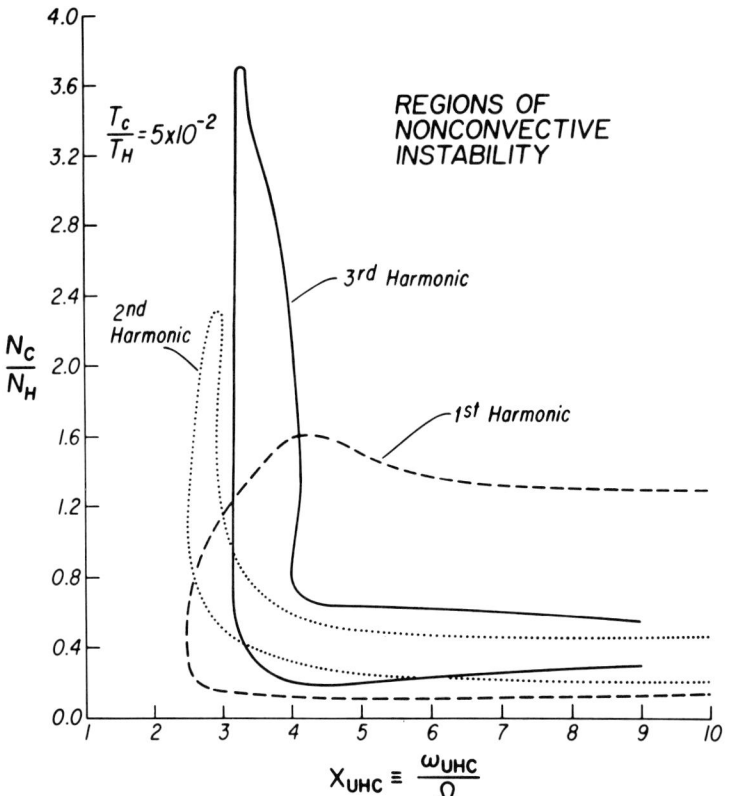

Figure 5. Regions of nonconvective instability.

In this figure, we superpose the boundaries of the regions of nonconvective instability for the three harmonics. Since the contours of constant convective spatial growth rate tend to follow these boundaries, these curves give a visual impression of when growth in each harmonic is significant. Nonconvective growth in the 2nd or 3rd harmonic is significant. Nonconvective growth in the 2nd or 3rd harmonics alone can occur if $N_c/N_H > 1.6$ and X_{UHC} lies in or near the appropriate frequency band. A simultaneous nonconvective instability is possible in all three bands if X_{UHC} is sufficiently large, and N_c/N_H sufficiently small. The much larger volume of NCI for the first harmonic suggests that the 1st harmonic should be the most often observed, and should normally, except in the special circumstances detailed above, accompany the higher harmonics.

spatial growth rates, exceeds those for the higher harmonics, the first harmonic should be the most commonly observed, and the higher harmonics should usually be accompanied by the first. Only in the rather special circumstances outlined above would this linear stability analysis

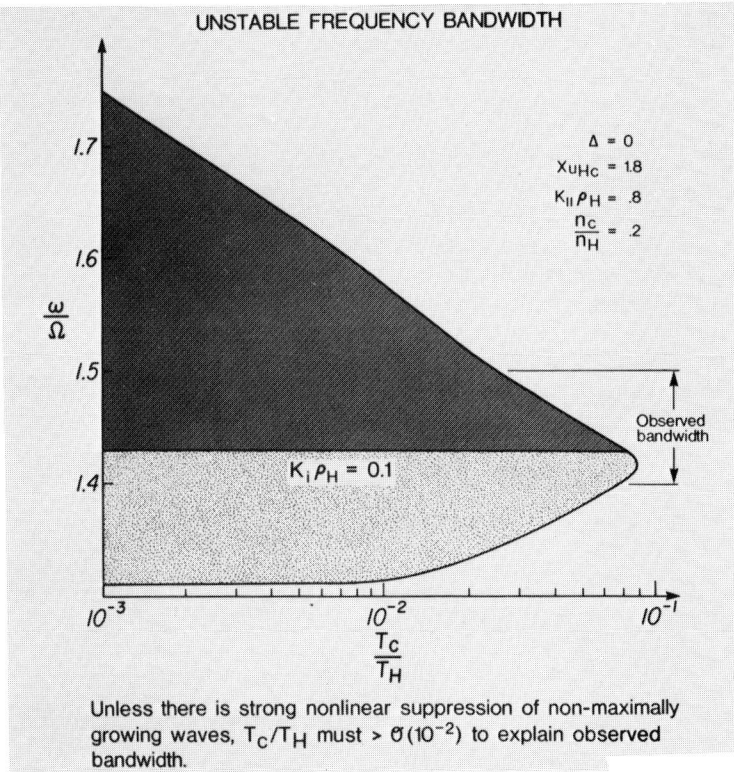

Figure 6. Unstable frequency bandwidth

In this figure we plot the unstable frequency bandwidth against T_c/T_H for waves with a given $k_{\|}$, but variable $k_{\perp}$ for an $X_{UHC} = 1.8$. $k_i\rho_H$ exceeds 0.1 in the dark shaded region, and is positive everywhere within the two shaded regions.

suggest that higher harmonics could be observed separately.

In figure 6, we provide a further argument for our belief that T_c/T_H is not small for the majority of observations of odd half-harmonic emissions. Here we plot the unstable frequency bandwidth against T_c/T_H for waves with a given $k_{\|}$, but variable $k_{\perp}$ for an $X_{UHC} = 1.8$ such that only the first harmonic can be unstable. $k_i\rho_H$ exceeds 0.1 in the dark shaded region, and is positive everywhere within the two shaded regions. Increasing T_c/T_H effectively decreases the maximum unstable frequency from X_{UHC} at T_c/T_H very small to about $3\Omega/2$ at $T_c/T_H \simeq 5 \times 10^{-2}$

The frequency bandwidth of the observed emissions is at most about 10% of the cyclotron frequency. (Those in figure 1 were 5% or less.) Thus to account theoretically for the observed bandwidth, we have a spectrum of choices to make which spans two extremes. At one extreme, we may say that electrons of ionospheric origin remain cold, the instability is nonconvective, and we must rely upon a nonlinear saturation mechanism to strongly suppress the growth of waves outside a narrow band of frequencies. At the other extreme, we may say that ionospheric electrons have been heated (or the the "cold" electrons come from a more energetic source), the instability has turned convective, and amplification is limited to a narrow frequency band. This second line of investigation has another advantage. While it is certainly not true that the odd-half harmonic emissions are observed exactly centered at $(n + \frac{1}{2})\Omega$, it is also true that waves with frequencies of 1.1 or 1.9 or 2.1 or 2.9, at the extremes of the bands are observed less frequently. As we showed in figure 4 the frequencies of the most rapidly growing waves shift to the center of the bands with increasing T_c/T_H. To our way of thought it is simpler to assume that the waves have a narrow band spectrum peaked near the middle of a harmonic band because T_c/T_H is at least $1 - 5 \times 10^{-2}$. Thus, if $T_H \simeq 1 - 10$ keV, T_c will be in the range 10 - 100 eV.

Our instability analysis suggests that odd-half harmonic emissions, when observed, will have been generated locally. If the instability is non-convective, it must nonlinearly saturate locally; if it is strongly and convectively growing, it will still nonlinearly saturate more or less locally; if it is more weakly convectively growing its amplification bandwidth will be small, and it can saturate by propagating out of a small region of spatial growth.

In figure 7, we show an old result from OGO-5 (Kennel *et al.*, 1970), a frequency time spectrogram which shows 3/2 emissions superposed upon convenient interference lines from the rubidium vapor magnetometer at $3\Omega/4$, $5\Omega/4$, $7\Omega/4$. The magnetic field strength varied on a time scale of tens of seconds and the center frequency of the emission tended to follow the magnetic field variations. In this case at least, one may say that the emissions were locally generated.

In the case of convective amplification and saturation, we may estimate the convective saturation length by the following simple argument. It is an expensive matter to perform ray-path propagation calculations for these modes because the dispersion relation depends on details of the hot and cold electron distributions and must be solved numerically. Thus, we have little detailed information on how they propagate. However, the most rapidly growing modes have perpendicular group velocities $\partial \omega/\partial k_\perp$ small and a parallel group velocity that is proportional to the cold electron thermal speed. Thus, we expect the wave energy to propagate parallel to the magnetic field. If the waves propagate towards a decreasing magnetic field strength, they will soon hit the maximum $X = \omega/\Omega$ for which propagation is possible and presumably be reflected. If they propagate towards increasing magnetic field strength, X will decrease and they will pass out of the region of amplification. The amplification length ℓ is thus

MULTIHARMONIC INSTABILITIES

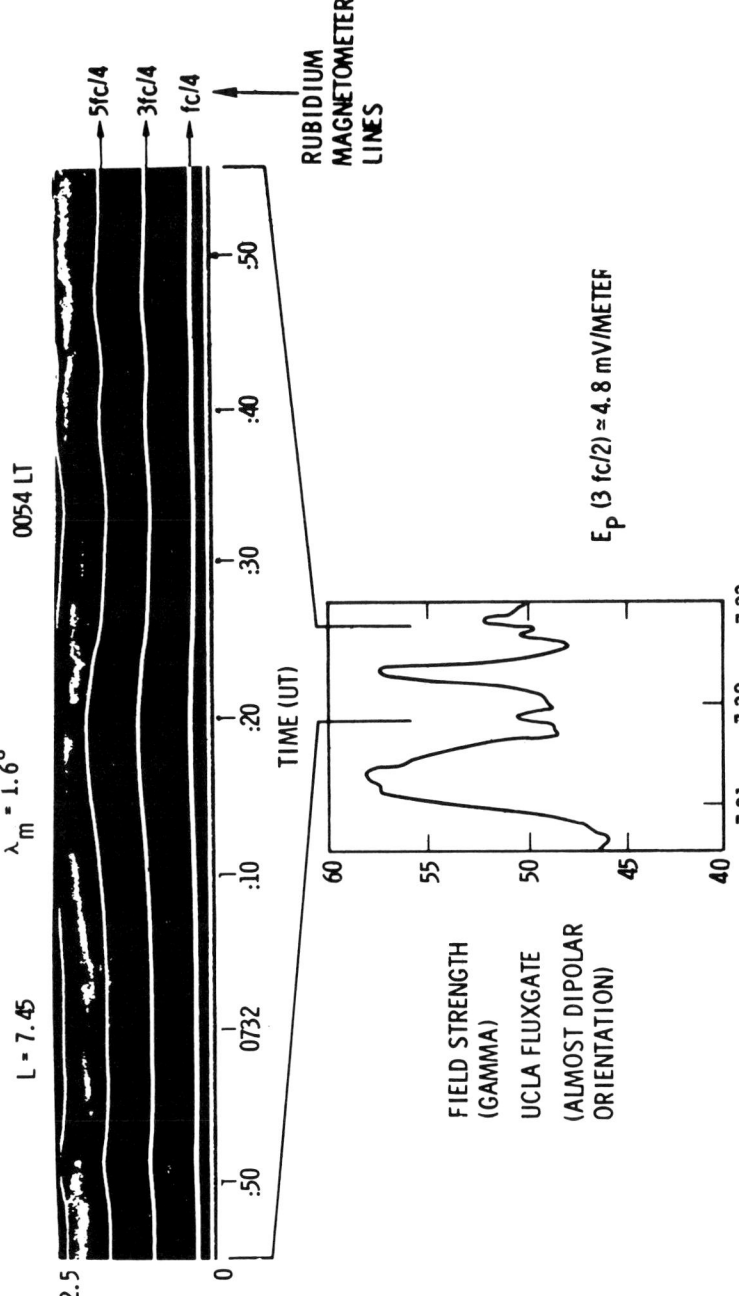

Figure 7. Control of emission frequency by local magnetic field. At the top we show a 0- to 2.5 kHz spectrogram, observed August 15, 1978, which contains Rb magnetometer lines (labeled at the right) and a variable strength emission with $f = 3f_c/2$. The bottom inset contours simultaneous UCLA fluxgate magnetometer data. The above data were obtained during a ~10 sec pulsation of the geomagnetic field, as determined by either magnetometer output. The emission tended to stay near $3f_c/2$ even in the changing magnetic field. The peak amplitude Ep of the f & $3f_c/2$ emission was 4.8 mV/m during a 27.6-sec data cycle of the 3-kHz filter channel (after Kennel et al., 1970).

$$\ell \approx \frac{\Delta\omega}{\Omega} L_B \simeq \left(\frac{\Delta\omega}{\Omega}\right)\frac{LR_E}{3}$$

where $\Delta\omega$ is the unstable band width and L_B is the scale length of the field magnitude in the direction of the field. For a dipole field $L_B \simeq LR_E/3$, so that if $(\Delta\omega/\Omega) \simeq 0.1$ we expect the amplification length to be the order 200 L Km. Note, however, that the parallel gradient of the magnetic field strength is zero at the magnetic equator, so that the amplification length could be considerably longer, and possibly, the amplitudes higher. We have already noted that $k_i\rho_H$ in the range $10^{-2} - 10^{-1}$ will produce sufficient amplification over this small region (Ashour-Abdalla and Kennel, 1978a).

III. SUMMARY OF LINEAR THEORY

We may summarize our investigations of the linear instability of odd-half harmonic waves as follows.
1) Very small wrinkles in the distribution function of hot electrons can produce waves with large spatial growth rates.
2) The cold electron density controls the frequency bands in which nonconvective or large spatial amplification can occur.
3) T_c/T_H determines the convective or non-convective nature of the instabilities. When T_c/T_H is small enough, the instabilities are non-convective.
4) When $N_c/N_H < 1$, the first few harmonic bands, at least, behave similarly. They can be simultaneously non-convective, or can have comparable spatial growth rates. Their convective-non-convective transition takes place at roughly the same T_c/T_H.
5) All theoretical evidence suggests that the waves grow and saturate locally. The fact that the waves are observed to have frequencies which correspond to the local magnetic field strength is consistent with this hypothesis.
6) The most straightforward interpretation, based on linear theory, of the facts that the observed waves have a narrow bandwidth centered near the middle of the harmonic bands and are locally generated, requires that T_c/T_H considerably exceed that expected from ionospheric injection of cold electrons.

We have thus far computed the instabilities of electron cyclotron harmonic waves using a "loss-cone" free-energy source in the hot electrons. The computations of Ashour-Abdalla and Cowley (1974) showed that the combination of convective electric field and magnetic gradient and curvature drifts lead naturally to the development of a positive slope region in the perpendicular velocity distribution. Also such distributions have been observed (De Forest and McIlwain, 1971). However, it is a general feature of cylcotron harmonic modes that their group velocities can be small, especially in the presence of cold electrons. Thus, other sources of free energy could also provide significant spatial growth. For example, Nambu (1975) has shown that an anti-loss cone can produce instability of the first harmonic mode at least, Young *et al*. (1973) showed that a bi-Maxwellian thermal anisotropy ($T_\perp > T_\parallel$) could produce instability but only above the half-harmonic frequency, and a laboratory experiment of Bernstein (1976) showed that electron beams also produce instability of electron cyclotron harmonics. Thus far, there seem to have been no calculations of spatial growth rates for these other types of instability. Of the above three, sources of free energy electron beams seem most pertinent to the magnetosphere, since they have been observed on field lines connecting to the aurora at disturbed times (McIlwain, 1975; Mauk and McIlwain, 1975; Parks *et al.*, 1977; Lin *et al.*, 1978).

The above remarks make it incumbent upon the experimentalists to determine the hot electron distribution function, and therefrom, the source of free energy, when cyclotron harmonic waves are present. We feel that techniques of numerical evaluation of the dispersion relations are sufficiently flexible that given a distribution function, we could determine its stability with relative ease.

IV. NONLINEAR SATURATION AND DIFFUSION

The aim of nonlinear theory in this problem is to predict the saturation amplitude of the multiharmonic instability and therefrom, the effect of the instability upon the distribution of hot and cold electrons. As mentioned previously, there have been great improvements in the threshold sensitivities of the electric field detectors flown on recent spacecraft. For example, the electric field detector on OGO-5 had a threshold sensitivity such that the amplitudes of the multiharmonic waves observed were typically in the estimated range of 1 - 10 mV/m (Kennel *et al.*, 1970), and during a substorm, the amplitude could have been 100 mV/m (Scarf *et al.*, 1973). Waves of this amplitude were observed primarily when the spacecraft crossed within ±10° of the geomagnetic equator in the local midnight to dawn sector of the magnetosphere. By contrast, the GEOS and ISEE spacecraft have electric field sensitivity thresholds of 1μV/m or less. We would expect these more recent spacecraft to observe a higher occurrence of cyclotron harmonic waves. In this regard, we have not been disappointed: as figure 1 indicates, multiharmonic waves were detectable between the time ISEE crossed the magnetopause and the time it entered the plasmapause. Waves were observed at magnetic latitudes above the ±10° band quoted in the original OGO-5 paper and very near local noon, where OGO-5 essentially found nothing. On the other hand, their amplitudes were much lower. It is natural that the "typical" amplitude quoted for a high threshold experiment will be closer to the "peak" amplitude on a more sensitive experiment. Our experimental view of the typical behavior of the cyclotron harmonic emissions has changed: consequently we have modified our beliefs concerning their possible saturation mechanisms.

We believe that discussions of nonlinear saturation should take into account at least the following four points. First, even given the possiblity that OGO-5 experimentalists may have over-estimated their measured wave amplitudes it still seems that a wide range of amplitudes exists, from the order of $\sim 1\mu V/m$ to $\sim 1 mV/m$ and perhaps beyond. Second, the wide range of observed amplitudes suggests that perhaps different saturation mechanisms apply to different magnetospheric conditions. Third, models of the saturation process and amplitude depend not only on the properties of the hot electron distribution, but also the cold electrons. For example, if $T_C/T_H >$ few $\times\ 10^{-2}$ one can consider simple linear convective saturation. Fourth and finally, one should consider the effects of the waves upon both hot and cold electrons in s self-consistent model. The quasilinear diffusion of the hot electrons is well understood; with it Lyons (1974) has shown that an integrated amplitude of 1 - 10 mV/m could account for the observed strong diffusion precipitation of 1 - 10 keV electrons from the plasma sheet into the diffuse aurora (see also Ashour-Abdalla and Kennel, 1978b). Careful studies are needed to ascertain whether the amplitudes of cyclotron harmonic waves are typically large enough near the equator inside the electron plasma sheet to account for the quasi-permanent diffuse aurora.

The reasons for expecting the turbulence to affect the cold electrons significantly are more subtle. First of all, cold electrons escaping from the ionosphere have very long bounce times - tens to hundreds of seconds. Even a small turbulent diffusion coefficient can scatter them out of their ionospheric loss cose in space onto trapped orbits in a bounce time, thereby putting them effectively in a "strong diffusion" regime (Kennel, 1969). Furthermore, a small turbulent diffusion coefficient can reconstruct the cold electron distribution, relative to itself, on time-scales comparable to or even shorter than the energetic electron diffusion times. Let us define the hot and cold diffusion times, τ_H and τ_C respectively, by $\tau_H = a_H^2/D_H$, $\tau_C \sim a_C^2/D_C$, where a_H and a_C and D_H and D_C are the hot and cold electron thermal speeds and diffusion coefficients. Since $\tau_C/\tau_H \sim (T_C/T_H)(D_H/D_C)$, even a small cold diffusion coefficient, $D_C/D_H \sim T_C/T_H \sim$ few $\times\ 10^{-2}$ will significantly modify the cold electrons on the time scale on which hot electrons are diffused and precipitated. The nonlinear heating of cold particles is not a common problem in theoretical plasma physics, but our linear instability analysis suggests that it is an important one for multiharmonic emissions in the magnetosphere.

We have found it convenient to order our conceptions concerning possible saturation mechanisms using the following idealized picture of the time development of a multiharmonic instability. Suppose at time zero, electron cyclotron instabilities first start to grow in a plasma consisting of hot electrons, with a source of free energy such as a "loss cone" region of positive $\partial F/\partial v_\perp$, and cold electrons confined to the ionospheric loss cone in space. The unstable waves grow non-convectively and must saturate nonlinearly. At present, we have only one nonlinear saturation theory in the literature due to Young (1975). While Young's calculated saturated amplitudes are much higher than typically observed, it is likely that the saturation ampltidues in a nonconvective regime could still be quite large. Suppose there is some scattering of cold electrons out of the loss cone and some cold electron perpendicular heating. In the absence of further losses, cold electrons continue to heat, possibly until the nonconvective-convective boundary is passed. At this point, the instability could turn convective, and saturate by propagation. We have seen that T_c/T_H can be chosen in the linear theory to make this a consistent possible theory. This might yield an intermediate range of amplitudes. Finally, the system could evolve to be stable, at which point we might expect a non-negligible spectrum of thermal fluctuations, which would in turn represent the lowest observable wave amplitudes.

The scenario described above can be misleading in one important sense: the probable diffusion and growth times are such that we must consider a steady-state -- one in which sources of energetic and cold electrons are balance with precipitation losses -- to compute the distribution function and from it the wave growth rate. Cold electron heating, if it occurs, should be balanced with heat losses to determine a steady state cold electron temperature, and from it whether or not waves turn convective. Given the crucial role the cold electrons play in determining the choice saturation mechanism, measurement of cold electron temperature would be useful or even vital to the further development of the theory.

Acknowledgements. We are indebted to D. Gurnett and W. Kurth for providing data from the University of Iowa Plasma Wave experiment on board the ISEE satellite. We would like to give special thanks to W. Kurth for participating in many illuminating discussions concerning the observations. We would also like to acknowledge useful discussions with F. V. Coroniti, R. C. Elphic, W. S. Kurth, C. S. Lin, W. A. Livesey, W. Lotko, K. B. Quest, C. T. Russell, and R. J. Walker. The work was supported by NASA NGL-05-007-190 and NSF ATM-76-13792.

REFERENCES

Anderson, R. R., and Maeda, K.: 1977, *J. Geophys. Res.*, 82, p. 135.
Ashour-Abdalla, M., and Cowley, S. W. H.: 1974, Wave particle interactions near the geostationary orbit, in *Magnetospheric Physics*, B. M. McCormac, ed., D. Reidel Publishing Co., Dordrecht, p.270.
Ashour-Abdalla, M., and Kennel, C. F.: VLF electrostatic waves in the magnetosphere, in *Physics of the Hot Plasma in the Magnetosphere*, B. Hultquist and L. Stenflow, eds., Plenum, New York, pp. 201-227.
Ashour-Abdalla, M., and Kennel, C. F.: Convective cold upper hybrid instabilities, in *Magnetospheric Particles and Fields*, B. M. McCormac, ed., D. Reidel Publishing Co., p. 181.
Ashour-Abdalla, M., and Kennel, C. F.: 1978a, *J. Geophys. Res.*, 83, p. 1531.
Ashour-Abdalla, M., and Kennel, C. F.: 1978b, *J. Geomagnetism and*

Geoelectricity, 30, p. 239.

Bernstein, W., Leinbach, H., Cohen, H., Wilson, P. S., Davis, T. N., Hallinan, T., Baker, B., Martz, J., and Ziemke, R.: 1975, *J. Geophys. Res.*, 80, p. 4375.

DeForest, S. E., and McIlwain, C. E.: 1971, *J. Geophys. Res.*, 76, p. 3587.

Fredricks, R. W.: 1971, *J. Geophys. Res.*, 76, p. 5344.

Fredricks, R. W., and Scarf, F. L.: 1973, *J. Geophys. Res.*, 78, p. 310.

Gurnett, D. A.: 1975, *J. Geophys. Res.*, 80, p. 2751.

Gurnett, D. A.: 1978, Initial results from the ISEE-1 and -2 plasma wave investigation, to be published in *Space Science Reviews*.

Hubbard, R. F., and Birmingham, T. J.: 1978, *J. Geophys. Res.*, 83, p. 4837.

Karpman, V. I., Alekhin, Ju. K., Borisov, N. D., and Rjabova, N. A.: 1973, *Astrophys. Space Sci.*, 22, p. 267.

Karpman, V. R., Alekhin, Ju. K., Borisov, N. D., and Rjabova, N. A.: 1975, *Plasma Phys.*, 17, p. 361.

Kennel, C. F.: 1969, *Rev. Geophys.*, 7, p. 379.

Kennel, C. F., Scarf, F. L., Fredricks, R. W., McGehee, J. G., and Coroniti, F. V.: 1970, *J. Geophys. Res.*, 75, 6136.

Lin, C. S., Parks, G. K., Mauk, B., DeForest, S., and McIlwain, C. E.: Electron beam spreading, University of Washington preprint, 1978.

Lyons, L. R.: 1974, *J. Geophys. Res.*, 79, p. 575.

Mauk, B., and McIlwain, C. E.: 1978, *IEEE Trans.*, AES-11, p. 1125.

McIlwain, C. E.: Auroral electron beams near the magnetic equator, in *Physics of the Hot Plasma in the Magnetosphere*, B. Hultquist and L. Stenflo, eds., Plenum Press, New York, pp. 91-109, 1975.

Nambu, M.: 1975, *Geophys. Rev. Lett.*, 2,

Parks, G. K., Lin, C. F., Mauk, B., DeForest, S., and McIlwain, C. E.: 1977, *J. Geophys. Res.*, 82, p. 5208.

Scarf, F. L., Fredricks, R. W., Kennel, C. F., and Coroniti, F. V.: 1973, *J. Geophys. Res.*, 78, p. 3119.

Shaw, R. R., and Gurnett, D. A.: 1973, *J. Geophys. Res.*, 80, p. 4259.

Young, T. S., Callen, J. D., and McCune, J. E.: 1973, *J. Geophys. Res.*, 78, p. 1082.

Young, T. S.: 1975, *J. Geophys. Res.*, 80, p. 3995.

SOME THEORETICAL ASPECTS OF ELECTROSTATIC DOUBLE LAYERS

Per Carlqvist
Department of Plasma Physics, Royal Institute of Technology,
S-100 44 Stockholm, Sweden

ABSTRACT

A review is presented of the main results of the theoretical work on electrostatic double layers. The general properties of double layers are first considered. Then the time-independent double layer is discussed. The discussion deals with the potential drop, the thickness, and some necessary criteria for the existence and stability of the layer. As a complement to the study of the time-independent double layer a few remarks are also made upon the time-dependent double layer. Finally the question of how double layers are formed and maintained is treated. Several possible formation mechanisms are considered.

1. INTRODUCTION

In recent years electrostatic double layers have attracted an increasing interest within the field of cosmic plasma physics. One important reason for this is, that space experiments have yielded results which strongly indicate that double layers are often present in the Earth's magnetosphere above the auroral zones. Observations of the drift velocity of artificially injected Ba^+-ions show, for instance, that strong electric fields, which might be due to double layers, exist at altitudes of several thousand kilometers (Wescott et al., 1976; Haerendel et al., 1976). Another interesting evidence of the presence of double layers in the magnetosphere comes from direct measurements of the electric field made onboard the S3-3 satellite (Mozer et al., 1977). When the satellite crossed the auroral zones at altitudes in the range $2 \times 10^3 - 8 \times 10^3$ km, both perpendicular and parallel electric field components of the order of $10^2 - 10^3$ mV m^{-1} were measured. The strong electric fields were often found to be concentrated in narrow regions in which the perpendicular component switched from one direction to the reverse in accordance with double layer theory (Shawhan et al., 1978).

Double layers have also been suggested to have other cosmic applications. Thus a model for solar flares has been proposed in which double layers constitute the basic mechanism for the energy release in flares (Alfvén and Carlqvist, 1967; Carlqvist, 1969). In this model one or several double layers are formed in current filaments penetrating the solar atmosphere when the current density in them exceeds a certain critical value. Stored magnetic energy is then rapidly released as kinetic energy of energetic particles and plasma. The final result is expected to be a solar flare.

Another region where double layers have been suggested to occur is in the Jovian magnetosphere (e.g. see Shawhan et al., 1975; Shawhan, 1976; Smith and Goertz, 1978). These double layers, which among other things may give rise to the decametric radiation observed from Jupiter, are assumed to form in an electric current system generated by the Jovian satellite Io.

In addition to the above mentioned applications of double layers to objects within our solar system, Alfvén (1978) has considered the possibility that double layers are also formed in radio galaxies where they might account for the large release of power observed.

The application of double layers on the magnetosphere of the Earth and on other cosmic phenomena has led to intensified experimental and theoretical investigations of such layers. Great progress in the understanding of the physics of the double layers has been made in the last few years. However, much work remains to be done before a general and complete description of double layers can be said to exist.

In the present article we shall discuss some of the most important results which have been obtained up to now in the theoretical work on double layers. On the basis of experimental investigations we first consider the general properties of double layers (Section 2). Next (Section 3) we discuss the time-independent double layer which is the type of double layer most frequently analysed. The discussion deals with the potential drop, the thickness, and some necessary criteria for the existence and stability of this kind of layer. Various models of the time-independent layer are considered as well. As a complement to the discussion on the time-independent layer a few remarks are also made upon the type of double layer that changes with time (Section 4). Finally the important question of how double layers are formed and maintained is treated (Section 5).

Reviews on double layers have earlier been presented by Block (1975, 1978).

2. GENERAL PROPERTIES OF DOUBLE LAYERS

For many decades double layers have been studied in a considerable number of laboratory experiments (for a review see e.g. Torvén, 1978). In some of the experiments external effects of the double layers such as current limitation, potential drop, voltage surges, and accelerated particles have been investigated (e.g. see Langmuir and Mott-Smith Jr., 1924; Hull and Elder, 1942; Schönhuber, 1958; Crawford and Freeston, 1963; Jacobsen et al., 1968; Torvén and Babić, 1975, 1976) while in others rather the internal structure of the layers has been studied (Quon and Wong, 1976; Coakley et al., 1978; Torvén and Andersson, 1978). From the experimental investigations combined with theoretical considerations the following general picture of the double layer has come out:

1. The double layer is a local region which is capable of sustaining a high potential drop, ϕ_{DL}, and which is surrounded by plasma. Generally ϕ_{DL} is larger than the equivalent thermal potential, kT/e, of the plasma. Potential drops of the order $10-10^5$ V have been experimentally found over structures interpreted to be double layers.

2. The potential within the double layer, ϕ, is usually assumed to vary monotonically (see Figure 1a). Such a potential variation has been experimentally observed by Quon and Wong (1976) and Torvén and Andersson (1978). However, double layers may also very well exist which have potential distributions similar to that shown in Figure 1a but with a superimposed fine structure consisting of several maxima and minima.

3. The electric field, E (see Figure 1b), which gives rise to the potential drop across the double layer is generated by space charges, ρ, according to Poisson's equation. The space charges are primarily concentrated to two adjacent layers of opposite polarity (see Figure 1c) – hence the term double layer. Inside the double layer the positive charge density produced by the ions may drastically differ from the negative charge density produced by the electrons. Quasi-neutrality does therefore not prevail in the double layer.

4. The electric field is much weaker in the plasma that surrounds the double layer than it is in the layer itself. This implies that the double layer, taken as a whole, is electrically neutral. It therefore contains approximately equal numbers of positive and negative charges.

5. The thickness of the double layers is generally much smaller than the mean free path of the ions and electrons. For layers with a fairly small potential drop ($\phi_{DL} = 10 - 10^2$ V) in low pressure discharges (density $n_e = n_i = 10^{15} - 10^{16}$ m^{-3}) the thickness is a fraction of a millimetre, which corresponds to some ten Debye lengths in the surrounding plasma, while the mean free path is of the order of a metre.

6. Inside the double layer the ions and electrons, which account for the space charge, are acted upon by the electric field. Some of the particles are accelerated and form beams on both sides of the layer (Crawford and Freeston, 1963) while the rest are either decelerated or reflected. The conditions in the double layer are in several respects similar to those in cathode sheaths. An important difference is, however, that the double layer is surrounded by plasma while the cathode sheath borders on plasma on one side only.

7. Most double layers which have been experimentally investigated have carried an electric current, I_{DL}. This does not mean that the current is a necessary condition for the formation of all types of double layers. However, in case electric energy is released in the double layer a current must flow through it, since the power developed is $P_{DL} = I_{DL} \phi_{DL}$.

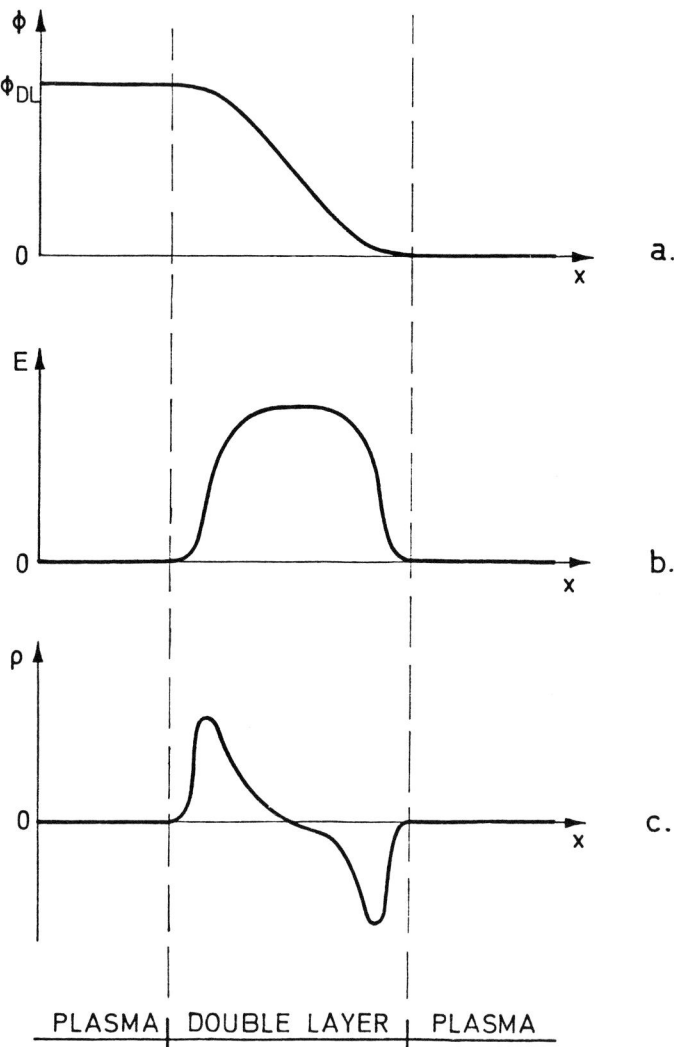

Figure 1. Schematic picture of the variation of a) the potential ϕ, b) the electric field E, and c) the net charge density ρ with the distance x in a double layer surrounded by plasma.

From the experiments it is well-known that the formation and development of a double layer does not only depend on the plasma in which it occurs but also on the whole circuit in which it is an integral part. As a consequence of this, double layers show a great variation of their appearance.

Double layers can be classified according to their properties in several different ways. We can for instance distinguish between strong layers and weak layers. The strong layers are characterized by a potential drop that is much larger than the equivalent thermal potential of the surrounding plasma, $\phi_{DL} >> kT/e$. The temperature, T, is here adopted to be a measure of the mean energy coupled with the random motion of the particles and does therefore not necessarily presuppose a Maxwellian velocity distribution. In the weak layers on the other hand, ϕ_{DL}, is of the same order of magnitude as kT/e.

Another way to classify double layers is by means of their behaviour in time. There are layers that vary strongly with time, but there are also layers that are essentially time-independent.

As was pointed out above (point 5) the mean free path of the charged particles is generally much larger than the thickness of the double layer. Collisions can therefore generally be assumed to play a negligible role within the layer. The motions of the ions and electrons in the layer are instead mainly governed by the electrostatic force and the inertia force. This means that if the boundary conditions were known, it would, in principle, be possible to obtain a complete description of the double layer through a combined analysis of the Vlasov equations

$$\frac{\partial f_\alpha}{\partial t} + \underline{v} \cdot \nabla_{\underline{r}} f_\alpha + \frac{q_\alpha (\underline{E} + \underline{v} \times \underline{B})}{m_\alpha} \cdot \nabla_{\underline{v}} f_\alpha = 0 \tag{2.1}$$

valid for the ions and electrons and the Poisson equation

$$\nabla^2 \phi = - \nabla \cdot \underline{E} = - \sum_\alpha q_\alpha n_\alpha \tag{2.2}$$

in which $f_\alpha = f_\alpha(\underline{r}, \underline{v}, t)$, $n_\alpha = \int f_\alpha d\underline{v}$, m_α, and q_α denote the distribution function, number density, mass and charge of the particles respectively, and where the subscript α is i for the ions and e for the electrons. Such a complete description of the layer does not yet, however, exist owing to the complexity of the analysis and unknown boundary values of f_α.

The theoretical studies on double layers performed up to now have mainly been limited to time-independent and one-dimensional models with particular distribution functions for the ions and electrons (see e.g. Block, 1972; Knorr and Goertz, 1974; Lee et al., 1977; Hasan and ter Haar, 1978). In most models studied the magnetic field has been assumed to be either zero or parallel to the motion of the charged particles which implies that no magnetic force acts upon the particles. Exceptions in this respect are the self-consistent models of oblique double layers considered by Swift (1975, 1976) in which the magnetic field plays an important role.

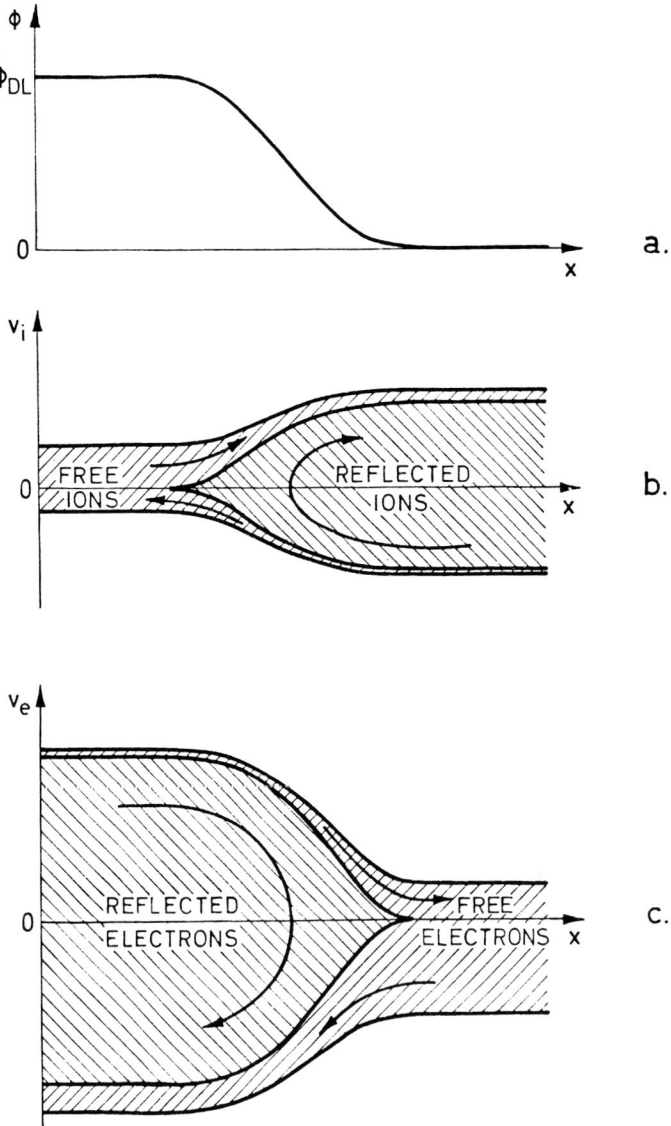

Figure 2. Schematic picture of a) the potential distribution, b) phase space for the ions, and c) phase space for the electrons in a double layer. Both the ion and the electron populations consist of free particles which can pass the double layer and of reflected particles which cannot penetrate the layer because of the potential barrier.

The general structure of phase space for the type of double layer that do not include any perpendicular component of the magnetic field is shown in Figure 2. As is evident from the Figure the particles can be divided into two categories - the free particles and the reflected (trapped) particles. The free particles, which can pass through the double layer, consist partly of particles that are accelerated in the electric field of the layer, partly of particles that are decelerated. The accelerated particles gain energy when passing the layer and form energetic ion and electron beams on opposite sides of the layer. The decelerated particles again lose energy but possess a sufficiently high initial energy to overcome the potential drop, ϕ_{DL}. In strong double layers the fluxes of the free decelerated particles are generally so small that they can be neglected.

Contrary to the free decelerated particles the reflected particles have not energy enough to penetrate the whole potential barrier of the double layer. In point 4 above we found that considerable space charges may exist in the double layer, while approximate charge neutrality must prevail in the surrounding plasma. For charge neutrality to be maintained in the plasma, reflected particles necessarily have to exist. In principle, charge neutrality can be maintained by means of only one kind of reflected particles - ions or electrons - if suitably chosen (Block, 1978), but usually both kinds of reflected particles are present in the plasma.

3. THE TIME-INDEPENDENT DOUBLE LAYER

3.1. The particle distribution

The type of double layer that is most easy to study theoretically is the time-independent layer. In this type of layer all physical quantities such as the particle velocities, the potential drop, the electric field, and the thickness remain constant in time. An important contribution to our knowledge about the time-independent layer was given by Bernstein et al. (1957) who studied self-consistent and time-independent solutions of the potential in a one-dimensional and collision-free plasma. They showed that essentially arbitrary potential distributions can be constructed by choosing trapped particle populations in an appropriate way.

On the basis of this result Montgomery and Joyce (1969) could demonstrate the existence of shock-like solutions of the combined Vlasov-Poisson equations. The conditions in such shocks are in several respects similar to those in the double layers.

Later, time-independent and one-dimensional double layers possessing particular particle or potential distributions have been investigated in several works. Thus Andrews and Allen (1971) have investigated a double layer model which contains both reflected and free accelerated ions and electrons. They first study some conditions for the existence of double layers with Maxwellian distributions of the reflected ions and electrons and arbitrary distributions of the free particles. Then they treat numerically a special layer model in which the free accelerated particles are mono-energetic and the reflected ones are Maxwellian.

For ionospheric applications Block (1972) has considered a model of a strong double layer in which the free accelerated particles are described by means of the fluid equations. According to Bertrand and Feix (1968) such a fluid approach is equivalent to a water-bag distribution of the particles. The reflected particles on the other hand are assumed to maintain a constant temperature in the layer which is consistent with a Maxwellian distribution of the particles.

A similar model of the double layer, which is also based on the fluid equations, has been treated by Lee et al. (1977). However, in this model the temperatures of the reflected particles are allowed to vary with their density depending on the value that is adopted of the adiabatic constant of the particles.

As a complement to the two fluid models mentioned above Hasan and ter Haar (1978) have studied a model of the double layer using a kinetic approach. The model, which describes both weak and strong layers, includes free accelerated particles as well as reflected particles. The free particles are assumed either to be monoenergetic or to have a power-law distribution with a low-energy cut-off, while the reflected particles possess a water-bag distribution. On the basis of the distribution functions of the particles it is possible to calculate a relation between the potential drop and the thickness of the double layer. Similarly, certain conditions for the existence of the layer can be found. This will be discussed more in detail in Sections 3.2, 3.3, and 3.4.

Another approach to the study of the time-independent double layer, which more follows the methods developed by Bernstein et al. (1957) have been made by Knorr and Goertz (1974). They assume a well-defined potential distribution of the layer ($\phi \propto \tanh(x/x_0)$) and consider the distribution functions of the free ions and electrons and of the reflected ions as given. By means of the potential distribution and the three particle distribution functions they calculate the remaining distribution function of the reflected electrons.

3.2 The potential drop

Half a century ago Langmuir (1929) studied a simple time-independent and one-dimensional model of a strong double layer. In this model electrons and singly ionized ions are accelerated in opposite directions through the potential drop of the layer, ϕ_{DL}. The ions and electrons are emitted with zero velocity from a plane anode and a plane cathode respectively, which are parallel to each other and separated by the distance, d. The electric field is zero at both the electrodes (in accordance with Figure 1a) implying that both the ion current density, i_i, and the electron current density, i_e, are space charge limited. No surface charges do then exist on the electrodes. The electric field is completely formed by the space charges which the accelerated ions and electrons give rise to. The only function of the electrodes is therefore to emit particles.

Using Poisson's equation and the non-relativistic equations of motion for the ions and electrons Langmuir found the potential drop across the double layer

$$\phi_{DL} = C_1 \left\{ \frac{m_e i^2 d^4}{\varepsilon_o^2 e \left[1+(m_e/m_i)^{1/2}\right]} \right\}^{1/3} \quad (3.1)$$

where $i = i_e + i_i$ is the total current density and $C_1 \approx 0.90$. The electrode separation, d, constitutes the thickness of the layer. It is interesting to note that the expression (3.1) is of the same form as the expression for the potential drop of a vacuum diode, but the former is a factor 1.5 smaller than the latter (Child, 1911; Langmuir, 1913).

The densities of the ions and electrons which are accelerated in the double layer are shown in Figure 3a. Because of the assumption of zero initial velocities of the particles, the densities tend to infinity at the emitting electrodes. This property of the model is obviously non-physical, but still the model must be considered interesting since it may help us understand much of what is going on in real double layers.

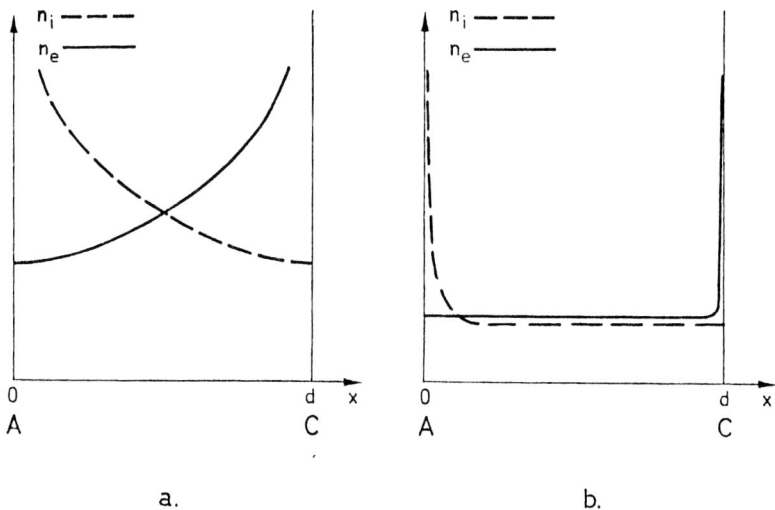

a.   b.

Figure 3. Densities of ions, n_i, and of electrons, n_e, schematically shown as functions of the x-coordinate in a double layer of thickness d. A is the anode boundary and C is the cathode boundary. a) Non-relativistic double layer according to Langmuir (1929) - b) Relativistic double layer where both the ions and electrons are accelerated to velocities close to the velocity of light (Carlqvist, 1969).

Although Langmuir's model only includes accelerated particles inside the double layer, it can be considered equivalent to a more complete model of the type shown in Figure 2 which also contains free and reflected particles outside the layer. These latter particles must then be cold and of infinite density.

The expression for the potential drop (3.1) is valid only for strong double layers. In the case the double layer is weak, the reflected particles will have a neutralizing influence on the layer, leading to a potential drop that is smaller than that given by equation (3.1). Using a model with monoenergetic free particles and water-bag distributed reflected particles (see Section 3.1) Hasan and ter Haar (1978) have calculated the potential drop when the double layer is weak. Figure 4 shows how the potential drop varies with the thickness of the layer when the current density is kept constant. For the sake of comparison Langmuir's relation (3.1) is also included in the Figure. As one would expect the

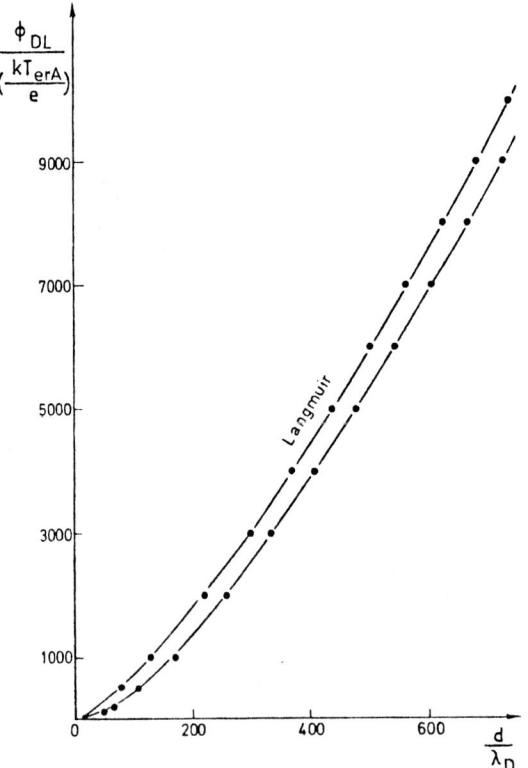

Figure 4. Normalized potential drop of a one-dimensional double layer, $\phi_{DL}/(kT_{erA}/e)$, shown as a function of its normalized thickness, d/λ_D, according to Hasan and ter Haar (1978). The current density in the layer is assumed to be constant, $i = 2.20\ en_{efC}(kT_{erA}/m_e)^{1/2}$, and further $n_{efC} = n_{ifA}$ and $T_{erA} = T_{irC}$. For the sake of comparison Langmuir's (1929) relation for the potential drop, valid for a strong double layer, is shown as well.

relative difference between the Langmuir curve and the curve derived by Hasan and ter Haar decreases as the layer becomes stronger.

It is important to stress that the potential relations derived by Langmuir and by Hasan and ter Haar are not generally valid, but are merely applicable to double layers in which the free particles are only accelerated. As has been pointed out by Block (1978) it is possible to reverse the direction of motion of an arbitrary number of the free particles in a double layer without changing either the charge distribution or the potential distribution in the layer. By choosing the particle fluxes in the forward and backward directions in an appropriate way one can, for instance, construct double layers which carry no current, but have a finite potential drop.

The validity of Langmuir's relation for the potential drop of the double layer (3.1) is limited to non-relativistic energies of the electrons, i.e. when $\phi_{DL} \ll \phi_e = m_e c^2/e = 5.1 \times 10^5$ V. However, astrophysical applications of double layers, e.g. on solar flares, have made it necessary to consider layers of larger potential drops where both the electrons and the ions are accelerated to relativistic energies. A one-dimensional double layer similar to that treated by Langmuir but having a potential drop $\phi_{DL} \gg \phi_i = m_i c^2/e \gtrsim 10^9$ V has been studied by Carlqvist (1969). The charge distribution in such a relativistic double layer is shown in Figure 3b. Both the electrons and the ions entering the layer are rapidly accelerated to velocities close to the velocity of light, c, whereupon the densities of the particles remain almost constant. The potential drop of this type of double layer has been found to be

$$\phi_{DL} = \left(\frac{m_i c i d^2}{4\varepsilon_o e}\right)^{1/2}, \qquad \phi_{DL} \gg \phi_i \qquad (3.2)$$

where, i, and, d, as before represent the current density and the thickness of the layer respectively. The variation of the potential drop, ϕ_{DL}, with id^2 according to equations (3.1) and (3.2) is shown in Figure 5 for both non-relativistic and relativistic double layers accelerating electrons and protons.

Strong arguments have been advanced for the view that electric currents which flow in cosmic plasmas have a pronounced tendency to contract into filaments (Murty, 1962; Alfvén and Fälthammar, 1963; Marklund, 1978). If a double layer is formed in such a filament - e.g. as a result of some current instability (see Section 5) - it will get a limited extension perpendicular to the filament which we can define by the radius, r_o. As long as the thickness of the layer, d, is much smaller than r_o, the potential drop of the layer, ϕ_{DL}, can be described by one-dimensional theory. For a relativistic double layer ϕ_{DL}, then grows in proportion to d according to equation (3.2). On the other hand, in case the thickness, d, becomes much larger than r_o a saturation of the relativistic potential drop can be expected to occur, in about the same

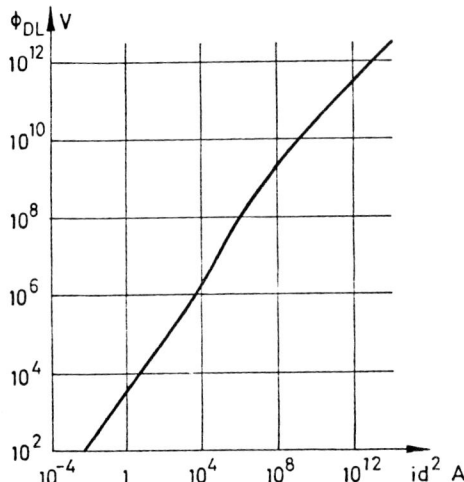

Figure 5. The potential drop across a strong one-dimensional double layer, ϕ_{DL}, is shown as a function of id^2. For $\phi_{DL} \ll m_e c^2/e$ the curve tends towards that given by Langmuir (1929), while for $\phi_{DL} \gg m_i c^2/e$ it tends towards the asymptotic relation $\phi_{DL} = (m_i c \, i \, d^2/4 \, \varepsilon_o e)^{1/2}$. The current is supposed to be carried by electrons and protons.

way as the potential drop between two circular condensor plates of constant charge reaches a saturation level when the distance between the plates is increased indefinitely. The maximum potential drop across the double layer, ϕ_{DLm}, can be roughly estimated if we insert $d = r_o$ into equation (3.2) and notice that the total current through the layer is $I_{DL} \approx \pi r_o^2 i$ (Carlqvist, 1969). We then obtain

$$\phi_{DLm} \approx C_2 \, I_{DL}^{1/2} \tag{3.3}$$

with $C_2 = (m_i c/4\pi \varepsilon_o e)^{1/2}$. The maximum potential drop across a time-independent filamentary double layer is consequently a function only of the total current passing the layer.

3.3 The Langmuir condition

In the strong double layer studied by Langmuir (1929) the ratio of the ion current density and the electron current density, measured in the frame of the layer, is

$$\frac{i_i}{i_e} = \left(\frac{m_e}{m_i}\right)^{1/2} . \tag{3.4}$$

This specific ratio between the current components is necessary for the momentum balance to be maintained in the layer.

Conditions similar to equation (3.4) regulating the conservation of total momentum have also been derived for other models of double layers. Usually these conditions are referred to as "Langmuir conditions". In the relativistic double layer ($\phi_{DL} \gg \phi_i$), for instance, both the electrons and the ions are accelerated to such high energies that their masses become approximately equal. In order to conserve the momentum in this kind of layer the ion flux must be about equal to the electron flux. This can for singly ionized ions be written as

$$\frac{i_i}{i_e} \approx 1 \quad , \quad (\phi_{DL} \gg \phi_i) \tag{3.5}$$

(Carlqvist, 1969). Hence, in the relativistic double layer the ion current is approximately the same as the electron current.

Various Langmuir conditions have also been derived for more complex models of double layers containing both free particles and reflected

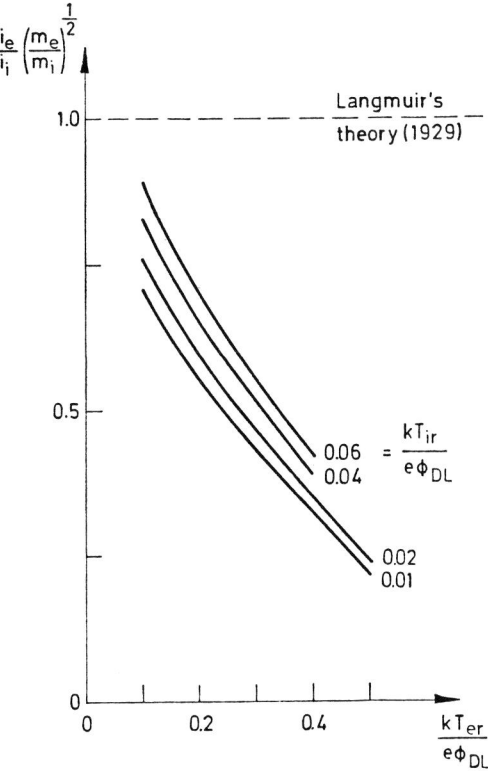

Figure 6. Variation of $(i_e/i_i)(m_e/m_i)^{1/2}$ with $kT_{er}/e\phi_{DL}$ in a weak double layer for some different values of $kT_{ir}/e\phi_{DL}$ according to Andrews and Allen (1971). When the double layer is strong, $(i_e/i_i)(m_e/m_i)^{1/2}$ approaches unity as prescribed by Langmuir (1929).

particles of finite temperatures. Thus Andrews and Allen (1971) found from a numerical treatment of their model (see Section 3.1) that large deviations from the condition (3.4) can occur when the temperatures of the reflected particles differ from zero and the layer is weak (see Figure 6). We find from Figure 6 that the current ratio i_i/i_e tends to the Langmuir value, defined by equation (3.4), as the layer becomes strong. The Langmuir value should also be reached in weak layers when the densities and temperatures of the reflected ions and electrons on both sides of the layer are equal. Similar results have later been obtained analytically by Hasan and ter Haar (1978) by means of the model treated by them (cf. Section 3.1).

It should be noted, that no general Langmuir condition can be given for weak double layers since the current components in these layers are not uniquely defined (Block, 1978; cf. Section 3.2).

For strong double layers, including both free and reflected particles of finite temperatures, Block (1972) and Lee et al. (1977) (cf. Section 3.1) have derived an equation for the conservation of momentum throughout the layers, which in a one-dimensional geometry can be written as

$$m_e n_{ef} u_{ef}^2 + m_i n_{if} u_{if}^2 + p_{ef} + p_{if} + p_{er} + p_{ir} - \frac{\varepsilon_o E^2}{2} = \text{const.} \quad (3.6)$$

Here n is the number density, u the mean velocity, and p the pressure of the particles while the subscripts f and r refer to free and reflected particles respectively. Equation (3.6) can be interpreted as an indirect Langmuir condition.

3.4 The Bohm condition

When Bohm (1949) many years ago studied the conditions in stable wall sheaths he found that the ions must impinge on the sheath with a velocity

$$u_i \geq \left(\frac{kT_e}{m_i}\right)^{1/2} \quad (3.7)$$

in order for a self-consistent space charge to be maintained in the sheath. Bohm's discussion on wall sheaths can also be applied to the anode and cathode boundaries of time-independent double layers (Persson, 1968, 1969).

We can easily see this by considering for instance the anode boundary of a strong and one-dimensional double layer. The free ions which impinge on the layer with the velocity u_{ifA} are for the sake of simplicity assumed to have zero temperature while the reflected electrons are Maxwell distributed with the temperature T_{erA}. The densities of the ions and electrons are approximately equal, $n_{ifA} = n_{erA} = n_A$, at the boundary of the layer since the density of the free electrons accelerated through the layer can be neglected compared to n_A in strong layers. When accelerated in the layer the free ions reach the velocity

$$u_{if} = \left(u_{ifA}^2 + \frac{2e\Delta\phi}{m_i}\right)^{1/2} \tag{3.8}$$

where $\Delta\phi = \phi_{DL} - \phi$ is the potential drop from the anode boundary. From the constancy of the ion current $i_i = en_{if} u_{if} = en_A u_{ifA}$ throughout the layer and from equation (3.8) the ion density is found to vary with the potential drop as

$$n_{if} = n_A \left(1 + \frac{2e\Delta\phi}{m_i u_{ifA}^2}\right)^{-1/2} \tag{3.9}$$

The density of the reflected electrons, on the other hand, varies according to Boltzmann's law as

$$n_{er} = n_A \exp\left(-\frac{e\Delta\phi}{kT_{erA}}\right) \tag{3.10}$$

Both n_{if} and n_{er} thus decrease with increasing potential drop $\Delta\phi$. For a positive potential drop to occur it is according to Poisson's equation necessary that n_{er} decreases more rapidly than n_{if} so that a positive space charge is formed. This condition can by means of equations (3.9) and (3.10) be expressed as

$$m_i u_{ifA}^2 > kT_{erA} \tag{3.11a}$$

which is identical with the condition (3.7) found by Bohm for stable wall sheaths. A condition similar to (3.11a).

$$m_e u_{efC}^2 > kT_{irC} \tag{3.11b}$$

can in the same way be shown to ve valid for the free electrons at the cathode boundary of the layer provided the free electrons there are of zero temperature while the reflected ions are thermal with the temperature T_{irC}.

Conditions like (3.11) which are necessary for self-consistent space charge and potential distributions are usually termed "Bohm conditions". Such Bohm conditions have been derived for several of the double layer models treated in Section 3.1.

Thus Andrews and Allen (1971) found from numerical calculations on a layer model similar to that described above that the systematic velocities u_{ifA} and u_{efC} of the impinging ions and electrons must be larger if the layer is weak compared to if it is strong. This enhancement of u_{ifA} and u_{efC} in weak layers is caused by the accelerated beam particles which have a finite density when leaving the layer.

For strong double layers with finite temperatures of both the reflected and the free particles (cf. Section 3.1) Lee et al. (1977) have derived the following Bohm conditions

$$m_i u^2_{ifA} > \gamma k T_{ifA} + \gamma_e k T_{erA}$$
$$m_e u^2_{efC} > \gamma k T_{efC} + \gamma_i k T_{irC}$$
(3.12)

where γ, γ_e, and γ_i are the adiabatic constants of the free particles, the reflected electrons, and the reflected ions respectively. The conditions (3.12) can be said to include the Bohm conditions earlier derived by Block (1972) for his model (see Section 3.1) if we put $\gamma_e = \gamma_i = 1$ corresponding to a constant temperature of the reflected particles. Also the Bohm conditions found by Hasan and ter Haar (1978) for their kinetic model (see Section 3.1) can be obtained from (3.12) by putting $\gamma_e = \gamma_i = 3$ and $T_{ifA} = T_{efC} = 0$.

3.5 The stability problem

In order to decide under what conditions time-independent double layers can exist it is of decisive importance to learn when these layers are stable and when they are unstable. Unfortunately only a few investigations have been performed in this field, and for that reason our understanding is here still strongly limited.

Knorr and Goertz (1974) have studied some aspects of the stability of their specific double layer model (cf. Section 3.1). They find that it is possible to construct plasmas containing the double layer which are Penrose stable everywhere. It should be noted, however, that the stability criterion derived by Penrose (1960) is strictly valid only for homogeneous plasmas. The criterion can therefore be applied to double layer models only to the extent that they include scale lengths that are large compared to the wave lengths of the modes considered.

Recently Wahlberg (1977) has studied electron oscillations of a collisionless plasma in the presence of a static electric field. The configuration considered has certain features in common with a double layer. A cold electron population (the beam component) is accelerated in a local and time-independent electric field, having the same principal structure as that shown in Figure 1b, while a hot electron component (the thermal component) is partly reflected by the same electric field. Hence, the phase space distribution of the electrons is similar to that shown in Figure 2c. The ions, on the other hand, form a fixed background of positive charge contrary to what is the case in ordinary double layers.

Wahlberg has shown that the amplitude of the uncoupled oscillations in the cold electron beam decreases as the beam is accelerated in the electric field. This result suggests that the acceleration of charged particles in a double layer might have a stabilizing effect on certain modes in the layer.

On the other hand Wahlberg has also given arguments which indicate that, under certain conditions, a new type of instability may be present in the configuration considered, even if the plasma is everywhere Penrose stable. The instability is caused by a mutual coupling and reflection of various modes in both the beam plasma and the thermal plasma.

There are, however, not only waves that may become unstable but also the whole structure of the double layer (Tonks, 1937; Jacobsen and Carlqvist, 1964; Carlqvist, 1972). In the specific models described in Section 3.1 collisions were completely neglected as well in the double layer as in the plasma outside it. In reality collisions (ordinary or collective) between the accelerated beam particles and the plasma surrounding the double layer are always present.

In order to demonstrate the importance of such collisions we consider a time-independent and one-dimensional double layer which is surrounded by a plasma having an extension that is much larger than the mean free path of the beam particles. By means of collisions the beam particles gradually transfer their momentum to the plasma. In the steady state the pressure of the plasma must then decrease towards the double layer. A necessary but not sufficient condition for the double layer to be stable is that the dynamic pressure, which the beam particles exert on the plasma, must be smaller than the total pressure of the plasma. If the opposite situation is prevailing the plasma will tend to be pushed away from the double layer. This means that the layer widens and that the potential drop across the layer increases provided the current is constant. The dynamic pressure of the beam particles then increases further leading to an unstable situation of the whole double layer. In the case the double layer has a two- or three-dimensional geometry, the efficiency of the beam pressure may be reduced because of spreading of the beam particles after the acceleration.

One may ask whether the beams necessarily must possess such a large dynamic pressure that they are capable of pushing away the plasma adjacent to the double layer for an unstable situation to occur. From the discussion in Section 3.4 it follows that if a time-independent double layer shall be able to exist, the plasma pressure must be reduced so much at the boundaries of the layer that some kind of Bohm condition becomes fulfilled there. If on the other hand the beams tend to reduce the plasma pressure to a value which is still smaller, this may lead to an inconsistent and possibly unstable situation since the locations of the boundaries of the double layer are roughly defined by the sites where the Bohm conditions are fullfilled.

From what has been said above it is clear that the stability of the double layers represents a difficult but important problem where much remains to be done.

4. THE TIME-DEPENDENT DOUBLE LAYER

Most double layers which have been studied experimentally in the laboratory show variations with time. Both fluctuating layers and transient layers have been observed. Although the time-dependent double layers are thus very common in the laboratory experiments, they have been studied theoretically only to a small degree due to their complex behaviour. In a few cases time-variable structures with the character of double layers have been investigated by means of numerical particle simulations (Goertz and Joyce, 1975; De Groot et al., 1977; Joyce and Hubbard, 1978).

There may be several reasons for the time-variations of the double layers. For instance, various instabilities may give rise to a fluctuating or transient appearance of the layers. Another cause of the time-variations may be a lack of the momentum balance in the plasma surrounding the layer in accordance with the discussion in Section 3.5.

As was pointed out above the whole electric circuit containing the double layer is generally of decisive importance for the behaviour of the layer. It is therefore often illuminating to divide the circuit into its separate elements like resistances, inductances, capacitances, and EMF (see Figure 7) (Boström, 1974; Alfvén, 1977). The double layer itself can be considered as a complex non-linear and time-variable element.

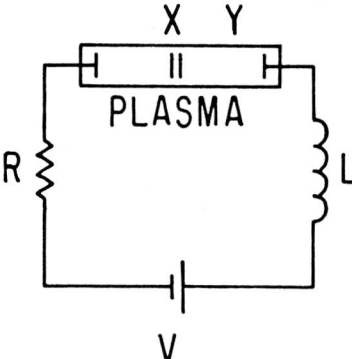

Figure 7. Electric circuit containing a plasma in which a double layer is formed (Alfvén, 1977). The plasma is represented by the two circuit elements X and Y. X refers to the double layer while Y refers to the surrounding plasma. The external circuit is represented by the resistance R, the inductance L, and the voltage source V. It is important to note that the properties of the double layer does not only depend on the properties of the local plasma but also on the whole circuit in which the plasma current flows.

Although extremely complicated, the analysis of the time-development of the double layer circuit can, in principle, be performed in the same way as of an ordinary electric circuit.

In a double layer which varies with time, extra inertia forces (positive or negative) exist which are not present in a time-independent layer of similar structure. As long as the time constant for the variation of the double layer is long compared to the times needed for the ions and electrons to pass a scale-length of the layer, the extra forces are small in comparison with other forces acting in the layer. If on the other hand the opposite situation is prevailing the extra inertia forces can be expected to be of relatively great importance. As a result of this, the conditions for the existance of the time-dependent layer may differ appreciably from those of the time-independent layer treated in Sections 3.3 and 3.4.

5. MECHANISMS FOR THE FORMATION AND MAINTENANCE OF DOUBLE LAYERS

5.1 The double layer as an adapting agent

Experiments show that double layers can behave in many different ways. On one hand there are double layers which show a transient appearance or which fluctuate strongly. This type of layer is mostly formed out of a fairly homogeneous plasma when some plasma parameter such as the current, the density, or the temperature attains a certain critical value. On the other hand there are also more stable and time-independent double layers which can exist for long periods. Noise and small fluctuations do, however, generally occur also in these layers.

Several mechanisms have been suggested to explain how these different types of double layers are formed and maintained. Since the appearance of double layers is so variable it is likely that several mechanisms may be active, although under different conditions.

First we shall devote our interest to how double layers can be maintained. From experiments it is well-known that persistent double layers tend to occur in low-pressure discharges when there is an abrupt constriction of the anode end of the discharge tube (Schönhuber, 1958; Crawford and Freeston, 1963; Andersson, 1977). The layer, which has an electric field that accelerates electrons towards the anode and ions towards the cathode, generally forms at the constriction where it bulges out more or less into the cathode plasma (see Figure 8).

Crawford and Freeston (1963) have constructed a simple theory of this kind of double layer. This theory suggests that the double layer has two functions. In the first place the layer serves as an adapting agent between two plasmas of different character - the electron gas on the anode side of the layer is hotter and denser than that on the cathode side. Crawford and Freeston assume that the electron gas on both sides of the double layer has a Maxwellian velocity distribution. If the number

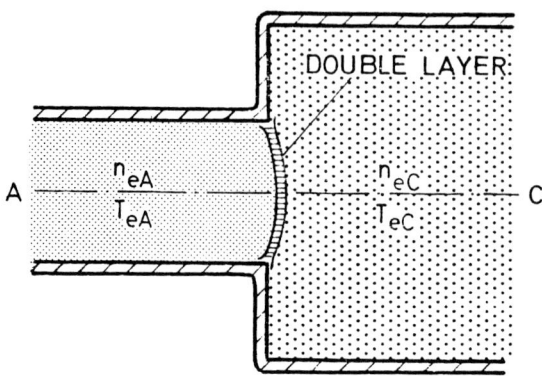

Figure 8. Double layer appearing at the constriction of a discharge tube. Both the temperature and the density of the electrons are larger in the narrow anode tube than in the wider cathode tube.

densities and temperatures of the electrons on the anode and cathode sides of the layer are denoted by n_{eA}, n_{eC}, T_{eA}, and T_{eC} respectively, the random electron flux densities in the two sections are $\Phi_{eA} = n_{eA}(kT_{eA}/2\pi m_e)^{1/2}$ and $\Phi_{eC} = n_{eC}(kT_{eC}/2\pi m_e)^{1/2}$. Since $T_{eA} > T_{eC}$ and $n_{eA} > n_{eC}$ we have $\Phi_{eA} > \Phi_{eC}$. In the case there were no double layer at the constriction there would consequently be a net electron flux from the plasma in the more narrow anode tube to the plasma in the wider cathode tube. Such a net electron flux corresponds to a current which flows in opposite direction to the normal discharge current. Since the ion current is not capable of influencing the total current very much this opposing electron current cannot be allowed to exist. In reality a double layer is formed at the constriction which adjusts the current crossing it, I_{DL}, to the applied discharge current. If A_{DL} is the effective area of the double layer this condition can be expressed by the equation

$$A_{DL}\, en_{eC}\left(\frac{kT_{eC}}{2\pi m_e}\right)^{1/2} - A_{DL}\, en_{eA}\left(\frac{kT_{eA}}{2\pi m_e}\right)^{1/2} \exp\left(-\frac{e\phi_{DL}}{kT_{eA}}\right) = I_{DL} \quad . \quad (5.1)$$

Here the first term on the left hand side represents the random current from the cathode plasma while the second term represents the random current from the anode plasma diminished to a first approximation by the Boltzmann factor. Reorganizing equation (5.1) and putting $I_{eC} = A_{DL}\, en_{eC}(kT_{eC}/2\pi m_e)^{1/2}$ we find the potential drop across the double layer to be

$$\phi_{DL} = \frac{kT_{eA}}{e} \ln\left[\frac{n_{eA}}{n_{eC}}\left(\frac{T_{eA}}{T_{eC}}\right)^{1/2} \frac{I_{eC}}{I_{eC}-I_{DL}}\right] \quad (5.2)$$

The second function of the double layer consists in that it can capture electrons from the cathode plasma and focus them into the narrow anode tube. We see from equation (5.2) that a plane double layer covering the constriction and having the area $A_{DL} = \pi r_A^2$ cannot carry a current larger than $I_{eC} = \pi r_A^2 en(kT_{eC}/2\pi m_e)^{1/2}$. If a current larger than this is forced to flow through the discharge tube and consequently also through the double layer, the layer compensates this by bulging out towards the cathode. In that way the cathode plasma electrons experience a larger surface of capture at the same time as they are focused towards the more narrow anode tube.

The idea that double layers may be formed at the boundary of two plasmas of different character has also been applied to the ionosphere and magnetosphere. Alfvén (1958) early realized that shocklike structures similar to double layers might be formed when a hot plasma encounters a cold.

Recently Lennartsson (1978) has in more detail studied how double layers may be formed in the magnetosphere. He considers a current loop which passes through both the hot magnetospheric plasma and the cold ionospheric plasma. In those parts of the circuit which connect the ionosphere with the magnetosphere the current is field-aligned. In order for this field-aligned current to flow from the ionosphere to the magnetosphere a considerable potential drop is required along the magnetic field lines. This is a consequence of the magnetic mirroring of the electrons which mainly carry the current. The plasma in which the field-aligned current flows must for the main part be quasi-neutral ($n_e \approx n_i$), otherwise excessively high electric fields would be produced. Lennartsson finds that quasi-neutrality is generally possible only if the potential makes a jump at some point in the field-aligned current path. He identifies this potential jump with a double layer. The role of the double layer is to fit two plasmas of different character to each other. One can therefore say that in principle it has the same function as the constriction layer discussed earlier.

5.2 Current-driven instabilities and double layers

As mentioned above time-dependent double layers have been observed to occur in fairly homogeneous discharge plasmas when for instance the current is increased above a certain critical value. High over-voltages of the order of $10^3 - 10^5$ V, leading to current limitation or current disruption, have then sometimes been measured across the discharge (e.g. see Hull and Elder, 1942; Jacobsen et al., 1968; Lutsenko et al., 1975). A necessary condition for these overvoltages to occur is that the discharge circuit contains an inductance or a sufficiently large voltage source. Immediately before the current reaches the critical value at which the double layer is formed, disturbances in the form of propagating solitary pulses appear in the discharge plasma (Torvén and Babić, 1975; Quon and Wong, 1976). Thus, there are reasons to assume that some kind of instability in the plasma gives rise to this type of dynamic double layer.

It has been argued that double layers might be formed as a result of a local evacuation of the plasma (Alfvén and Carlqvist, 1967; Carlqvist, 1969). The principal idea is, that if the plasma is forced to leave a local region at the same time as the current is kept constant - e.g. by means of an inductance - charges of opposite polarity will collect at the boundaries of the evacuated region until the potential drop across the region has grown so large that a space-charge limited current, similar to that described by Langmuir (1929) (see Section 3.2), can start to flow.

Several instabilities may give rise to a local evacuation of the plasma. One possibility is that the evacuation is caused by the two-stream instability (Carlqvist, 1973) which is active as soon as the current density exceeds a certain critical value which depends on the density and the temperature of the electrons and ions. This evacuation process can easily be explained if we consider a fully ionized plasma of temperature $T_e = T_i = T$ carrying the current density, i. The plasma is initially (t = 0) homogeneous (density = n_o) except for a small and local disturbance in the form of a dip in the density (see Figure 9a, the solid curve). For the sake of simplicity the disturbance is assumed to be at rest with respect to normal wave modes and to have a scale-length, ℓ, which is much larger than the Debye length. By means of Fourier analysis we can then divide the disturbance into a set of waves, the amplitude spectrum of which is shown by the solid curve in Figure 9b. If the current density is sufficiently large for the two-stream instability to occur ($i > 0.926 \; ne(2kT/m_e)^{1/2}\left[1 + (m_e/m_i)^{1/2}\right]$) all the mentioned waves will grow in amplitude as described by well-established linear

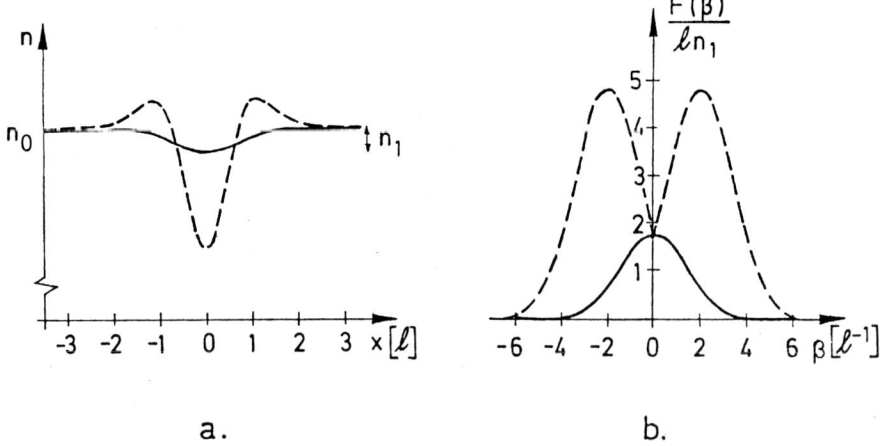

a.   b.

Figure 9. The development of a small disturbance of the density, n(x), and its associated Fourier spectrum $F(\beta)$ normalized by ℓn_1 is illustrated in a) and b) respectively. The solid curves show the initial state at t = 0 when the density disturbance is assumed to have a Gaussian shape with the amplitude, $n_1 (\ll n_o)$, while the dashed curves show the state at a later time $t = t_1$.

theory (Buneman, 1959; Jackson, 1960; Stringer, 1964). For the long wave lengths considered, the growth is approximately proportional to the wave number, β. The growth of the waves causes the spectral distribution of the disturbance to change with time (see Figure 9b, the dashed curve). By putting together the Fourier components again at a later time, $t = t_1$, we can find how the density disturbance develops in the linear phase. As is clear from Figure 9a the density dip becomes deeper and more narrow as time passes. Preliminary numerical calculations show that this trend seems to persist also in the non-linear phase of the development (Raadu and Carlqvist, 1979).

In order to demonstrate how the evacuation process works we have above studied a particular example, comprising the two-stream instability for equal ion and electron temperatures. However, nothing seems to prevent that the process may work also for other temperature ratios or even by means of other instabilities.

Another evacuation mechanism that might lead to the formation of double layers is the caviton as suggested by Torvén (1976). This mechanism also presupposes that a local disturbance in the form of a density dip is initially present in the plasma. When the current density exceeds a certain critical limit, plasma oscillations are supposed to be produced by some current-driven instability and trapped in the dip. When bouncing back and forth in the dip, the oscillations give rise to a <u>ponderomotive force</u> which acts outwards on the slopes of the dip. If the energy density of the electric field of the trapped oscillations becomes sufficiently large the ponderomotive force may give rise to a local evacuation of the plasma (see e.g. Zakharov, 1972; Morales and Lee, 1974, 1975). Whether this evacuation mechanism will be more effective than the mechanism discussed above or not, depends on the efficiency of the generation and trapping of oscillations in the dip.

Smith and Goertz (1978) have also suggested that the ponderomotive force might be important for the generation of double layers. They consider a plasma having a density that decreases in space. At those places where the density of the plasma is below a certain critical value, a current driven instability is assumed to occur and to give rise to plasma oscillations. Owing to the ponderomotive force, resulting from the oscillations, the density gradient steepens in time so that it finally becomes arbitrarily large. Since the ponderomotive force mainly acts on the electrons a charge separation with accompanying dc electric field arises at the position of the steep density gradient. This situation, Smith and Goertz expect, will lead to a double layer.

Besides the above discussed theoretical investigations, numerical particle simulations have also been used at the study of how double layers are formed and maintained (Goertz and Joyce, 1975; DeGroot et al., 1977; Joyce and Hubbard, 1978). The calculations, which often have been performed with somewhat artificial boundary conditions, have revealed structures similar to the double layers considered above. In addition to this, it has been possible to show that the bulk drift between ions and electrons must exceed a certain critical value for these structures to be formed. In most cases treated the minimum drift needed has turned out to be of the order of the electron thermal velocity.

For the future we may expect that particle simulations and other numerical methods will be increasingly used in investigations on the behaviour of double layers. Analytical studies of the double layer phenomenon should not for that reason be neglected, since they are of decisive importance for the understanding of the physics behind the layers. However, it is most important that both the numerical calculations and the theoretical analyses are performed in close contact with experimental investigations of real double layers, in the laboratory and in space.

ACKNOWLEDGEMENTS

I wish to thank Prof. H. Alfvén, Prof. C.-G. Fälthammar, and Drs M. Raadu och S. Torvén for valuable comments.

REFERENCES

Alfvén, H.: 1958, Tellus 10, pp. 104-116.
Alfvén, H.: 1977, Rev. Geophys. Space Phys. 15, pp. 271-284.
Alfvén, H.: 1978, Astrophys. Space Sci. 54, pp. 279-292.
Alfvén, H. and Carlqvist, P.: 1967, Solar Phys. 1, pp. 220-228.
Alfvén, H. and Fälthammar, C.-G.: 1963, "Cosmical Electrodynamics", 2nd ed., Clarendon Press, Oxford, pp. 195-199.
Andersson, D.: 1977, J. Phys. D.: Appl. Phys. 10, pp. 1549-1556.
Andrews, J.G. and Allen, J.E.: 1971, Proc. Roy. Soc. Lond. A. 320, pp. 459-472.
Bernstein, I.B., Greene, J.M., and Kruskal, M.D.: 1957, Phys. Rev. 108, pp. 546-550.
Bertrand, P. and Feix, M.R.: 1968, Phys. Letters 28A, pp. 68-69.
Block, L.P.: 1972, Cosmic Electrodynamics 3, pp. 349-376.
Block, L.P.: 1975, B. Hultqvist and L. Stenflo (eds.),"Physics of the Hot Plasma in the Magnetosphere", Plenum Press, New York, London, pp. 229-249.
Block, L.P.: 1978, Astrophys. Space Sci. 55, pp. 59-83.
Bohm, D.: 1949, A. Guthrie and R.K. Wakerling (eds.),"The Characteristics of Electrical Discharges in Magnetic Fields", McGraw-Hill, New York, Toronto, London, pp. 77-86.
Boström, R.: 1974, B.M. McCormac (ed.), "Magnetospheric Physics", Reidel Publ. Comp., Dordrecht-Holland, pp. 45-59.
Buneman, O.: 1959, Phys. Rev. 115, pp. 503-517.
Carlqvist, P.: 1969, Solar Phys. 7, pp. 377-392.
Carlqvist, P.: 1972, Cosmic Electrodynamics 3, pp. 377-388.
Carlqvist, P.: 1973, Tech. Rep. TRITA-EPP-73-05, Dept of Plasma Phys., Royal Inst. of Tech., Stockholm, Sweden.
Child, C.D.: 1911, Phys. Rev. 32, pp. 492-511.
Coakley, P., Hershkowitz, N., Hubbard, R., and Joyce, G.: 1978, Phys. Rev.

Letters 40, pp. 230-233.
Crawford, F.W. and Freeston, I.L.: 1963, VIe Conférence "Internationale sur les Phénomènes d'Ionisation dans les Gas", Paris, Vol. I, pp. 461-464.
DeGroot, J.S., Barnes, C., Walstead, A.E., and Buneman, O.: 1977, Phys. Rev. Letters 38, pp. 1283-1286.
Goertz, C.K. and Joyce, G.: 1975, Astrophys. Space Sci. 32, pp. 165-173.
Haerendel, G., Rieger, E., Valenzuela, A., Föppl, H., Stenbaek-Nielsen, H.C., and Wescott, E.M.: 1976, in "European Programmes on Sounding-Rocket and Balloon Research in the Auroral Zone", Rep. ESA-SP115, European Space Agency, Neuilly, France.
Hasan, S.S. and ter Haar, D.: 1978, Astrophys. Space Sci. 56, pp. 89-107.
Hull, A.W. and Elder, F.R.: 1942, J. Appl. Phys. 13, pp. 372-377.
Jackson, E.A.: 1960, Phys. Fluids 3, pp. 786-792.
Jacobsen, C.T. and Carlqvist, P.: 1964, Icarus 3, pp. 270-272.
Jacobsen, C.T., Robinson, T., Sandahl, S., and Terlouw, J.C.: 1968, Arkiv för Fysik 38, pp. 113-126.
Joyce, G. and Hubbard, R.F.: 1978, Rep. 78-10, Dept of Phys. and Astron., U. of Iowa, Iowa City.
Knorr, G. and Goertz, C.K.: 1974, Astrophys. Space Sci. 31, pp. 209-223.
Langmuir, I.: 1913, Phys. Rev., Ser.2 2, pp. 450-486.
Langmuir, I.: 1929, Phys. Rev. 33, pp. 954-989.
Langmuir, I. and Mott-Smith, H.M., Jr.: 1924, Gen. Elec. Rev. 27, pp. 762-771.
Lee, H.-J., McKenzie, J.F., and Axford, W.I.: 1977, Astrophys. Space Sci. 51, pp. 3-32.
Lennartsson, W.: 1978, Tech. Rep. TRITA-EPP-78-08, Dept of Plasma Phys., Royal Inst. of Tech., Stockholm, Sweden. Submitted for publication to Planet. Space Sci.
Lutsenko, E.I., Sereda, N.D., and Kontsevoi, L.M.: 1975, Zh. Tekhn. Fiz. 45, pp. 789-796 (Sov. Phys. Tech. Phys. 20, pp. 498-502).
Marklund, G.: 1978, Tech. Rep. TRITA-EPP-78-09, Dept of Plasma Phys., Royal Inst. of Tech., Stockholm, Sweden.
Montgomery, D. and Joyce, G.: 1969, J. Plasma Phys. 3, pp. 1-11.
Morales, G.J. and Lee, Y.C.: 1974, Phys. Rev. Letters 33, pp. 1016-1019.
Morales, G.J. and Lee, Y.C.: 1975, Rep. PPG-211, Dept of Physics, U. of Calif., Los Angeles.
Mozer, F.S., Carlson, C.N., Hudson, M.K., Torbert, R.B., Parady, B., Yatteau, J., and Kelley, M.C.: 1977, Phys. Rev. Letters 38, pp. 292-295.
Murty, G.S.: 1962, Arkiv för Fysik 21, pp. 203-211.
Penrose, O.: 1960, Phys. Fluids 3, pp. 258-265.
Persson, H.: 1968, Rep. 68-18, Depts of Electron and Plasma Phys., Royal Inst. of Tech., Stockholm, Sweden.
Persson, H.: 1969, Rep. 69-05, Depts of Electron and Plasma Phys., Royal Inst. of Tech., Stockholm, Sweden.
Quon, B.H. and Wong, A.Y.: 1976, Phys. Rev. Letters 37, pp. 1393-1396.
Raadu, M.A. and Carlqvist, P.: 1979, in preparation.
Schönhuber, M.J.: 1958, "Quecksilber-Niederdruck-Gasentladungen", Lachner, München.
Shawhan, S.D.: 1976, J. Geophys. Res. 81, pp. 3373-3379.
Shawhan, S.D., Goertz, C.K., Hubbard, R.F., Gurnett, D.A., and Joyce, G.: 1975, V. Formisano (ed.), "The Magnetospheres of Earth and Jupiter",

Reidel Publ. Comp., Dordrecht-Holland, pp. 375-389.
Shawhan, S.D., Fälthammar, C.-G., and Block, L.P.: 1978, J. Geophys. Res. 83, pp. 1049-1054.
Smith, R.A. and Goertz, C.K.: 1978, J. Geophys. Res. 83, pp. 2617-2627.
Stringer, T.E.: 1964, Plasma Phys. 6, pp. 267-279.
Swift, D.W.: 1975, J. Geophys. Res. 80, pp. 2096-2108.
Swift, D.W.: 1976, J. Geophys. Res. 81, pp. 3935-3943.
Tonks, L.: 1937, Trans. Electrochem. Soc. 72, pp. 167-182.
Torvén, S.: 1976, private communication.
Torvén, S.: 1978, Tech. Rep. TRITA-EPP-78-13, Dept of Plasma Phys., Royal Inst. of Tech., Stockholm, Sweden (Astrophysics and Space Science Library - see this volume).
Torvén, S. and Andersson, D.: 1978, Tech. Rep. TRITA-EPP-78-12, Dept of Plasma Phys., Royal Inst. of Tech., Stockholm, Sweden.
Torvén, S. and Babić, M.: 1975, J. G. A. Hölscher and D.C. Schram (eds.), Proc. 12th Int. Conf. on Phenomena in Ionized Gases, American Elsevier Publ. Comp., New York, pp. 124-125.
Torvén, S. and Babić, M.: 1976, IEE 4th Int. Conf. on Gas Discharges, Swansea, Conf. Publ. No 143, p. 323.
Wahlberg, C.: 1977, J. Plasma Phys. 18, pp. 415-431.
Wescott, E.M., Stenbaek-Nielsen, H.C., Hallinan, T.J., and Davis, T.N.: 1976, J. Geophys. Res. 81, pp. 4495-4502.
Zakharov, V.E.: 1972, Zh. Eksp. Teor. Fiz. 62, pp. 1745-1759 (Sov. Phys. - JETP 35, pp. 908-914).

FORMATION OF DOUBLE LAYERS IN LABORATORY PLASMAS

Staffan Torvén
Department of Plasma Physics, Royal Institute of Technology,
S-100 44 Stockholm, Sweden

ABSTRACT

A review of experimental observations of electric double layers in different types of laboratory plasmas is presented. Particular investigations relating to the formation of a double layer and the layer stability are also discussed, and some typical plasma parameters in the different experiments are summarized. The existence of double layers in unmagnetized as well as magnetized plasma is now well established. However, problems concerning the layer formation, the layer stability, and the origin and significance of observed high frequency fluctuations remain to be solved.

1. INTRODUCTION

Space charge regions, which are spatially confined within distances of some tens of Debye lengths, can exist in low density plasmas. They give rise to potential drops much larger than kT_e/e, where T_e is the electron temperature on the low potential side of the region.

In the unmagnetized plasma of wall-confined low pressure arcs such regions have been observed as potential discontinuities existing at a sharp constriction of the diameter of the confining tube (Langmuir and Mott-Smith Jr 1924). This type of discontinuity, which is of importance in technical applications, has since been subject to many investigations (Schönhuber 1963, Crawford and Freeston 1963, Andrews and Allen 1971, Sandahl 1972, Andersson 1977, Torvén 1978).

Another type of potential discontinuity, which may be called a free double layer, has also been observed (Torvén and Babić 1975, Levine, Ilić, and Crawford 1977). It can form in an axially uniform column without any external disturbance from a tube constriction, and its formation is triggered by increasing the discharge current to sufficiently large values.

Recently it has become possible to resolve the internal structure of such space charge regions through the development of double plasma devices (Taylor, MacKenzie and Ikezi 1972). In such devices stationary electric double layers have been produced (Quon and Wong 1976, Coakley, Hershkowitz, Hubbard, and Joyce 1978), and investigations of the layer formation process and the layer stability have been presented.

A question of particular interest is the excitation of double layers in magnetized plasma since wall effects are suppressed by the magnetic confinement. Indication that double layers exist in that case too has been reported by Lutsenko, Sereda and Kontsevoi (1975) who observed localized potential drops on the discharge tube wall. Recently Torvén and Andersson (1978), by means of direct observations within a magnetized plasma column, have found conclusive evidence of the existence of double layers.

In this review observations of double layers and potential discontinuities in different devices are described in Section 2. Particular investigations relating to the formation of a double layer and the layer stability are discussed in Sections 3 and 4, and some typical plasma parameters in the different experiments are summarized in Section 5.

2. OBSERVATIONS OF DOUBLE LAYERS AND POTENTIAL DISCONTINUITIES IN DIFFERENT DISCHARGE DEVICES

2.1 Double plasma devices

The double plasma devices contain a large volume of collisionfree and quiescent plasma at such low densities that the Debye length is of the order of 1 mm. This means that the particle motion can be controlled by grids. Too, the long Debye length implies that the internal structure of localized space charge regions can be resolved.

The basic design of a double plasma device is shown in Fig. 1. The vacuum chamber is divided into two electrically insulated chambers. Plasma is produced in each chamber by electron bombardment of the neutral gas from a multiplicity of filaments placed symmetrically at the chamber walls. A flow of electrons from one chamber (the source or driver chamber) into the other (the target chamber) can be produced by applying a voltage between the chamber walls.

Quon and Wong (1976) used a modified version of this device containing two closely spaced grids which separated the two chambers, and plasmas were produced by filaments in the source chamber. When a voltage is applied between the chambers, the potential drop between the two plasmas usually occurs between the grids, which normally are biased at the same potentials as their neighbouring chambers. Quon and Wong observed that the potential jump instead occurred in the plasma in the target chamber when the electron drift speed exceeded $(kT_e/m_e)^{1/2}$. Here m_e is the electron mass. An electric double layer with a monotonically increasing potential

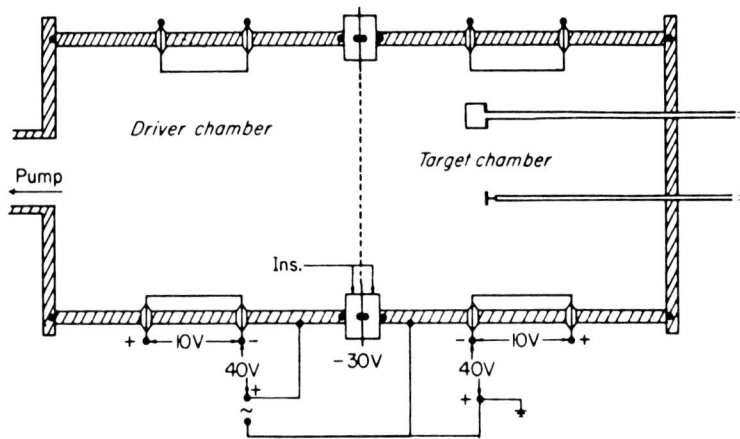

Figure 1. A double plasma device with two electrically insulated chambers separated by a grid (Taylor, MacKenzie, and Ikezi 1972). Quon and Wong (1976) used a modified version of this device containing two closely spaced grids. Plasma was produced by electron bombardment of neutral gas from filaments in the driver chamber. A voltage applied between the chamber walls produced an electron flow into the target chamber where a double layer was formed when the electron drift speed exceeded $(kT_e/m_e)^{1/2}$.

was formed, and stable layers were observed for values of $e\phi/kT_e$ between 1 and 4.5. Here ϕ is the layer voltage. The existence of trapped and free electrons and ions was confirmed by measurements with probes and energy analyzers. Time resolved potential profiles during the layer formation process were also measured (see Section 3). The ions accelerated across the layer towards the source chamber were reflected by one of the grids which had to be biased positive with respect to the source plasma in order to obtain layer formation.

The importance of this grid bias, giving counterstreaming ion beams on the low potential side of the layer, was confirmed in the experiments by Coakley, Hershkowitz, Hubbard, and Joyce (1978). These authors observed stable double layers (Fig. 2) with such large values of $e\phi/kT_e$ as 14. In this experiment trapped electrons were provided by a separate plasma source on the high potential side of the layer. This seems to have important consequences for the layer stability as is further discussed in Section 4.

2.2 Positive columns in low pressure discharges

In low pressure arcs, plasma columns can be maintained in glass tubes under conditions such that the mean free paths are larger than the column diameter while the Debye length usually is much smaller (typically 0.1 mm). Under these conditions potential discontinuities have

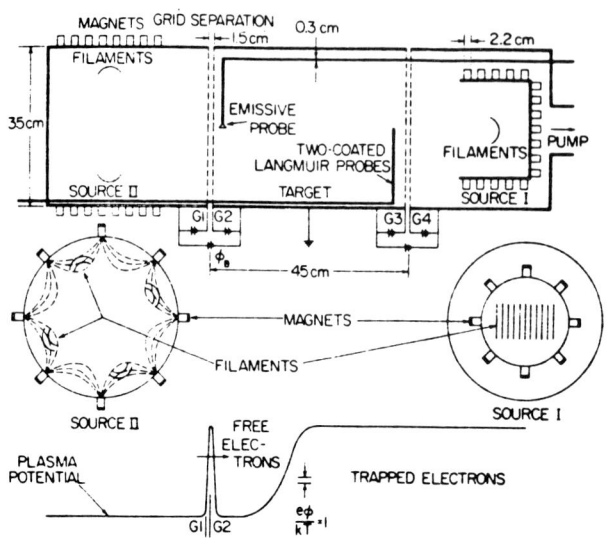

Figure 2. Device used by Coakley, Hershkowitz, Hubbard and Joyce (1978). In this device an additional plasma source (source I) provides trapped electrons on the high potential side of the double layer. The double layer formed in the central target chamber as shown in the diagram below the apparatus. As also was noticed by Quon and Wong (1976) it was possible to obtain a layer only when the grid G_2 was biased positive with respect to the plasma. A large number of small permanent magnets, stacked at the walls of the sources, give cusp fields which increase the source densities.

been observed with $e\phi/kT_e \gg 1$.

One type of layer has been observed in front of the cathode from which it is separated by a "dark space" (analogous to the Faraday dark space in glow discharges) containing strongly inhomogeneous plasma with a large electron drift speed (Schönhuber 1963). Another type of layer exists at a constriction in the discharge tube wall as shown in Fig. 3. Typical values of $e\phi/kT_e$ are found to be between 2 and 5 (Schönhuber 1963). Measurements of electron energy distributions (Section 4) have confirmed the existence of free and trapped electrons (Crawford and Freeston 1963, Andersson 1977).

Layers can also form in axially uniform arc plasma columns without any external disturbance from a constricted tube. Torvén and Babić (1975) produced potential discontinuities in an initially axially uniform mercury plasma column with a length of 100 cm, a diameter of 2 cm and a Debye length of 0.005 cm. In this case the electron number density was found to saturate at sufficiently large currents, and the electron drift speed began to increase with increasing current. When the drift speed

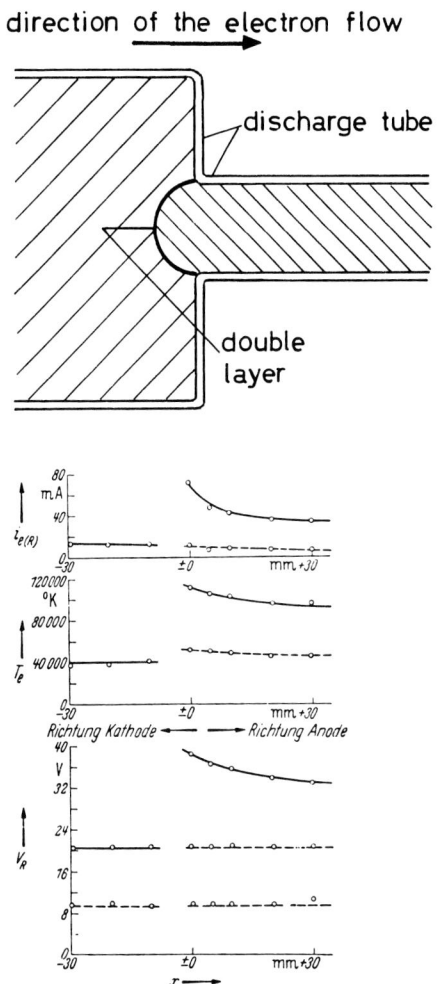

Figure 3. Double layer at a tube constriction in a low pressure mercury arc (Schönhuber 1963). The full lines in the diagrams show the axial variation of the random electron current (i_e), the electron temperature (T_e), and the plasma potential (V_R).

reached values between 0.9 à 1.1 times $(kT_e/m_e)^{1/2}$, a layer with a stationary position was formed. However, the formed layer was unstable as shown in Fig. 4. Pulses with a duration of about 2 µs, corresponding to about 10 ion plasma periods ($2\pi/\omega_{pi}$), were superimposed on the layer voltage, and pulse amplitudes up to 200 V ($e\phi/kT_e \simeq 20$) were measured. Between the pulses, the layer voltage assumed only small values (10 à 30 V) as indicated by the zero line inserted in the oscillogram. The voltage fluctuations were measured by external capacitive probes relative to the quiescent plasma on the cathode side of the layer. The layer vol-

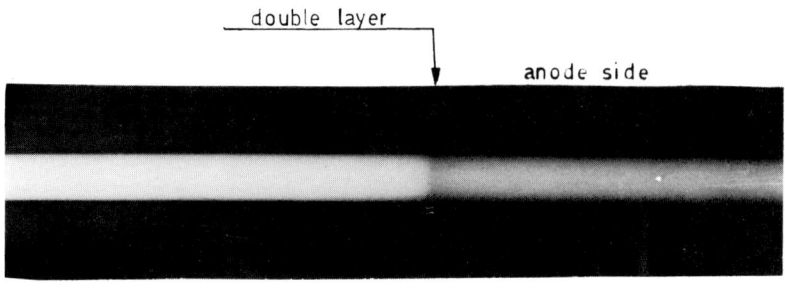

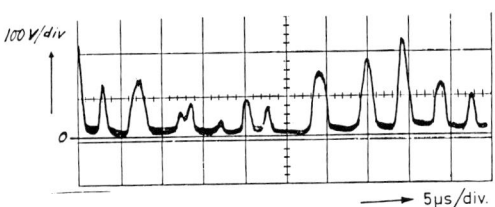

Figure 4. Double layer formed in an axially uniform plasma column in a mercury arc when the electron drift speed exceeds $(kT_e/m_e)^{1/2}$ (Torvén and Babić 1975). The layer position was stationary, but the layer voltage was strongly fluctuating as shown by the oscillogram where the inserted zero line refers to the quiescent plasma on the cathode side of the layer. The fluctuations were measured by external capacitive probes. The layer voltage between the pulses was inferred from separate measurements with Langmuir probes.

tage between the pulses was inferred from separate measurements with Langmuir probes.

Voltage fluctuations almost identical with the layer voltage fluctuations existed between any point along the plasma on the anode side of the layer and the quiescent cathode plasma. Attempts to observe any time delay between pulses measured in the anode plasma by axially separated probes showed that the voltage changes propagate with a velocity $v > 10^6 ms^{-1}$ which is of the order of the thermal electron speed.

Transient layers with life times up to 30 μs were also observed. They usually appeared about periodically in time. The period between two layers was determined by the time the discharge current needed to increase to the threshold value for layer formation after the current decrease which occurred when a layer existed. The layers formed at different positions at different times, and the layer voltage was superimposed by pulses similar to those in Fig. 4.

The plasma on the anode side of the layer also emitted high frequency radiation which was detected outside the tube. The time-averaged frequency spectrum had maxima at the plasma frequency and at whole number multiples of this frequency.

Potential discontinuities have also been observed under similar conditions in a hydrogen discharge (Armstrong and Torvén 1974, Armstrong 1975).

Levine, Ilić, and Crawford (1977) produced potential discontinuities in an argon plasma column by pulsing the discharge current periodically from a smaller to a higher value. The layer formation was accompanied by rf emission. In another experiment (Levine, Ilić, and Crawford 1978) the break-up of an electron beam by beam-plasma interaction was studied. It was found that the beam rf break-up was followed by a dc potential step. The double layer formation was interpreted in terms of a non-linear model predicting a rectified dc potential proportional to the square of the rf field. This work is presented in another paper in this volume (Crawford, Levine, and Ilić 1979).

Striations, i.e. a marked spatial variation of the light output from the plasma column, have also been produced in low density plasma columns (Foulds 1956, Schönhuber 1963). However, since no measurements of potential profiles were made in these experiments, it is not known whether these striations are associated with potential discontinuities. For considerably higher gas pressures such that the plasma approaches the diffusion regime, striations have been observed to be accompanied by potential discontinuities with potential drops of the order of the ionization potential for the neutral gas (Boyd and Twiddy 1960, Twiddy 1961).

2.3 Double layers in magnetized plasmas

Lutsenko, Sereda, and Kontsevoi (1975) studied a plasma column (Fig. 5a) confined by an axial magnetic field (0.2 T). A voltage from a capacitor C (15 kV) was applied between a cold cathode (2) and an anode (3). The space between the electrodes was filled with a pregenerated plasma with an ionization degree of nearly 100%. After an initial high current pulse (10^3A) with a duration of 1 à 2 µs, the current decreased to small values (10-100 A). This "current pause" could last for times up to about 10 µs, when usually a new high current pulse appeared. During these current variations, the discharge voltage was kept approximately constant.

Measurements with external capacitive probes showed an axial potential distribution according to Fig. 5b. The figure shows that most of the discharge voltage is concentrated to a cathode sheath at a time of 2 µs,

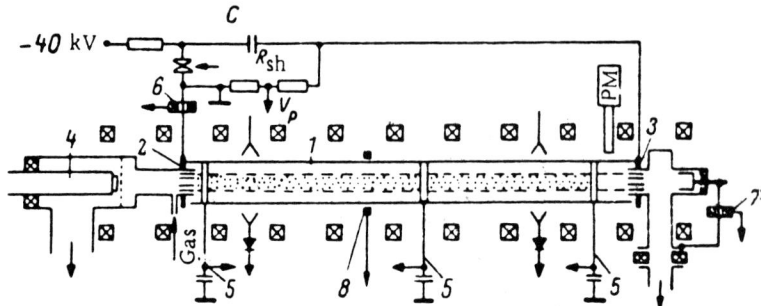

Figure 5a. Device used by Lutsenko, Sereda and Kontsevoi (1975). A nearly fully ionized plasma column with a diameter of 3 cm is produced in the glass chamber (1) (length 60 cm, diameter 8 cm). This was done by injecting a 10 keV electron beam from a source (4) into the chamber at a background pressure of $3 \cdot 10^{-5}$ T. The axial magnetic field is 0.2 T. After the electron beam is shut off, the main discharge is started by applying a voltage from a capacitor C between the grounded cathode (2) and the anode (3). Potential jumps of 10 à 15 kV over distances of 2 à 5 cm were found to evolve from a cathode sheath as shown in Fig. 5b.

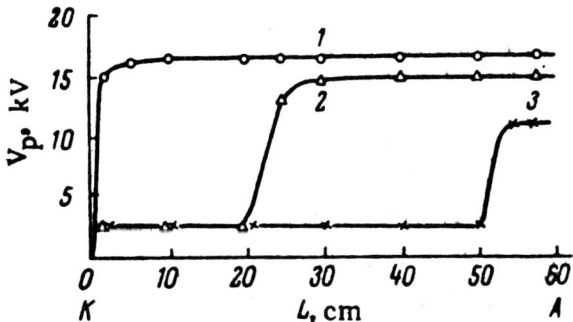

Figure 5b. Axial potential profiles measured by external capacitive probes. Curves 1, 2, and 3 corresponds to the times 2, 5, and 15 μs after the capacitor discharge has been switched on. K and A denote the positions of the cathode and the anode (Lutsenko, Sereda, and Kontsevoi 1975).

and the potential of the whole plasma column is the same as the anode potential. When the "current pause" begins, the cathode sheath is converted into a double layer which moves towards the anode with a velocity between 10^3 and 10^5 m/s. The moving layer was also associated with a large density jump. When the layer began to move, a dense plasma ($10^{19} m^{-3}$) had formed between the layer and the cathode. The density in the rest of the plasma column, which in the beginning of the discharge was $3 \cdot 10^{18} m^{-3}$, had decreased to $10^{17} m^{-3}$. This density decrease was associated with a radial expansion of the column.

During the high current pulses an electron beam was formed in the column, and radiation in a broad band (0.05-10 GHz) was emitted from the column.

In recent experiments Torvén and Andersson (1978) observed double layers in a magnetized plasma column, confined by an axial magnetic field variable between 0.005 and 0.15 T. The plasma was produced in the device shown in Fig. 6a. A source of plasma is obtained from a dc arc discharge between the mercury pool cathode C and the hollow electrode E_1. The plasma penetrates along the magnetic field lines through E_1 and through another aperture in the plate E_2 and into the vacuum chamber. There a quiescent and currentfree plasma column is obtained when the anode A is left floating.

When the anode potential V_A relative to the plasma was increased from floating potential, the anode acted essentially as a Langmuir probe.

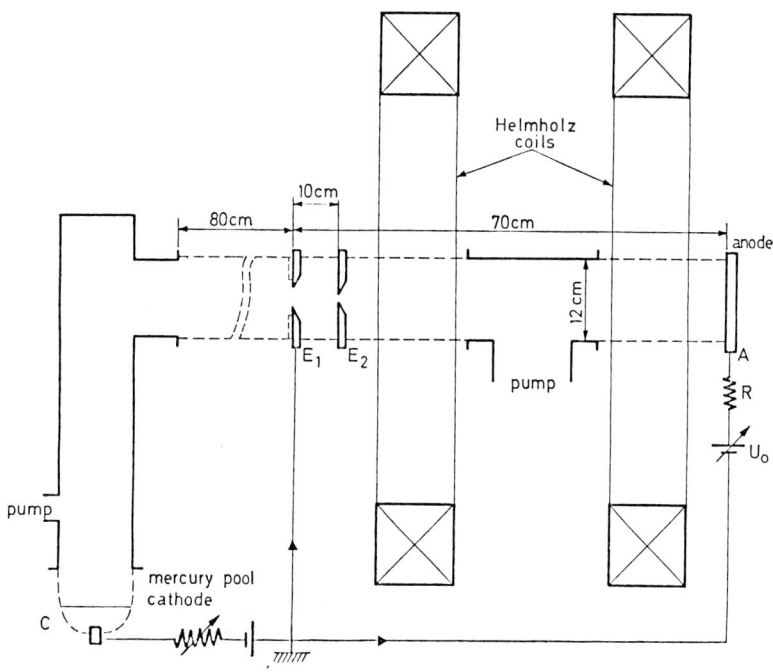

Figure 6a. Device used by Torvén and Andersson (1978) (dashed lines: glass, full lines: stainless steel). A plasma column confined by an axial magnetic field (0.005-0.15 T), is obtained between E_2 and A by inflow of plasma through the aperture in E_2. The plasma source is a dc arc between C and E_1. A double layer evolving from an anode sheath is formed when the potential of the anode A is increased to sufficiently large values.

The current increased exponentially until V_A reached the plasma potential. For larger values of V_A the current saturated, and an electron-rich anode sheath was formed. During this increase of V_A the potential along the plasma column remained the same as in the currentfree column because the electric field did not penetrate into the plasma, but the potential difference between the plasma and the anode was concentrated to the anode sheath. The anode sheath was converted into a double layer when V_A reached values of about 25 V. This value refers to a background density of $5 \cdot 10^{18} m^{-3}$ of mercury atoms, and larger values of V_A were required to bring about the conversion at lower background densities.

In Fig. 6 b typical axial potential profiles are given. Diagram 1 shows a profile before the layer formation, when the externally applied voltage is concentrated to the anode sheath. Diagram 2 shows a profile after the double layer formation when a plasma has formed between the double layer and the anode. By increasing the discharge current further (through an increase of the external EMF U_o) the layer could be moved to any new position between A and E_2, and diagram 3 shows a profile for another layer position at a discharge current of about 60 mA. The plasma on the cathode side of the layer was maintained by the plasma source. On the anode side of the layer the plasma was maintained entirely by ionization through fast electrons accelerated across the layer because positive ions originating from the cathode plasma did not have sufficient energy to move into the anode plasma.

The profiles were measured by three electron emitting Langmuir probes

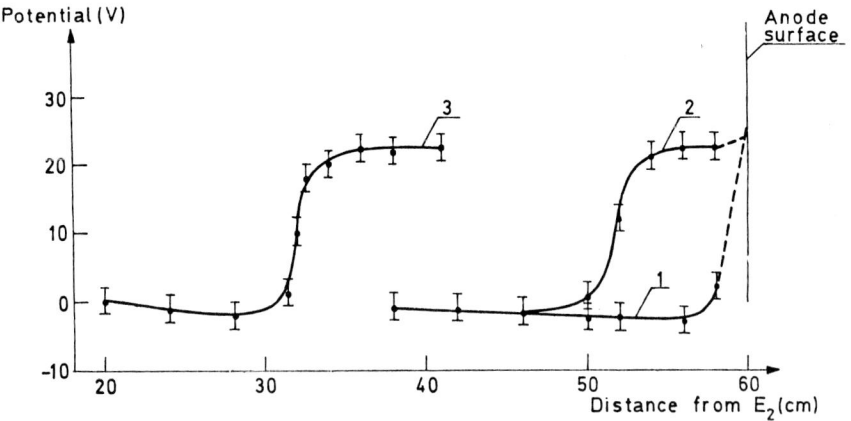

Figure 6b. Diagram 1 shows the potential profile along the magnetized plasma column before the double layer formation when the applied voltage is concentrated to an anode sheath. The probes could not be moved closer to the anode than 2 cm. Diagrams 2 and 3 show typical double layer profiles at two different positions corresponding to two different discharge currents. The profiles were measured by electron emitting probes (bandwidth 0-200 kHz), and they were obtained by a sampling technique because fluctuations (10-200 kHz) were present (Torvén and Andersson 1978).

(bandwidth 0-200 kHz), and they represent instantaneous profiles obtained by a sampling technique because fluctuations (10-200 kHz) were present in the layer voltage ($\delta\phi/\phi < 30\%$). These fluctuations propagated in the anode plasma with a velocity $v > 10^5 ms^{-1}$. Too, the layer exhibited an axial motion back and forth with amplitudes somewhat larger than the layer thickness. The velocity in this motion was about 300 ms^{-1}. However, the potential within the layer was always monotonically increasing, and a typical value of $e\phi/kT_e$ was 12.

Before the layer formation the electron drift speed always assumed its maximum value close to the anode. This property is inherently connected with the single ended plasma inflow in this device.

Fluctuations with frequencies close to the plasma frequency existed in the plasma on the anode side of the layer.

3. THE FORMATION OF A DOUBLE LAYER

Some experiments (Quon and Wong 1976, Torvén and Babič 1975) have related the layer formation to a threshold value for the electron drift speed equal to $(kT_e/m_e)^{1/2}$. Such a threshold value has also been observed in numerical simulations (Goertz and Joyce 1975, DeGroot, Barnes, Walstead, and Buneman 1977), but it has not yet been possible to relate the initial phase of the layer formation to the linear growth rate of some instability.

However, there is another characteristic phenomenon preceding the layer formation. This is the excitation of solitary pulses which propagate in the direction of the electron drift simultaneously as their amplitudes grow to non-linear values. These precursors of double layers have been observed both in laboratory experiments and numerical simulations.

Quon and Wong (1976) studied the formation of a double layer by pulsing the electron drift speed, v_d, on and off for $v_d = 1.5 \, (kT_e/m_e)^{1/2}$. The plasma potentials were measured by Langmuir probes at successive stages of the temporal evolution. The obtained profiles are shown in Fig. 7a. The electron drift was switched on at $t = 0$, and x is the distance from the grids. The figure shows that a potential pulse is excited close to grid and propagates in the direction of the electron drift with a speed of $10^4 ms^{-1}$. Its amplitude grows to $e\phi/kT_e \gtrsim 1$, and the pulse evolves into a moving potential step which leaves the considered region. After the passage of this structure a double layer develops. It was reported that this layer remained essentially stationary in later times.

Torvén and Babič (1975) studied the layer formation by investigating the different quasi-stationary states succeding each other, when the electron drift speed was increased in small steps. The first evidence of the forthcoming formation of the layer was the development of a

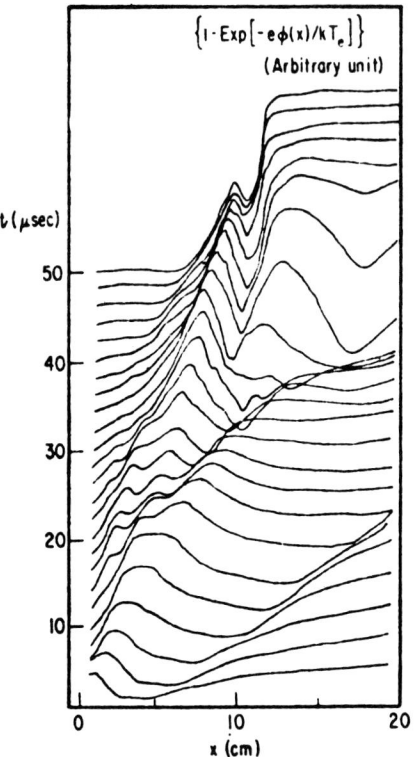

Figure 7a. Time resolved Langmuir probe measurements of axial profiles showing the formation of a double layer and propagating solitary pulses preceding the layer formation (Quon and Wong 1976). x is the distance from the grids. The electron drift was switched on at t = 0, and the curves were measured by pulsing the drift on and off.

relaxation region for the axial dc field. Over a distance of 10 à 20 cm (in a 100 cm long column with a Debye length of 0.05 mm) the magnitude of the dc field decreased from a higher value on the cathode side of the region to a lower value on the anode side. On the cathode side of this relaxation region solitary pulses were excited when v_d was 0.8 à 0.9 times $(kT_e/m_e)^{1/2}$. In Fig. 7b a small pulse is seen to be spontaneously excited. It propagates with a velocity of $10^3 ms^{-1}$ in the direction of the electron drift. The amplitude grows and saturates at a level of $e\phi/kT_e \lesssim 1$ whereafter the pulse disintegrates leaving only a wave trail in the downstream region. This wave trail had a much larger propagation velocity ($v > 10^6 ms^{-1}$), and probably it represents a disturbance of the electron flow introduced by the pulses and carried downstream by the electrons. For a constant value of the drift the pulse formation was repeated quasi-periodically.

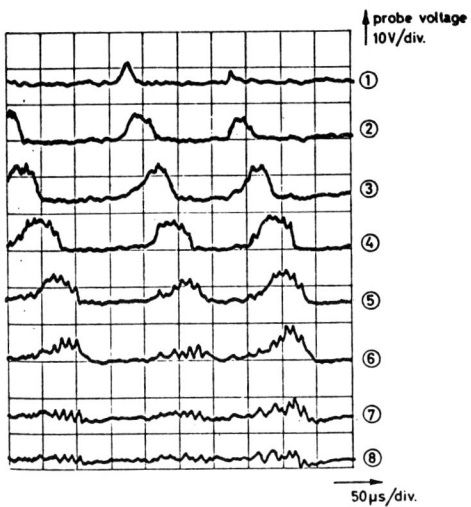

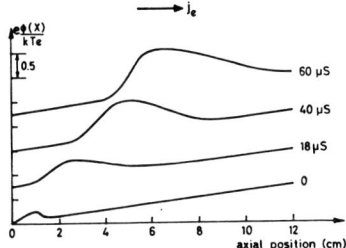

Figure 7b. Propagating potential pulses observed as precursors of double layers (Torvén and Babić 1975). The potentials were measured simultaneously by eight capacitive probes with a spacing of 2 cm. The electron drift speed was kept at a constant value slightly smaller than the threshold value for layer formation, and the formation of the pulses was repeated quasi-periodically. Below the oscillogram, axial potential profiles are depicted. The dc potential drop was inferred from Langmuir probe measurements.

In a numerical simulation Joyce and Hubbard (1978) observed similar pulses in the temporal evolution of the plasma during the initial stage of the layer formation (Fig. 7c). A well defined pulse speed was obtained, and the results led the authors to suggest, that the pulse velocity is proportional to $(m_e/m_i)^{1/2}$ and $(\Delta\phi_o)^{1/2}$. Here $\Delta\phi_o$ is the dc potential drop over the considered region. Using the suggested proportionality it is found that the simulated pulse velocity ($m_i/m_e = 64$, $\Delta\phi_o = 50$ kT_e/e) agrees roughly (within a factor of three) with the velocities determined

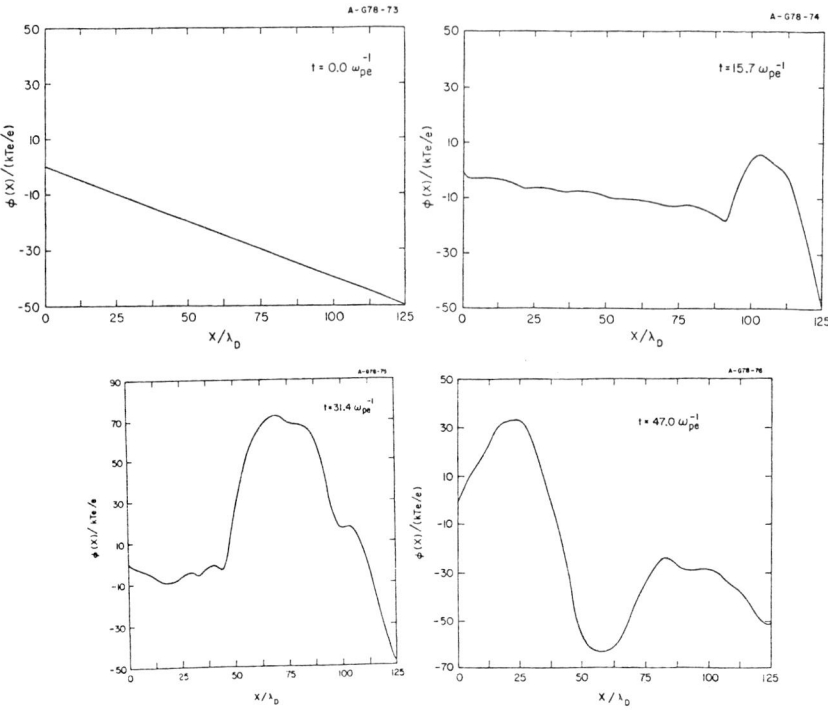

Figure 7c. Example of a propagating pulse observed in a numerical simulation during the initial stage of the layer formation (Joyce and Hubbard 1978). A time independent voltage drop $\Delta\phi_o = 50 \, kT_e/e$ was applied over the simulation region.

in the laboratory experiments.

In magnetized plasmas double layers have so far been observed to evolve from electrode sheaths. Further experiments are required to explore the layer formation under more general conditions.

4. STABILITY OF THE LAYER

The electron beam injected into the plasma on the high potential side of the layer as well as the ion beam on the low potential side represent potentially unstable systems. Nevertheless double layers with a time-independent voltage drop have been observed over a wide range of values of $e\phi/kT_e$.

Stable double layers were observed by Quon and Wong (1976) in the interval $1 < e\phi/kT_e < 4.5$. The authors report that for layer voltages in this interval the electron beam was thermalized through beam plasma interactions, which produced hot electrons, moving both in the direction of the beam and in the opposite direction, in a downstream region 35 à 70 Debye lengths beyond the layer. These electrons were the only source of trapped electrons in this experiment. For $e\phi/kT_e > 4.5$ no double layers were observed.

In contrast to this result Coakley et al. (1978) observed stable double layers with $e\phi/kT_e = 14$. These authors, who used a separate plasma source to produce the trapped electrons, suggested that the apparent discrepancy depends on the fact that the only source of trapped electrons was thermalized beam electrons in Quon's and Wong's experiment. They also estimated electron energy distribution functions from differentiated Langmuir probe characteristics and determinations of plasma potentials by electron emitting probes. The result was presented as distribution functions of the magnitude of the velocity (Fig. 8a). It appears from the figure that the electrons accelerated across the layer do not thermalize over a distance of 30 cm (200 Debye lengths). Similar energy distribution functions were also measured at a layer formed at a tube constriction in a low pressure arc (Crawford and Freeston 1963, Andersson 1977). In this case (Fig. 8b) relaxation lengths up to 1000 Debye lengths were required to damp the high energy hump. However, this long relaxation length still seems to be quite too short to be accounted for by inelastic binary collisions (Tendler and Arora 1978).

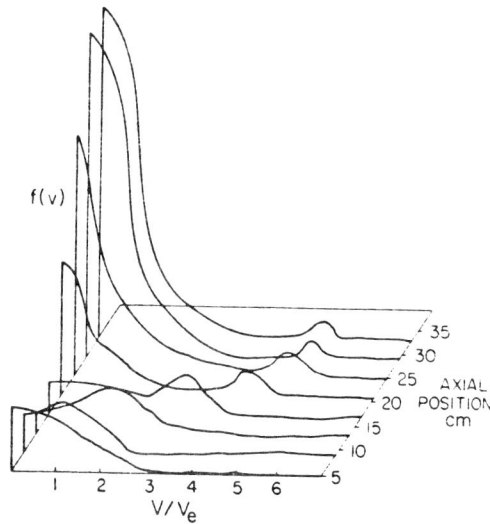

Figure 8a. Variation of the electron velocity distribution, f(v), along the target chamber when a double layer is present (Coakley, Hershkowitz, Hubbard, and Joyce 1978). Electrons accelerated at the double layer do not thermalize in the plasma on the high potential side of the layer. f(v) is the distribution of the magnitude of the electron velocity obtained from measurements of the electron energy distribution.

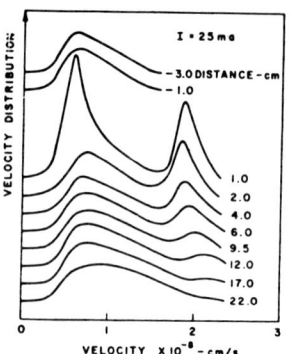 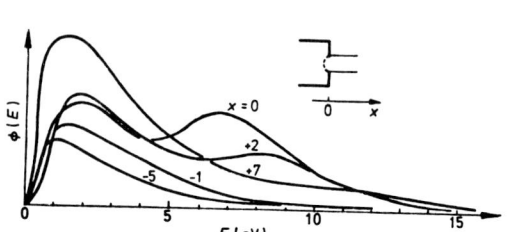

Figure 8b. Diagrams showing electron distribution functions at different distances from a layer at a tube constriction. Curves marked with a negative distance are distributions on the cathode side of the layer. The distributions on the anode side remain double humped up to a distance of 1000 Debye lengths beyond the layer. The diagram to the left (Crawford and Freeston 1963) shows the distribution function $v^2 f(v)$ (cf. Fig. 8a), and the other diagram (Andersson 1977) shows the electron energy distribution.

The fact, that the energy distribution is double humped, does not necessarily imply that the corresponding velocity distribution is unstable. The distribution of the axial velocity component along the plasma column was measured by Andersson (1976) at a tube constriction in a low pressure arc. This distribution, which is the relevant distribution for instability investigations, was found to be single humped in spite of the fact that the energy distribution was double humped (Fig. 9). Andersson suggested that this was due to the three-dimensional structure of the double layer at a discharge constriction. In this case the layer surface is approximately a hemisphere which bulges into the plasma on the cathode side of the layer. In contrast to this, the violently unstable layers observed by Torvén and Babić (1975) appeared to be plane over most of the column cross section. These observations give some support for the idea that the three-dimensional layer structure may be of importance for the layer stability.

Further experiments are required including estimations of the errors introduced when energy distributions are measured by the Druyvesteyn method in an anisotropic electron gas (Allen 1978).

5. SUMMARY OF SOME PROPERTIES OF THE OBSERVED LAYERS

The table below gives measured values of the layer voltage, ϕ, $e\phi/kT_e$ (T_e is the electron temperature on the low potential side of the

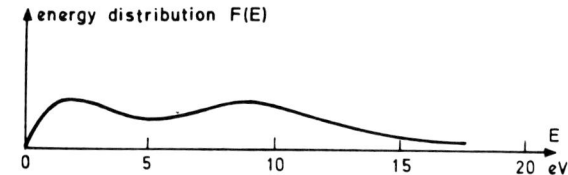

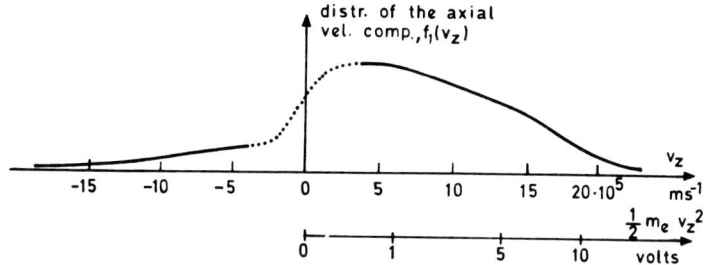

Figure 9. Comparison between the electron energy distribution, $F(E)$, and the distribution of the axial electron velocity component $f_1(v_z)$, measured at approximately the same positions in a plasma column on the anode side of a constriction layer (Andersson 1976). $f_1(v_z)$, which is the relevant distribution for instability investigations, is single humped whereas $F(E)$ is double humped. The detailed shape of f_1 around $v_z = 0$ could not be determined due to inaccurate determination of the plasma potential.

layer), the Debye length λ_D, and the layer thickness L. As a measure of the space charge strength within the layer $\gamma = (\lambda_D/L)(e\phi/kT_e)^{1/2}$ has been used. The typical space charge density in the layer is $\pm 4\gamma^2$ en if the densities n on each side of the layer are equal. (Shawhan, Fälthammar, and Block 1977). It can be shown from Poisson's equation that in general

$$\frac{\lambda_D}{L}\left(\frac{e\phi}{kT_e}\right)^{1/2} = \frac{1}{L}\left(\frac{1}{n}\int_{x_1}^{x_2} x \left[n_i(x) - n_e(x)\right] dx\right)^{1/2}$$

Here n is the electron number density, that is chosen to define the Debye length, n_i and n_e are the ion and electron number densities within the layer, and x_1 and x_2 are the points on each side of the layer at which the space charge density and the electric field assume negligible values ($|x_2 - x_1| = L$).

It should be observed that the evaluated values of γ give order of magnitude values only, because the given values of λ_D are based on rough estimations. No measurements of density profiles at a double layer seem to have been presented so far.

Experiment	ϕ (V)	$\frac{e\phi}{kT_e}$	λ_D (mm)	L (mm)	$\gamma = \frac{\lambda_D}{L} \left(\frac{e\phi}{kT_e}\right)^{1/2}$
Coakley et al.	14	14	1.5	150	0.04
Lutsenko et al.	10^4			40	
Quon and Wong	16	4.5	1.5	35	0.09
Torvén and Babić	200	20	0.05	<5	>0.03
Torvén and Andersson	25	12	0.35	12	0.1
Joyce and Hubbard (num. sim.)		50	$\frac{L}{\lambda_D} = 40$		0.17

6. CONCLUSIONS

Experiments in laboratory plasmas have unambigously demonstrated the existence of double layers in magnetized as well as unmagnetized plasmas. Measurements of distribution functions have also shown the existence of trapped and free ions and electrons, and stable double layers with $e\phi/kT_e > 10$ have been observed. In some experiments the layer formation has been related to a threshold value of the electron drift equal to $(kT_e/m_e)^{1/2}$. The early phase of the layer formation has been found to be characterized by solitary potential pulses which propagate in the direction of the electron drift with a velocity of the order of the ion sound speed. Simultaneously their amplitudes grow to non-linear values.

Rapid progress is foreseen, and problems concerning the layer formation, the layer stability, and the origin and significance of the observed high frequency fluctuations are expected to be attacked and solved within the next few years.

ACKNOWLEDGEMENTS

I wish to thank the American Institute of Physics, The Institute of Physics, North Holland Publishing Company, Springer-Verlag, and a large number of authors for their permission to publish figures and diagrams which have been published elsewhere. This work was supported by the Swedish Natural Science Research Council, contracts F 7014-104 and F 7014-107.

REFERENCES

Allen, J.E.: 1978, J. Phys. D.: Appl. Phys. 11, pp. L 35-38.
Andersson, D.: 1977, J. Phys. D.: Appl. Phys. 10, pp. 1549-1556.
Andersson, D.: 1976, Report 76-102, Dept of Electron Phys., Royal Inst. of Tech., Stockholm, Sweden.
Andrews, J.G. and Allen, J.E.: 1971, Proc. Roy. Soc. Lond. A. 320, pp. 459-472.
Armstrong, R.J. and Torvén, S.: 1974, Report 10-74, Auroral Observatory, University of Tromsø, Tromsø, Norway.
Armstrong, R.J.: 1975, Proc. 12th Int. Conf. on Phenomena in Ionized Gases (Ed. J.G.A. Hölscher and D.C. Schram, American Elsevier Publ. Comp., New York) p. 117.
Boyd, R.L.F. and Twiddy, N.D.: 1960, Proc. Roy. Soc. Lond. A. 259, pp. 145-158.
Coakley, P., Hershkowitz, N., Hubbard, R., and Joyce, G.: 1978, Phys. Rev. Letters 40, pp. 230-233.
Crawford, F.W. and Freeston, I.L.: 1963, 6th Int. Conf. on Phenomena in Ionized Gases (Ed. P. Hubert and E. Crémieu-Alcan, EURATOM-CEA, Paris), Tome I, pp. 461-464.
Crawford, F.W., Levine, J.S., and Ilić, D.B.: 1979, Astrophysics and Space Science Library (see this volume).
DeGroot, J.S., Barnes, C., Walstead, A.E., and Buneman, O.: 1977, Phys. Rev. Letters 38, pp. 1283-1286.
Foulds, K.W.H.: 1956, J. Electronics 2, pp. 270-278.
Goertz, C.K. and Joyce, G.: 1975, Astrophys. and Space Sci. 32, pp. 165-173.
Joyce, G. and Hubbard, R.: 1978, Report 78-10, Dept of Phys. and Astronomy, The University of Iowa, Iowa City, Iowa 52242.
Langmuir, I. and Mott Smith, Jr.H.: 1924, Gen. Electric Review 27, pp. 762-71 (see p. 768).
Levine, J.S., Ilić, D.B., and Crawford, F.W.: 1977, IAGA/IAMAP Joint Assembly, Seattle, Final Program, p. 125 (GA 539).
Levine, J.S., Crawford, F.W., and Ilić, D.B.: 1978, Phys. Lett. 65A, pp. 27-29.
Lutsenko, E.I., Sereda, N.D., and Kontsevoi, L.M.: 1975, Zh. Tekh. Fiz. 45, pp. 789-796 (Sov. Phys. Tech. Phys. 20, pp. 498-502).
Quon, B.H. and Wong, A.Y.: 1976, Phys. Rev. Letters 37, pp. 1393-1396.
Sandahl, S.: 1971, Phys. Scr. 3, pp. 275-278.
Schönhuber, M.J.: 1963, Zeitschrift für angewandte Physik 15, pp. 454-460.
Shawhan, S.D., Fälthammar, C.-G., and Block, L.P.: 1978, J. Geophys. Res. 83, pp. 1049-1054.
Taylor, R.J., MacKenzie, K.R., and Ikezi, H.: 1972, Rev. Sci. Instr. 43, pp. 1675-1678.
Tendler, M.B. and Arora, A.K.: 1978, J. Phys. D.: Appl. Phys. 11, pp. 1125-1132.
Torvén, S. and Babić, M.: 1975, Proc. 12th Int. Conf. on Phenomena in Ionized Gases (Ed. J.G.A. Hölscher and D.C. Schram, American Elsevier Publ. Comp., New York) p. 124.
Torvén, S. and Andersson, D.: 1978, Report TRITA-EPP-78-12, Dept of Plasma Phys., Royal Inst. of Tech., Stockholm, Sweden. (To be published

in J. Phys. D.: Appl. Phys.)
Torvén, S.: 1978, J. Appl. Phys. 49, pp. 2563-2565.
Twiddy, N.D.: 1961, Proc. Roy. Soc. Lond. A 262, pp. 379-394.

LABORATORY SIMULATION OF IONOSPHERIC DOUBLE-LAYERS

F. W. Crawford, J.S. Levine, and D.B. Ilić
Institute for Plasma Research
Stanford University
Stanford, California 94305, USA

ABSTRACT

A mechanism for double-layer formation in partially- or fully-ionized plasmas is described, founded on beam-plasma interaction: rf growth along the beam excites a rectified, ponderomotive electric field, which in turn causes charge separation. Laboratory studies of the mechanism are described.

1. INTRODUCTION

Formation of a double-layer (or double-sheath) — defined here as an essentially time-independent charge separation and potential step, typically tens of Debye lengths wide — may be invoked to explain a variety of laboratory and space plasma phenomena [1,2].

Laboratory experimentation is difficult, but such direct measurements as discharge parameters for double-layer formation, potential steps, rf generation near double-layers, and electron velocity distributions, have been made. Two of our experiments will be described in Section 4. Ionospheric double-layers remain a highly plausible inference, since comprehensive time- and space-resolved measurements are not yet available.

Several mechanisms may be capable of producing double-layers [2]; some possibilities will be discussed in Section 2. Section 3 describes an additional mechanism which is the subject of our current work.

2. ORIGINS OF DOUBLE-LAYERS

Double-layers at emitting electrodes are well known and adequately understood [3]. As shown in Fig. 1(a), the double-layer accelerates a beam of cathodic electrons into the plasma, prevents most thermal plasma electrons from reaching the cathode, and accelerates positive ions towards it. The double-layer is a mediating element between plasma and cathode; its profile and potential step depend on the discharge current.

The constriction double-layer plays a similar role [see Fig. 1(b)] [4]. The electron density, temperature, and random current towards the constriction are higher in the smaller section. The double-layer accelerates electrons from the larger to the smaller section, and confines thermal electrons in the smaller section. Its profile and potential step depend on the reduction in section, and discharge current.

Double-layers may form in a column of uniform section for reasons related to the ionization processes maintaining the discharge. For example, stationary or moving striations are observed in the rare gases. The potential distribution rises in steps of about the ionizing potential [see Fig. 1(c)], each serving to inject high-energy electrons into the region to its anode side. There they produce the ionization necessary to maintain the column. The next step (double-layer) forms when they have been depleted.

At low pressures (< 1 mTorr) and high current densities, arc starvation can occur, since relatively few neutrals are available to be ionized.

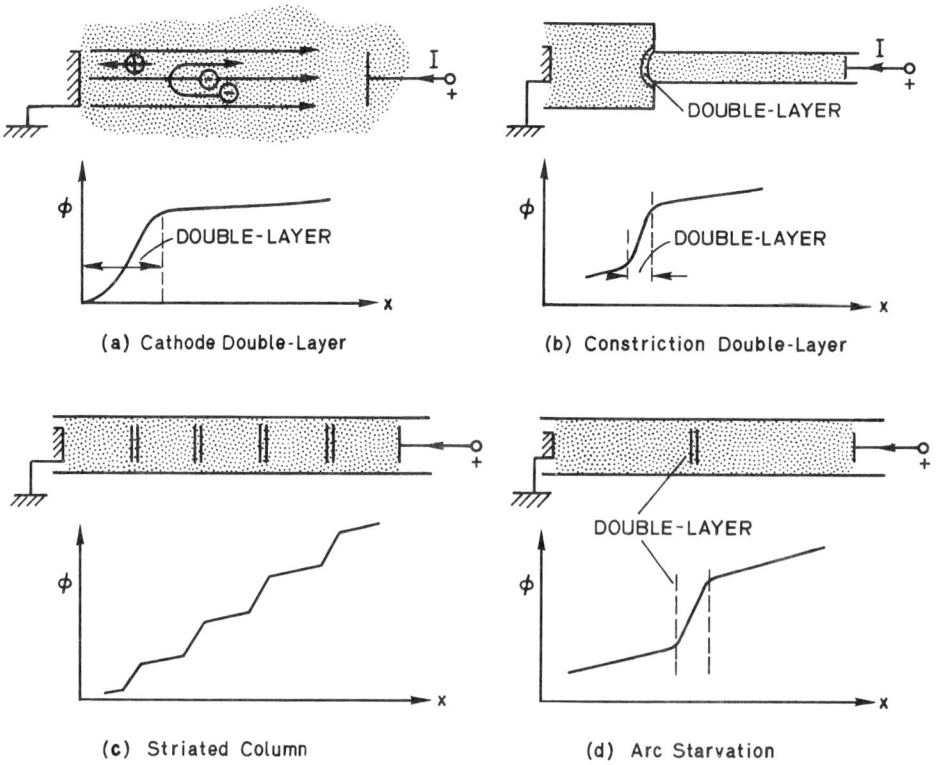

Fig. 1. Double-Layer Mechanisms.

Although theory supposes an infinite, axially uniform column [5], in practice high potential steps (analogous to striations) may form as the discharge struggles to maintain itself [see Fig. 1(d)].

The catalogue of double-layers extends further: laboratory experiments indicate potential steps from a few volts to kilovolts in columns whose sections (though not necessarily the plasma itself) are uniform, for parameters such that striations and arc starvation should not occur. Ionospheric double-layers may have a distinctly different origin, since the plasma where they are believed to occur is effectively fully-ionized. We shall discuss a mechanism which can operate in partially- or fully-ionized plasmas.

3. A MECHANISM FOR DOUBLE-LAYER FORMATION

We assume that an electron beam is present in the plasma (see Fig. 2). In the laboratory, the beam might enter from a cathode, an electron gun, or a constriction double-layer, for example. In space, it might be due to precipitation from the Van Allen belts.

Under such conditions, convective rf growth will occur in the direction of the beam velocity, near to the electron plasma frequency, as a result of beam-plasma interaction. Growth may proceed until the wave amplitude is sufficient to disperse the beam; thereafter, the wave will decay.

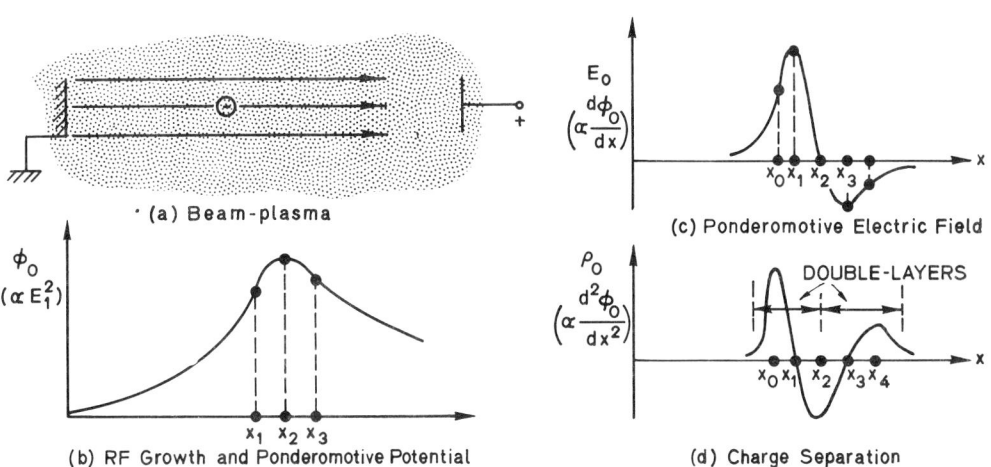

Fig. 2. Double-Layer Formation by Beam-plasma Interaction.

Nonlinear analysis predicts a rectified dc potential, $\emptyset_0 \propto E_1^2$, where E_1 is the rf electric field [see Fig. 2(b)]. Its derivatives are sketched in Fig. 2(c) and (d), demonstrating double-layer formation. Note that two double-layers form back-to-back; an electron emerges from them with its energy unchanged. If the spatial growth is more rapid than the decay, the positive step should be more conspicuous. In actual space or laboratory plasmas, the ponderomotive potentials and fields will be combined with any background dc potentials and fields of other origins.

The predicted magnitude of the potential step, $\emptyset_0$, is

$$\frac{\emptyset_0}{V_b} = \frac{1+\alpha^2}{32}\left(1 + \frac{3V_e}{2V_b}\left(1 - \frac{n_b}{\alpha^2 n_p}\right)\right) \quad , \quad \alpha = \frac{k_i}{k_r} = \frac{3^{1/6}}{2}\left(\frac{n_b}{n_p}\frac{V_b}{V_e}\right)^{1/3} . \quad (1)$$

Here, V_b and V_e are the beam voltage and plasma electron temperature, n_b and n_p the beam and plasma electron densities, and k_r and k_i the real and imaginary parts of the wavenumber, k, of the growing wave. The beam and plasma have been assumed infinite. For typical parameters in our experiments $(V_b = 50\ V, V_e = 7\ V, n_b/n_p = 0.02)$, $\emptyset_0 = 2\ V$.

4. EXPERIMENTS

Double-layers produced by cathodic beam-plasma interaction have been studied in the apparatus shown in Fig. 3(a). An argon discharge is main-

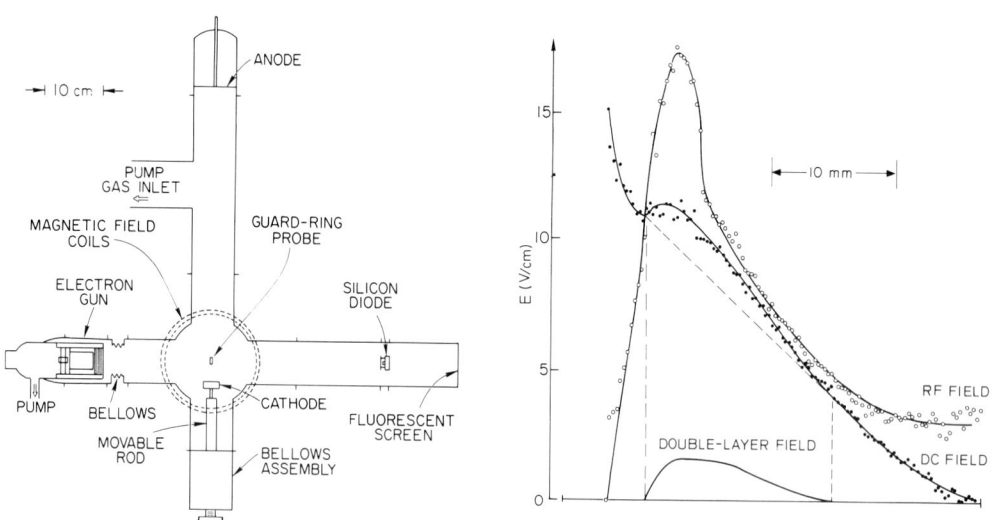

Fig. 3. Beam-plasma Interaction.

tained between the anode and a movable oxide-coated cathode. The dc and rf fields are measured by an electron beam probing technique [6].

Typical results are shown in Fig. 3(b). The lump on the dc electric field curve may be interpreted as due to a double-layer. To separate it from the background dc electric field, the variation of the latter has been assumed to be as shown dashed. The double-layer field can then be integrated to obtain the height of the potential step. The experimental and predicted values for Fig. 3(b) are 1.7 V and 1.5 V.

Double-layer parameters should also be obtainable from the rf electric field measurements. As currently calibrated, these are an order of magnitude lower than can be inferred from the occurrence of beam break-up. Our current work is directed towards correcting the calibration.

Double-layers are also being studied in a 60 cm long, 2.5 cm diameter, argon positive column, at about 1 mTorr pressure. A step of about 10 V occurs at current densities of ~ 1 A/cm^2 (see Fig. 4).

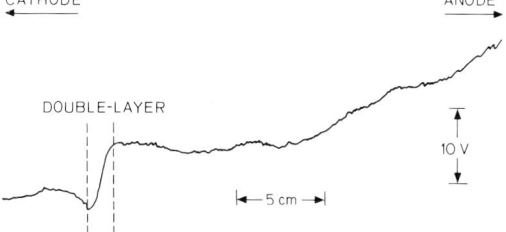

Fig. 4. Positive Column: Measurements of DC Potential Distribution.

To avoid wall heating due to the double-layer, the discharge is pulsed repetitively, for 300 μs periods, between a low current density (~ 0.4 A/cm^2) and that at which the double-sheath forms. Continuous operation is closely approached in 150 μs.

We wish to determine whether beam-plasma interaction influences this kind of double-layer. Pulsed Langmuir probe techniques are being developed to study the electron velocity distribution along the tube axis. If an electron beam is present (a cathodic beam, for example), it should be observed on the cathode side of the double-layer, and dispersed in its vicinity.

5. DISCUSSION

The rectification mechanism is just one means of ionospheric double-layer formation. It has the features of being able to occur in fully- or partially-ionized plasmas, with a potential step whose height depends on the beam energy and wave growth in the break-up region, rather than being limited to values of the order of the electron temperature or ionization potential.

Prediction of double-layer characteristics is complicated, in practice, by a variety of factors influencing the beam-plasma growth rate — beam and plasma electron velocity distributions and spatial inhomogeneities, the earth's magnetic field, and dc electric fields parallel to it — and involves a nonlinear theory to relate that growth to the beam break-up and ponderomotive fields causing double-layer formation; simplifications are necessary if the analysis is to be tractable. It is highly desirable that refined time- and space-resolved measurements of ionospheric double-layers be made in the near future for comparison with such analysis. Possible use of the Space Shuttle for this purpose commends itself for further study.

ACKNOWLEDGEMENTS

This work was supported by the National Aeronautics and Space Administration.

REFERENCES

[1] Torvén, S., Astrophysics and Space Science Library (see this volume).
[2] Carlqvist, P., Astrophysics and Space Science Library (see this volume).
[3] Crawford, F.W., and Cannara, A.B.: 1965, J. Appl. Phys., 36, pp. 3135-3141.
[4] Crawford, F.W., and Freeston, R.L.: 1964, Proc. Sixth International Conference on Ionization Phenomena in Gases, Paris, June 1963, 1, pp. 461-464.
[5] Stangeby, P.C. and Allen, J.E.: 1971, J. Phys. A., 4, pp. 108-119.
[6] Harp, R.S., Cannara, A.B., Crawford, F.W., and Kino, G.S.: 1965, Rev. Sci. Instrum. 36, pp. 960-968.

ANOMALOUS TRANSITION FROM BUNEMAN TO ION SOUND INSTABILITY

C.T. Dum[1] and R. Chodura[2]
(1) Max-Planck-Institut für Physik und Astrophysik, Institut für extraterrestrische Physik, 8046 Garching, Federal Republic of Germany
(2) Max-Planck-Institut für Plasmaphysik, 8046 Garching, Federal Republic of Germany

Using two- and one-dimensional particle simulation, we find for a constant current in an initially isothermal plasma that the initial and final stage of evolution are in accordance with quasi-linear theories for the Buneman and ion sound instability, respectively. Large deviations from Maxwellian distributions and resonant wave-particle interactions occur, however, already in the "hydrodynamic" phase and have to be accounted for. This phase is abruptly terminated by the emergence of long wavelength modes, a substantial drop in electric field energy and resistivity, and the formation of electron tails. In two dimensions, heating then proceeds as for ion sound turbulence.

1. Introduction

It appears that a current carrying plasma is the simplest configuration of practical importance which leads to instability. This importance stems from the macroscopic effects connected with the development of instabilities. Such effects have been discussed for many laboratory experiments and space plasmas, in particular for dynamic processes such as shocks, auroral disturbances and solar flares. The observations reveal, however, a great complexity and the evidence is perhaps not compelling enough for a generally accepted theory to emerge for some of the most basic instabilities. This is particularly true for instabilities due to currents aligned with the magnetic field and their macroscopic effects, as observed in turbulent heating experiments and along auroral field lines (for example, Gurnett and Frank 1977). Urgently needed are detailed measurements of particle distributions and of their dynamic response to wave spectra. Space observations of wave particle interactions may be particularly valuable. The failure of the many attempts at deriving quasi-stationary fluctuation levels and values of anomalous resistivity from the macroscopic parameters can be due to the dynamic evolution of the shape of particle and wave spectra. This was demonstrated very clearly by computer simulation of ion sound waves driven by a current across a weak magnetic field, thus eliminating complicated dynamic processes connected with electron runaway, and for a plasma entering a regime of constant ratios $T_e/T_i \gg 1$ between thermal energies (temperatures) of electrons and ions and $u/v_e < 1$ between the relative electron-ion drift and the thermal velocity $v_e = (T_e/m)^{1/2}$ (Dum et al. 1974, e.g. Fig. 1). We show in this paper that already in the regime $u/v_e > 1$, usually described by hydrodynamic equations (Buneman 1959), large changes in the shape of particle and wave spectra occur which have a significant effect on the macroscopic dynamics.

The complex interplay between evolution of the instability and macroscopic conditions may not only explain many of the apparent discrepancies between experiments, but must also be considered for computer simulations which allow the study of much simpler situations and with more detailed diagnostics than is possible in experiments. Usually, the case of large, constant electric field applied to a homogeneous plasma is considered ideal, and has been investigated theoretically and by one dimensional computer simulations in which drift and thermal velocity increase linearly with time, apart from oscillations about $u \approx v_e$, thus never entering the ion sound regime. (Boris et al. 1970, Morse and Nielson 1971). Numerical solutions of the corresponding one-dimensional quasi-linear equations for an initially isothermal plasma describe the first burst of wave activity (Chodura et al. 1971). Biskamp and Chodura (1973) show that the simulations may be visualized as a succession of wave bursts, each one doubling the wavelength of the dominant mode by coalescence of Bernstein-Greene-Kruskal-waves. At late times almost all of the electric energy resides in the mode of longest wavelength allowed by the size of the simulation system. In most experiments, however, the drift velocity, after reaching values above the initial electron thermal velocity which depend on the applied voltage, does not, as predicted, continue to increase at a sizable fraction of the free acceleration rate, and by electron heating also falls much below the electron thermal velocity. This behavior persists even if an effort is made to approximate the case of constant electric field (Gentle et al. 1978). An explanation, also inspired by some one-dimensional simulations, in terms of nonlinear formation of a single large amplitude Bernstein-Greene-Kruskal-wave which traps most of the electrons has been proposed (Drummond et al. 1971, Kan and Akasofu 1978).

From our more realistic two-dimensional simulations in a sufficiently large system (Sec. 2), actually the opposite picture emerges. The hydrodynamic phase is terminated abruptly by a spreading of the wave spectrum to all parallel wave numbers $k_\| \lesssim k_0$, where $k_0 = \omega_e/u$ corresponds to maximum hydrodynamic growth. The initially very wide range of wavenumbers $k_\perp$ perpendicular to the drift narrows only somewhat in the later ion sound regime, and electron trapping is unimportant. Instead of using a constant electric field or a drift imparted initially (Buneman 1959), the drift velocity is maintained constant by the applied electric field, which is then a measure for anomalous resistivity, $\nu^* = eE_0/mu$. The simulations, then much more closely resemble the experimental situation during the phase of rapid dissipation. Constancy of the current may also follow from a consideration of the equivalent electric circuit, e.g. for turbulent heating (Wharton et a. 1971, Mondelli and Ott, 1974),

Birkeland currents (Boström 1974) and the return current of a relativistic beam (Lovelace and Sudan 1972). Since the evolution of the system may depend on its past, the instability was followed from the hydrodynamic regime of large drifts and comparable electron and ion temperatures until well into the ion sound regime of drift velocities between the electron and ion thermal velocities and temperature ratios $T_e/T_i \gg 1$, studied previously (Biskamp and Chodura 1971).

Conclusions from the simulations about the instability mechanism and the macroscopic effects of current driven instabilities are presented in Sec. 3. A transition between hydrodynamic instability due to charge bunching and kinetic instability excited by resonant particles may also occur for a variety of beam plasma instabilities (O'Neil and Malmberg 1968, Mantei et al. 1976, Lampe et al. 1974). The discussion of the complex interplay between instability and macroscopic behavior strongly suggests, however, that no firm conclusions about the nonlinear behavior of an instability should be drawn from a mere resemblance of linear dispersion relations.

2. Computer simulation by the particle in cell method.

The two-dimensional simulations were carried out with 2×10^5 particles per species, 128×128 cells and periodic boundary conditions. The drift velocity was $u/v_{eo} = 20 (10)$ the system length 64 (128) u/ω_e and the initial temperatures were $T_{eo} = T_{io} = 2.5 \times 10^{-3}$ (10^{-2}). Mass ratios were either 100, 400, or 1600. Corresponding one-dimensional runs were also carried out. It was also verified that a weak perpendicular magnetic field ($\Omega_e/\omega_e = 0.04$), which is typical for shocks and by preventing runaway strongly affects the ion sound phase (Dum et al. 1974), has a negligible effect on the hydrodynamic phase. The electric field energy in the initial thermal fluctuations is very low, typically $W/nmu^2 = 2 \times 10^{-6}$.

The choice of simulation parameters is dictated by the properties of the linear dispersion relation and the interplay between nonlinear wave growth and macroscopic conditions, as discussed in Sec. 1. Hydrodynamic growth depends only on the wavenumber $k_\parallel$ parallel to the drift, having a strong peak at $(k_\parallel u/\omega_e) = 1$ and a cutoff at $(k_\parallel u/\omega_e) \approx 1 + (3/2)(m/M)^{1/3}$. The dependence of wave growth on the perpendicular wavenumber $k_\perp$ is determined by kinetic effects, for our initial Maxwellians by (kv_e/ω_e) or $v_e/u \cos\theta$ where θ is the angle between $\underline{k}$ and $\underline{u}$. For $T_e = T_i$ and $M/m = 1836$, thermal effects on the Buneman instability are negligible for $v_e/u \cos\theta \lesssim 0.2$, a transition to kinetic instability occurs for $v_e/u \cos\theta \approx 0.35$ and marginal stability corresponds to $v_e/u \cos\theta = 0.75$ (Jackson 1960). The system length L must be chosen such that at least several modes $k_\parallel = 2\pi n_\parallel/L$, $n_\parallel = 0, 1, 2,..$ near maximum growth can be excited and the later spread of the spectrum in the course of nonlinear evolution is possible. A reasonably large number of perpendicular wavenumbers $k_\perp = 2\pi n_\perp/L$ should also be excited, where $n_\perp = 0, 1, 2,..$ in the cold plasma is limited by the cell size Δx and only at later times by thermal effects. In this case we find no evidence for the often claimed nonlinear suppression of neighboring modes by the mode of largest linear growth rate. The wave

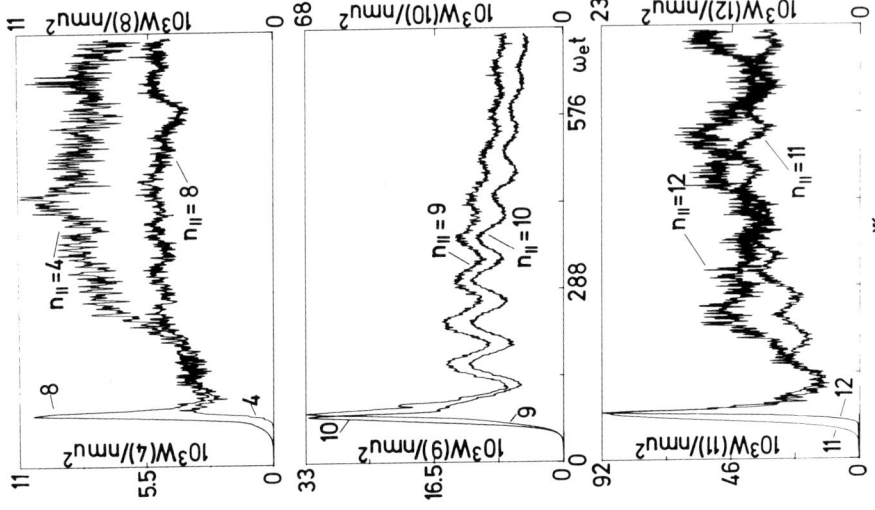
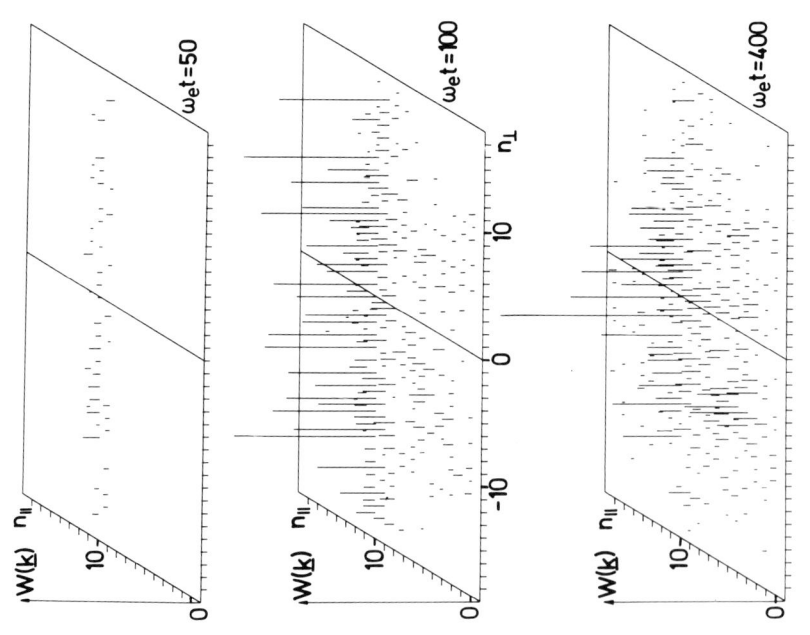

Figure 1. Wave spectrum $W(k_\|, k_\perp)$ before and after nonlinear transition at $\omega_e t^* = 80$ and $W(k_\|) = \Sigma W(k_\|, k_\perp)$ where $k_{\|,\perp} = (2\pi/L) n_{\|,\perp}$ are the wavenumber components parallel and perpendicular to the constant drift $\underline{u} = \underline{u}_e - \underline{u}_i$, and $n_\| = 9-11$ are the linearly most unstable modes. Simulation with $M/m = 400$, $u/v_{eo} = 20$, $\frac{u}{v_{eo}} = (T_{eo}^i/m)^{1/2}$ and $T_{eo} = T_{io}$.

spectrum corresponds to the features of the linear dispersion relation for $t < t^*$ (Fig. 1). The abrupt emergence of modes $k_{\parallel} < \omega_e/u$ is followed by slow growth while the energy in the linearly most unstable modes goes through several in phase oscillations at a much reduced level. If the transitions were due solely to thermal (quasi-linear) effects it should occur earlier for the more oblique modes for which $u \cos \theta/v_e$ is smaller (Lampe 1974). Fig. 2 shows, however, that the transition for different $k_{\perp}$ occurs at exactly the same time. In fact, a remarkably similar transition of the wave spectrum occurs in the one-dimensional simulations. Conservation of energy implies that the sudden drop in total electric field energy $W = \Sigma W_k$ is connected with an increase in electron energy T_e since in the early transition phase the ion (sloshing) energy T_i also drops and the external energy supply $\underline{E}_0 \cdot \underline{j} = \nu^* nmu^2$ continues, although at a much reduced rate (Fig. 3). A crescent shaped electron tail is formed just prior to the transition which later fills in and extends in height, but not much in velocity, $v_{\parallel} < 2u$. Eventually tail and bulk merge in a nearly isotropic distribution. (Fig. 4, $\omega_e t = 225$). During this time the ions maintain a nearly bi-Maxwellian distribution, $T_{i\parallel}/T_{i\perp} \lesssim 1.5$ (Fig. 5). The evolution of the ions may be described by nonresonant diffusion. The ion temperature at this stage is determined by the kinetic (sloshing) energy associated with the waves. In one dimension the electron tail formed at the time of transition of the wave spectrum develops into a plateau at $v \approx 0$ and then wave energy and thermal velocity level off, typically at $v_e/u \approx 1.7$ and $T_e/T_i \approx 12$. In two dimensions, however, electrons and ions at later times evolve in a way which is familiar from

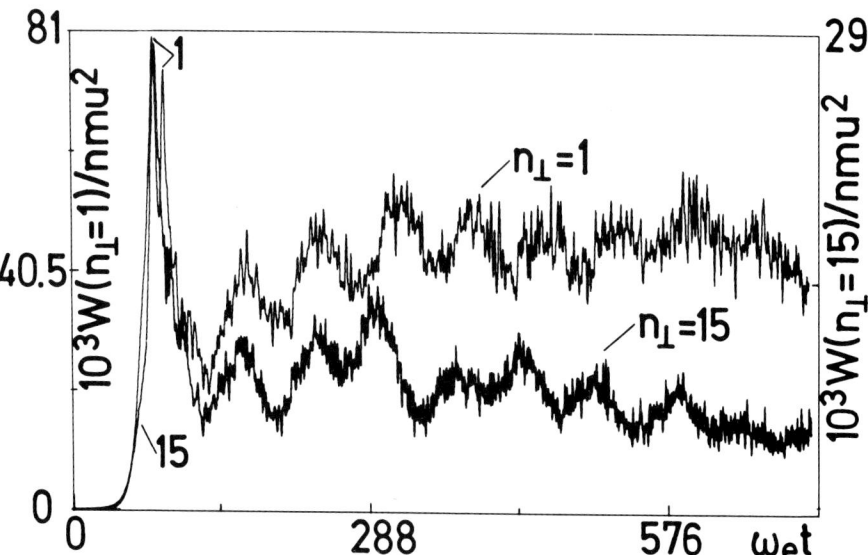

Figure 2. Time dependence of $W(k_{\perp}) = \Sigma W(k_{\parallel}, k_{\perp})$ for modes nearly parallel to the drift ($n_{\parallel} = 1$) and for oblique modes ($n_{\perp} = 15$).

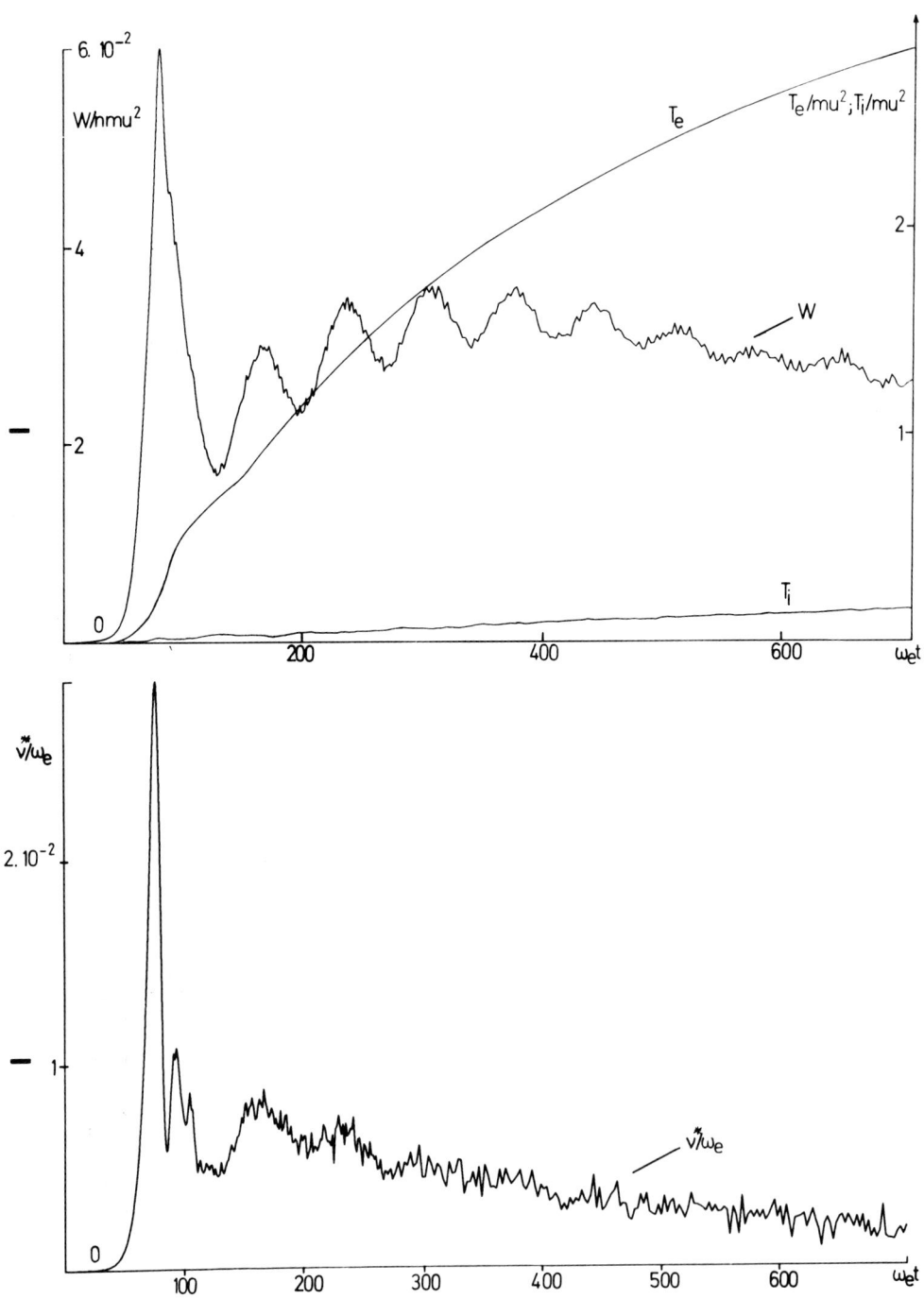

Figure 3. Total electric field energy $W = \Sigma W(\underline{k})$, thermal energies $T_j = m_j <(v-u_j)^2>$, $j = e, i$; and anomalous collision frequency $\nu^* = eE_o/mu$.

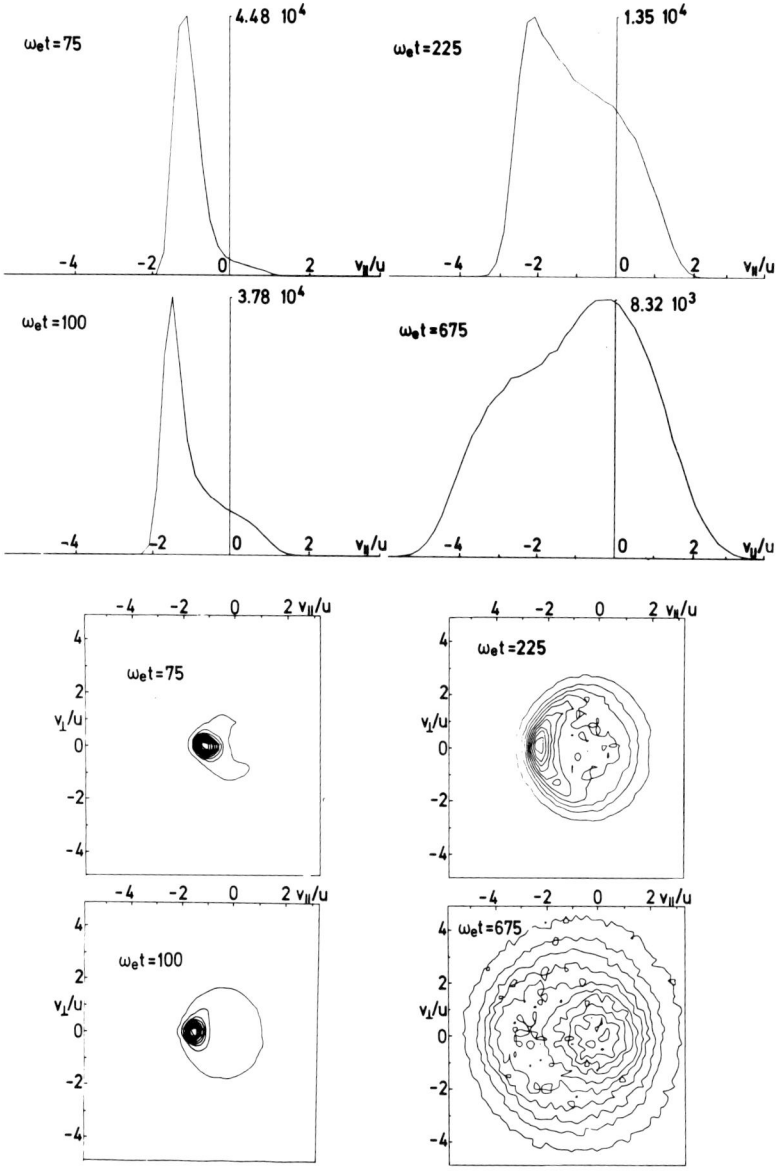

Figure 4. Reduced electron distribution $f_e(v_\parallel)$ and contours of $f_e(v_\parallel, v_\perp)$, showing tail formation in the hydrodynamic phase and runaway distortions in the ion sound phase.

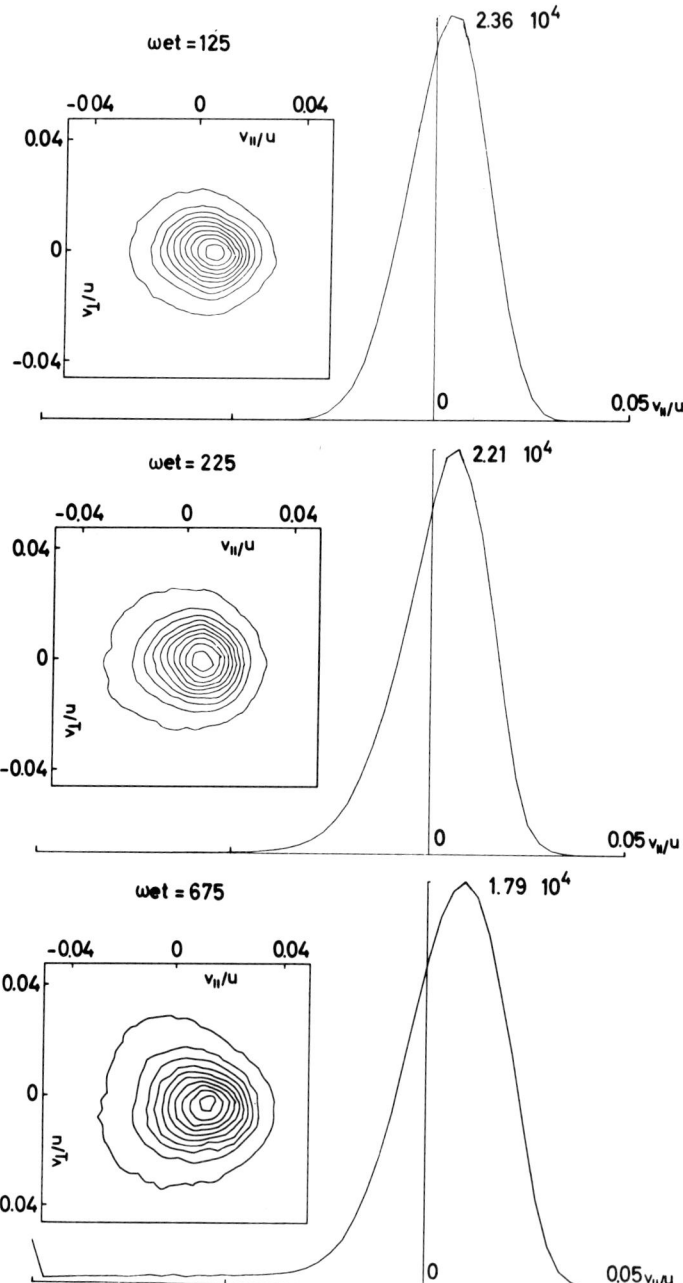

Figure 5. Reduced ion distributions $f_i(v_{\shortparallel})$ and contours of $f_i(v_{\shortparallel},v_{\perp})$, showing adiabatic heating in the hydrodynamic phase and formation of a high energy tail in the ion sound phase.

simulations of ion sound turbulence (Biskamp and Chodura 1971). The nearly isotropic electron distribution develops now a (runaway) tail in the direction of acceleration by the applied electric field. Irreversible ion heating by formation of a high energy tail sets in at a time and rate which depends strongly on the mass ratio. The reduction in the effective drift for wave growth by the runaway distortion, electron heating, and increased ion Landau damping must eventually combine in quenching the instability, after a period which lasts much longer than the hydrodynamic phase ($\omega_e t^* = 120$ for $M/m = 1600$). For a constant drift, $u/v_{eo} \lesssim 1$, across a weak magnetic field which prevents runaway, ion sound is quenched typically at $\omega_i t \approx 120$ (Dum et al. 1974).

3. Discussion and conclusions.

The simulations presented here show that current driven instabilities have a very complex dynamic behavior which sensitively depends on its entire past history and on the macroscopic conditions. Our results demonstrate the importance of using a sufficiently large two-dimensional system, sufficiently large ion to electron mass ratio and proper boundary conditions, in our case constant current, for a realistic simulation. The direction of the field aligned current will remain an axis of symmetry in the course of evolution and we expect no qualitative difference between two- and three-dimensional systems. In one dimension, however, the wave-particle interaction is fundamentally different.

The entire range from hydrodynamic instability by charge bunching, kinetic instability excited by resonant particles to stability is already covered initially in a multi-dimensional system, if we consider modes oblique to the drift for which $u \cos\theta /v_e$ is the effective drift to thermal velocity ratio. Electron diffusion may be described by pitch angle scattering, by choosing as rest frame the phase velocity of the most unstable mode in the hydrodynamic regime, and the mean ion velocity in the ion sound regime. Kinetic effects (interaction with the electron tail) are responsible for a reduction of the growth rate and an increase of resistivity and electron heating above the values connected with hydrodynamic sloshing, well before the transition time t^* of the wave spectrum. In the ion sound phase resonant pitch angle scattering completely dominates anomalous electron transport by maintaining a nearly isotropic distribution against the perturbing external forces and gradients (Dum 1978). In one dimension, however, electron diffusion is restricted to plateau formation and easily results in stabilization and termination of heating. Naturally, coherence and trapping effects are much more important in one dimension where they are observed for both electrons and ions. In two dimensions, only the ion phase space plots ($v_\parallel$, $x_\parallel$) show some modulation corresponding to the parallel wavelength of the most unstable modes.

For two ion beams counterstreaming with large drifts in a stationary electron background, which nearly independently also excite the Buneman instability, similar effects of dimensionality on wave-particle interactions are observed. The simulations (Lampe et al. 1974), in contrast to our work, cover only the "hydrodynamic" regime extending to $v_e \approx u$, where in corresponding one dimensional simulations wave growth and heating terminate (Davidson et al. 1970). In two dimensions exponential wave growth is terminated at different times, depending on the angle of wave propagation, but at roughly the same ratio $v_e/u \cos \theta \approx 0.35$, thus suggesting quasilinear saturation. This conclusion is in sharp contrast to our studies which show an abrupt termination of hydrodynamic wave growth at $v_e/u \approx 0.6$, $T_e/T_i \approx 25$, independent of the angle θ, and connected with the emergence of long wavelength modes. These observations and also the fact that a very similar transition occurs in our corresponding one-dimensional simulations, at nearly the same time, but $v_e/u \approx 0.9$, strongly suggest that a <u>nonlinear transition of the wave spectrum</u> takes place. The system must, of course, be large enough to allow the emergence of long wavelength modes and the excitation of at least several wavenumbers $k_\parallel$ near maximum growth $k_\parallel u/\omega_e = 1$, which may at least partly explain the discrepancy with the simulations mentioned above. Indeed, in a one-dimensional simulation which meets these requirements, but differs from our work by requiring constancy of the sum of current and displacement current $(1/4\pi) \partial E_o/\partial t$, the spectrum, after the hydrodynamic phase, shifts to longer wavelengths (Nishihara and Hasegawa 1972). This shift was attributed to a decay instability of plasma oscillations in the uniform electric field E_o. We have not observed such large amplitude plasma oscillations, in particular not for the two dimensional simulations where wave energy and applied electric field E_o have a comparatively much smoother time dependence, cf. Fig. 3. In this case we also have no evidence that the several oscillations in wave energy of the linearly most unstable modes, following the transition at t^*, are connected with particle trapping. At this time a smooth transition to resonant electron diffusion has occurred already. Resonant (quasi-linear) wave-ion interaction sets in much later, strongly dependent on the mass ratio. For mass ratio $M/m = 400$ (1600) the transition in the wave spectrum and resistivity takes place at $\omega_e t^* = 80$ (119), the oscillations in wave energy have the frequency $\omega = 1.7$ (1.5) ω_i, and resonant ion heating (tail formation) sets in at $\omega_e t \approx 250$ (450). For the smallest mass ratio $M/m = 100$, the wave ion interaction is more important, especially in one dimension. Simulations which use even smaller mass ratios and one dimensional simulations in general, must therefore be considered with great caution.

Apparently due to the varying importance of kinetic effects in the course of evolution of the instability, no simple scaling laws, such as $(m/M)^{1/3}$, emerge for our mass ratios $M/m = 100, 400, 1600$. The maximum wave energy W/nmu^2 decreases from 8×10^{-2} to 4.5×10^{-2} and the collision frequency ν^*/ω_e from 5×10^{-2} to 1.5×10^{-2} for M/m increasing from 100 to 1600. The meaning of such maxima is much reduced anyway by the fact that they result from a dynamic evolution which depends on the entire past of the system. For example, the fluctuation level W/nT_e decreases from a maximum of $0.6 - 0.5$ at $t < t^*$ to values in the ion sound phase which are still $2 - 3$ times larger than the maximum in simulations which start with parameters $u/v_e \lesssim 1$ and $T_e/T_i \gg 1$, corresponding to this phase. In the simulations with large drifts $u/v_{eo} \gg 1$ we may identify the start of the ion sound phase, e.g., with the dominance of resonant diffusion for the electrons, resulting in ν^*/ω_e proportional to W/nT_e (Dum et al. 1974), or with the cessation of oscillations in wave energy W which then increases or decreases, depending on the mass ratio. Decreasing the initial drift u/v_{eo} from 20 to 10 also decreases the maximum wave energy and resistivity, e.g. for $M/m = 1600$ from $W/nmu^2 = 4.5 \times 10^{-2}$ to 3.8×10^{-2} and from $\nu^*/\omega_e = 1.5 \times 10^{-2}$ to 1.2×10^{-2}. We do not know the limiting value of the drift u/v_{eo} for an abrupt transition in the wave spectrum, as observed for $u/v_{eo} = 20$, 10 and $T_{eo} = T_{io}$. As mentioned in Sec. 2, a transition in the linear dispersion relation from hydrodynamic to kinetic instability occurs for $u/v_e \approx 3$. No such transition is observed in simulations started with $u/v_{eo} = \sqrt{3}$ and $T_{eo}/T_{io} = 2$ (Biskamp and Chodura 1971, run not shown).

Wave frequencies and effective collision frequencies corresponding to the various stages of the current driven instability have been identified in turbulent heating experiments and shocks. From a recent experiment it has also been concluded, by determining moments of the electron distribution from the dispersion relation of a test plasma wave, that the reduced distribution is of the form shown in Fig. 4, $\omega_e t = 225$ (Clark and Hamberger 1978).

In many experiments the initial electron and ion temperatures are comparable and anomalous dissipation sets in only after the drift reaches values considerably above the initial electron thermal velocity, depending on the applied electric field (e.g. Gentle et al. 1977). The Buneman instability has the important function of pre-heating the electrons to $T_e/T_i \gg 1$ and $u/v_e \lesssim 1$, which is necessary for the much longer ion sound phase, but as shown by the simulation also affects the dynamics at least for the first burst of ion sound instability. The duration and number of such bursts depends on the equivalent electric circuit and other macroscopic conditions (Wharton et al. 1971, Mondelli and Ott 1974). It has often been pointed out that low frequency instabilities for a plasma in a magnetic field, such as the electrostatic ion cyclotron instability, require lower critical drift velocities for $T_e \approx T_i$ (Kindel and Kennel 1971). However, the growth rate and anomalous collision frequency are at most of the order of the ion cyclotron frequency Ω_i, although such waves can be excited to large amplitudes (Dum and Dupree 1970). Such waves are frequently observed, but for turbulent heating experiments and shocks at least, it has been shown that they can become important only after the phase of rapid dissipation by the Buneman and ion sound instabilities with characteristic frequencies $\omega_B = (\omega_e \omega_i^2)^{1/3}$ and $\omega_i \gg \Omega_i$, respectively, even in the case of prior excitation by some other mechanism (Wharton et al. 1971).

The simulations show that the hydrodynamic phase and the ion sound phase can be described by quasilinear theory, which must, however, include the large deviations from Maxwellian particle distributions, occurring already in the hydrodynamic phase. The transition between Buneman and ion sound instability requires a nonlinear theory. These questions will be considered in more detail in a forthcoming publication, for ion sound see Dum (1978).

Acknowledgements

We like to thank D. Biskamp for some stimulating discussions. The work of one of us (R. Chodura), and the work of the other (C.T. Dum) while at the Max-Planck-Institut für Plasmaphysik, was performed under the terms of the agreement on association between the Max-Planck-Institut für Plasmaphysik and EURATOM.

References

Biskamp, D. and Chodura, R.: 1973, Phys. Fluids, 16, pp. 888-892.

Biskamp, D. and Chodura, R.: 1971, Phys. Rev. Lett., 27, pp. 15553-1556.

Boris, J.P., Dawson, J.M., Orens, J.H., and Roberts, K.V.: 1970, Phys. Rev. Lett., 25, pp. 706-710.

Boström, R.: 1974, Magnetospheric Physics, B.M. McCormac (ed.), pp. 45-59.

Buneman, O.: 1959, Phys. Rev., 115, pp. 503-517.

Chodura, R., Bardotti, G., and Engelman, F.: 1971, Plasma Phys., 13, pp. 1099-1109.

Clark, W.H.M. and Hamberger, S.M.: 1978, Culham Laboratory Report, CLM-P531, Abingdon, Oxfordshire.

Davidson, R.C., Krall, N.A., Papadopoulos, K., and Shanny, R.: 1970, Phys. Rev. Lett., 24, pp. 579-582.

Drummond, W.E., Thompson, J.R., Sloan, M.L., and Wong, H.V.: 1971, Plasma Physics and Controlled Nuclear Fusion Research, Vol. 2, pp. 167-193, IAEA, Vienna.

Dum, C.T.: 1978, Phys. Fluids, 21, pp. 946-955, 956-969.

Dum, C.T., Chodura, R., Biskamp, D.: 1974, Phys. Rev. Lett., 32, pp. 1231-1234.

Dum, C.T. and Dupree, T.H.: 1970, Phys. Fluids, 13, pp. 2064-2081.

Gentle, K.W., Leifeste, G., and Richardson, R.: 1978, Phys. Rev. Lett., 40, pp. 317-320.

Gurnett, D.A. and Frank, L.A.: 1977, J. Geophys. Res., 82, pp. 1031-1050.

Jackson, J.D.: 1960, Plasma Phys., 1, pp. 179-189.

Kan, J.R. and Akasofu, S.I.: 1978, J. Geophys. Res., 82, pp. 735-738.

Kindel, J.M. and Kennel, C.F.: 1971, J. Geophys. Res., 76, pp. 3055-3078.

Lampe, M., Haber, I., Orens, J.H., and Boris, J.P.: 1974, Phys. Fluids, 17, pp. 428-439.

Lovelace, R.V. and Sudan, R.N.: 1972, Phys. Rev. Lett., 27, pp. 1256-1259.

Mantei, T.D., Doveil, F., and Gresillon, D.: 1976, Plasma Phys., 18, pp. 705-713.

Mondelli, A. and Ott, E.: 1974, Plasma Phys., 16, pp. 413-421.

Morse, R.L. and Nielson, C.W.: 1971, Phys. Rev. Lett., 26, pp. 3-6.

Nishihara, K. and Hasegawa, A.: 1972, Phys. Rev. Lett., 28, pp. 424-427.

O'Neil, T.M. and Malmberg, J.H.: 1968, Phys. Fluids, 11, pp. 1754-1760.

Wharton, C., Korn, P., Prono, D., Robertson, S., Auer, P., and Dum, C.T.: 1971, Plasma Physics and Controlled Nuclear Fusion Research, Vol. 2, pp. 25-36, IAEA Vienna.

MARGINAL PLASMA WAVES IN THE EQUATORIAL ELECTROJET OBSERVED BY HF COHERENT RADAR TECHNIQUES

C. HANUISE and M. CROCHET
Laboratoire de Sondages Electromagnétiques de l'Environnement Terrestre
Université de Toulon et du Var
"La Giponne" - Bd des Armaris - 83100 TOULON (France)

ABSTRACT. Multifrequency observations in the equatorial electrojet have shown that type 1 irregularities can be detected until frequencies as low as 4.6 MHz. Their phase velocity is equal to a threshold value which is influenced by the primary gradient in the E layer. During daytime the gradient is destabilizing and the threshold velocity is less than the ion acoustic velocity. Type 1 irregularities are then present for lower drift velocities and theories of plasma instabilities in the equatorial electrojet have to be reexamined at the light of these measurements.

1. INTRODUCTION

It is now well established that the two types of irregularities observed by coherent radar techniques in the equatorial electrojet are caused by the same plasma instability which has two main driving terms, one involving ion inertia and referred to as the two-stream term (FARLEY, 1963 ; WALDTEUFEL, 1965) and one involving the plasma density gradient and referred to as the cross-field term (ROGISTER and D'ANGELO, 1970 ; SUDAN et al, 1973). Nevertheless two types of completely different doppler spectra are obtained during analysis. Early observations having been performed at 50 MHz and the two-stream term prevailing at this frequency, it has become customary to associate type 1 spectra with the two-stream term and type 2 spectra with the cross-field one.

These two types of irregularities are distinguished by FARLEY and FEJER (1975) according to the wavelength at which the plasma is linearly unstable. The two-stream term is then supposed to be efficient mainly at short wavelengths (1-9 m) and the gradient drift term at considerably larger wavelengths. The type 2 which are observed at short wavelengths would be produced through some non linear cascading process (FARLEY and BALSLEY, 1973 ; SUDAN et al., 1973).

In the past most experimental studies were in fact performed at 50 MHz. Only one brief experiment combined three frequencies (BALSLEY and FARLEY, 1971) and measurements at 4.25 MHz and 49.92 MHz were compared during three days (BALSLEY et al., 1976). During these experiments, type 1 spectra were not observed below 16.25 MHz while type 2 spectra were present as low as 4.25 MHz. More recently a broad band HF coherent radar has been operating in Africa (CROCHET, 1977 ; HANUISE and CROCHET, 1978) and regular measurements of plasma instabilities at frequencies as low as 4.6 MHz have been performed from Addis-Ababa, Ethiopia (dip-0.4°). It was shown from the first experiments that strong type 1 were present at 14.2 MHz and probably below (HANUISE and CROCHET, 1977). This was confirmed by other experiments, the results of which are presented thereafter.

2. EXPERIMENTAL PROCEDURE

The radar has been operating from Addis-Ababa Geophysical Observatory with a peak power of 1.6 kw, a pulse width of 100 µs and a pulse repetition frequency of 50, 100 or 200 Hz. While it can be operated at any frequency between 3 and 30 MHz, the choice is limited by the antennas which are used. Measurements in the high part of the HF band are performed on a three frequencies Yagi antenna tuned on 14.2, 21.3 and 29.0 MHz. Forward gain is respectively 8, 8.5 and 9.5 db maintaining a front to back ratio of 25 db. Some measurements were performed at 7.3 MHz on another Yagi antenna of 4.9 db forward gain, and at 4.6 MHz on a dipole with a reflector element. For other frequencies a wide band antenna has been used. Directivity is then poor but echoes returned from east or west were distinguished by the sign of their doppler frequency. The analysis is performed in real time or after recording on magnetic tape using a minicomputer with hardware and software as described by GREENWALD (1972).

3. LINEAR THEORY

Dispersion equation for electrostatic irregularities of wave number K has been presented and discussed in numerous papers, including those listed in the introduction. Using fluid theory for waves propagating normal to the magnetic field and neglecting recombination effects, the real frequency ω and the growth rate γ can be written :

$$\omega = \frac{\vec{K}.\vec{V}e}{1+\Psi} \qquad (1)$$

$$\gamma = \frac{\Psi}{\nu_i(1+\Psi)} \left\{ \omega^2 + \frac{\nu_i \Omega e K y}{K^2 L n \nu e} \omega - K^2 C s^2 \right\} \qquad (2)$$

where Cs is the ion acoustic velocity, $\vec{V}e$ the electron drift velocity, $Ln = (1/N)(\partial N/\partial z)^{-1}$, Ky the wave vector component in the horizontal westward direction and $\Psi = \nu e \nu i / \Omega e \Omega i$ with the usual notations.

If we consider wavelengths for which (2) is valid and horizontally propagating waves, the threshold condition for instability is :

$$\omega^2 + \frac{\nu_i \Omega_e}{K L n \nu_e} \omega - K^2 C_s^2 = 0 \qquad (3)$$

During daytime electrons are drifting westwards and the electron density gradient is upwards. The second term in equation (3) is then positive (destabilizing gradient) and the threshold condition for instability can be written as in FARLEY and FEJER (1975) :

$$\omega = KC_s \left\{ (1 + F^2)^{1/2} - F \right\} \qquad (4)$$

where $F = \nu_i \Omega_e / 2K^2 \nu_e L n C_s$. Phase velocity Vph has then to be greater than a threshold velocity Vph = $C_s \left\{ (1+F^2)^{1/2} - F \right\}$.

Inserting typical values for the african equatorial ionosphere ($\nu_e \simeq 9 \times 10^4 S^{-1}$, $\nu_i \simeq 2 \times 10^3 S^{-1}$, $\Omega_e \simeq 6 \times 10^6 S^{-1}$) and $C_s = 360$ m/s we find $F = 4.6 \lambda^2 / L n$ with λ irregularity wavelength. Computed variation of the threshold drift velocity is shown on figure 1 for destabilizing gradients (daytime condition).

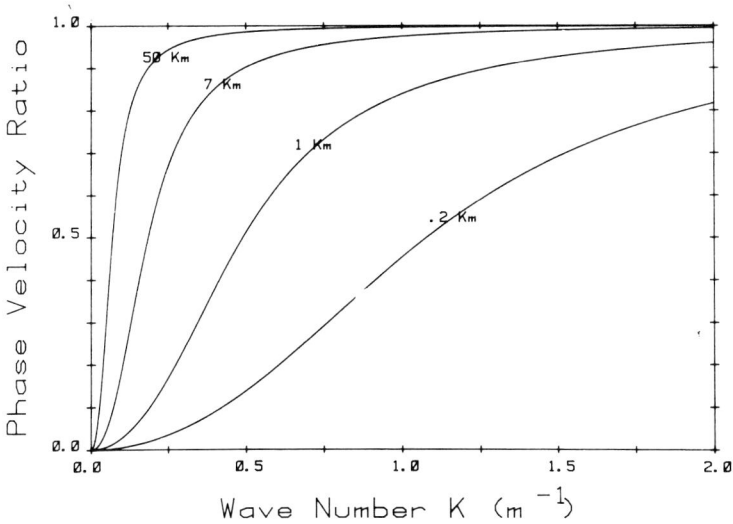

Figure 1. Ratio of the threshold value of the phase velocity of type 1 instability to the ion acoustic velocity for daytime conditions (destabilizing gradient). Lowest values of gradient length would correspond to secondary gradients generated by large scale type 2 irregularities.

When the effect of the gradient term is neglected, it is shown that unstable waves can be detected until an angle ϕ_c to the horizontal

for which they are marginally stable and which is defined as
cos φc = Cs (1 + Ψ)/Ve. With the gradient term, this critical angle
becomes φ'c, defined as :

$$\cos \phi'c = \frac{Cs\ (1 + \Psi)}{Ve\ \left(1 + \frac{2F(1+\Psi)Cs}{Ve}\right)^{1/2}} \quad (5a)$$

Around 105-107 Km, Ψ can be neglected and

$$\cos \phi'c = \frac{Cs}{Ve\ \left(1 + \frac{9.2\ \lambda^2 Cs}{Ln\ Ve}\right)^{1/2}} \quad (5b)$$

Values of the critical angle φ'c is shown on figure 2 for various
wavelengths. When the electron drift is near the threshold value for
small wavelength irregularities, aperture of the cone of unstable waves
is already important at higher wavelengths (for example 36° for 20 m
irregularities).

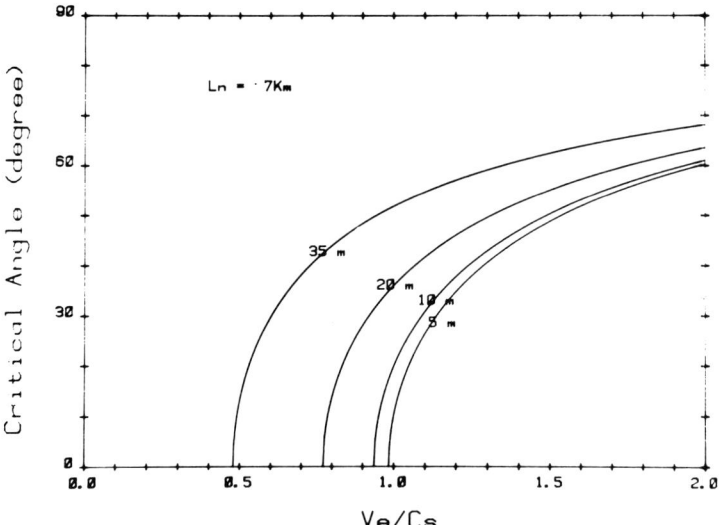

Figure 2. Variation of the critical angle φ'c for different
irregularity wavelengths with the electron
drift velocity when gradient is included.

It is then indicated by a linear theory that type 1 spectra should
be observed when the electron drift velocity is greater than a given
threshold velocity. Their aperture cone is depending on the ratio
between the drift velocity and this threshold velocity. Finally the
phase velocity of the unstable waves should be proportionnal to Ve.

4. MEASUREMENTS

4.1. Previous experiments in South America

The transmitted frequency during the first radar experiments in Jicamarca was 50 MHz and only type 1 spectra were identified at first as they are much more stronger than type 2 for high frequencies (BOWLES et al., 1963, COHEN and BOWLES, 1967). Distinctive features for type 1 spectra were :

- appearance above a given threshold,
- doppler shift constant with elevation and time,
- phase velocity independant from the electron drift velocity and always equal to the ion acoustic velocity,
- narrow width of the power spectrum.

These spectra were then associated with the two-stream instability, even if it is not explained by a linear theory that the phase velocity is independant from the electron drift velocity.

Type 1 spectra were later observed at lower frequencies (BALSLEY and FARLEY, 1971), the lowest being 16 MHz. It was indicated that the phase velocity and the basckscattered power are decreasing with frequency for these spectra.

At 16 MHz these spectra were always contaminated by a strong type 2 spectrum. It was then supposed that type 1 power spectrum was decreasing with frequency and that no such spectrum would be detected at lower frequencies. This was apparently confirmed by three days of measurements at 4.25 MHz (BALSLEY et al., 1976) : when type 1 were present at 50 MHz, only type 2 were identified at the lower frequency.

4.2. Recent measurements in Africa

Typical results showing the power spectra of echoes during strong electrojet conditions are presented on figure 3. The horizontal axis is scaled in doppler shift from transmitted frequency and the vertical axis represents the relative echo power density. All the maxima having been scaled to unity, comparison of power levels between two different spectra is not possible from this figure. All the measurements were taken around 10.45 LT on January 21, 1977. The three band Yagi antenna was looking westwards on 14.2 MHz and 29.0 MHz while returns are predominantly coming from the west on the wide band antenna at lower frequencies. Pulse rate frequency was 200 Hz, giving a doppler scale of ± 100 Hz at 29.0 MHz. It was divided by a factor of two at lower frequencies giving a scale of ± 50 Hz. The analysis was performed for the same time delay after the transmitted pulse on the different frequencies (in this case 1500 µs).

Type 1 spectra can be easily detected from 29.0 MHz to 5.5 MHz, with a phase velocity decreasing with the radar frequency. While the type 2 echoes are weak at 29 MHz, they are strong at lower frequencies especially at 14.2 MHz and 5.5 MHz. Spikes, appearing around 0 Hz are due to a direct coupling and the other ones are due to interferences from other transmissions.

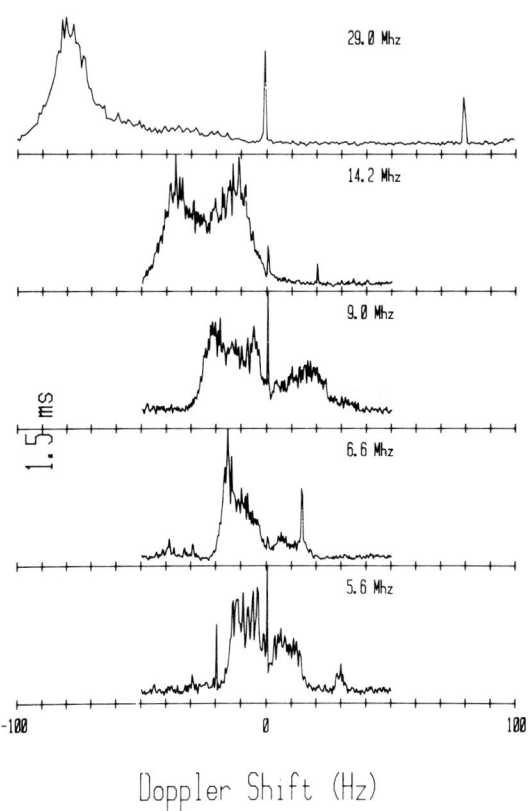

Figure 3. Type 1 doppler spectra observed on January 21, 1977 around 10.45 LT. The contamination at low frequencies is due to type 2 spectra. Returns are observed eastwards and westwards on the folded dipole antenna.

4.3. Type 1 phase velocity

Even if type 2 spectra are present simultaneously with type 1, it is nevertheless possible to perform doppler shift measurements by fitting gaussian curves to each type of irregularity spectrum. It has been seen that type 1 phase velocity is decreasing with frequency when the gradient term is introduced (equation 4).

Experimental ratios Vph/Vph29 of the phase velocity to the phase velocity at 29 MHz (the highest working frequency) are presented on figure 4. Also shown are theoretical curves when the effect of wave refraction is introduced. The irregularity wavelength selected by the radar is in this case modified according to.

$$\lambda irr = \lambda o / (1 - \omega^2 p/\omega^2)^{1/2} \qquad (6)$$

with λo irregularity wavelength selected by the radar if refraction is ignored. F in equation (4) has to be replaced by

$$F' = \frac{4.6 \; \lambda irr^2}{Ln} = \frac{4.6}{Ln} \cdot \frac{\lambda o^2}{(1 - \frac{\omega p^2}{\omega^2})} \qquad (7)$$

and $\quad \omega = KCs \left\{ (1 + F'^2)^{1/2} - F' \right\} \qquad (8)$

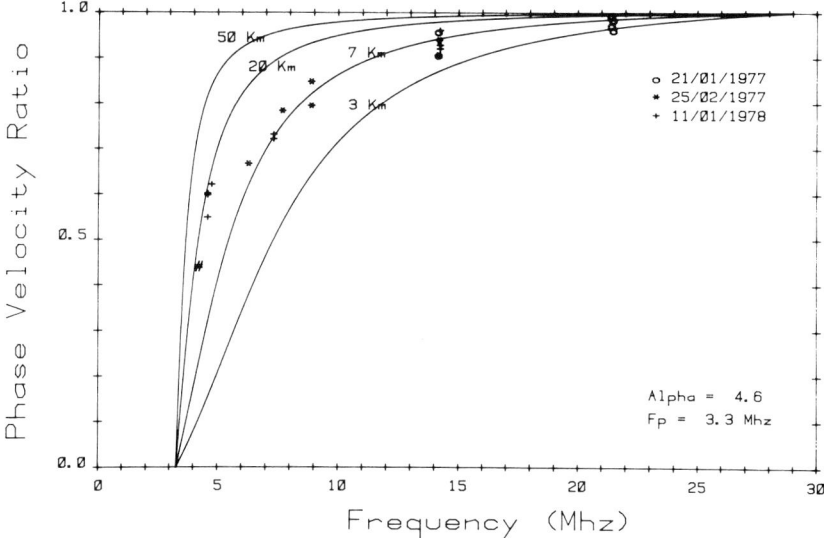

Figure 4. Three days experimental data of type 1 spectra phase velocity normalized to the phase velocity measured at 29 MHz (Also drawn as # is the point corresponding to BALSLEY et al. (1976) measurements). Theoretical curves have a better fit for a gradient length of 7 Km. Electromagnetic perturbation to the instability have to be accounted for at frequencies lower than 7 MHz.

A cut-off at the plasma frequency is introduced when refraction effects are included. The experimental data fit very well the theoritical curves for a gradient length Ln = 7 Km until about 7 MHz.

Consequently, it can be deduced from the experimental data that the influencing factor is the primary gradient in the layer and not the secondary gradients generated by large scale type 2 irregularities as it could be according to SUDAN et al. (1973). At lower frequencies, ratios would have a better fit with larger Ln values. As a matter of fact theory is only valid for electrostatic waves for $\lambda \lesssim 20$ m ; for lower frequencies electromagnetic perturbations to the instability have to be taken into account (KAW et al. , 1973).

DISCUSSION

The measurements presented here point out that the mechanism responsible for generation of type 1 irregularities is effective at frequencies as low as 4.6 MHz and probably lower even if their detection is difficult due to refraction phenomena. This conclusion differs from BALSLEY et al. (1976) one about the only previous published measurement near 4 MHz. It was then concluded that the two-stream instability was not effective at low frequencies : no type 1 spectra were detected at 4.25 MHz when they were present at 49.92 MHz (figure 5). It is suggested here that the type 1 was in fact present but not easily discernible due to refraction effects and type 2 mixing.

This is corroborated by two points :

- first, the mean phase velocity is the same at the two radar frequencies as long as only type 2 are present (this has been shown to be related to the electron drift velocity)(BALSLEY, 1969). But when type 1 are present this drift velocity is no more the same and is lower at the lowest frequency as if there was a threshold.

- secondly the type 2 spectra at 4.25 MHz have a good gaussian shape but this is no more true when type 1 is present at the other frequency. It is moreover possible to fit two gaussian curves for each spectra and this is done on figure 5c. The one with the highest doppler shift would be the type 1 spectra and its small phase velocity is due to the effect of the density gradient. The ratio between the two phase velocities is plotted on figure 4 and is seen to be perfectly coherent with our measurements and the theoretical curves.

Observations which are performed at low frequencies show then that type 1 spectra are present when the electron drift velocity is greater than the threshold value predicted by a linear theory when the electron density gradients are included. Contrary to the results of this linear theory the observed phase velocity is not proportional to Ve but is always equal to the threshold velocity. This can be interpreted in a convective theory as the detection of marginal waves with a K vector making an angle ϕc to the horizontal. These waves would

have the largest propagation path in the ionosphere and the strongest
amplitude (KAW, 1972 ; LEE and KENNEL, 1973 ; LEE et al., 1974).

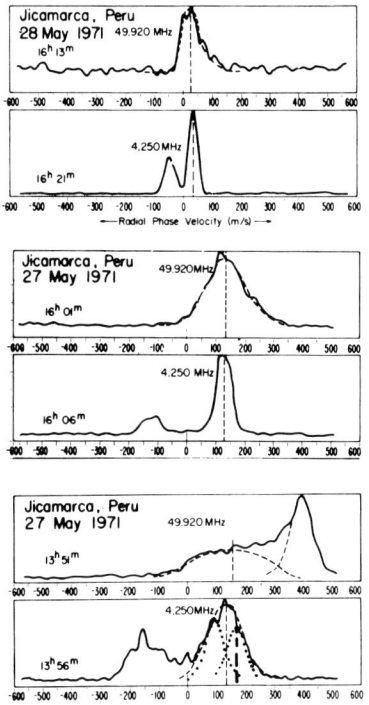

Figure 5. Power spectra obtained at Huancayo on
49.92 MHz and 4.25 MHz (from BALSLEY
et al., 1976). Only type 2 spectra are
present on figures 5a and 5b. On figure 5c
is shown the fitting of two gaussian
curves which would be type 1 and type 2 spectra.

The threshold drift velocity for instability has to replace the
ion acoustic velocity when gradients are included. It seems then
necessary to extend the quasi-linear or non-linear convective theories
by including these effects and get a possible explanation for low
frequency type 1 spectra. One such possibility is that, for a given
frequency, type 1 appears when electron drift threshold value is
reached. Spectra observed at higher frequencies for the same drift
velocity (type 2) would be in fact due to non linear cascading of the
type 1 spectra.

CONCLUSION

It has been shown unambiguously that type 1 radar echoes are present for frequencies as low as 4.6 MHz. Type 2 radar echoes also present at the lowest frequencies and the separation of the two spectral type may be difficult.

The phase velocity of type 1 instability is always equal to a threshold value which is influenced by the primary gradient in the E layer. During the day the gradient is destabilizing and this threshold velocity is less than the ion acoustic velocity C_s. Type 1 irregularities are then present for lower electron drift velocities at lower frequencies. For a given drift velocity, the critical angle for instability will be greater for these lower frequencies.

Electromagnetic effects have to be included in the theories for frequencies less than 7 MHz and the effectiveness of the two mechanisms proposed to account for radar observations in the equatorial electrojet have to be reexamined.

ACKNOWLEDGEMENTS - We want to acknowledge the help of Prof. P. GOUIN, GHEBREBRHAN OGUBAZGHI and all the staff of Addis-Ababa Geophysical Observatory. This work has been supported by the Centre National de la Recherche Scientifique (ERA 668), Université de Toulon et du Var, Institut de Physique du Globe de Paris, Ministère des Affaires Etrangères and D.G.R.S.T. (Contrat n° 75/7/1433).

REFERENCES

BALSLEY, B.B., 1969, J. Geophys. Res., 74, pp. 2333-2347.
BALSLEY, B.B., and FARLEY, D.T., 1971, J. Geophys. Res., 76, pp.8341-8351
BALSLEY, B.B., REY, A., and WOODMAN, R.F., 1976, J. Geophys. Res., 81, pp. 1391-1396.
BOWLES, K.L., BALSLEY, B.B., and COHEN, R., 1963, J. Geophys. Res., 68, pp. 2485-2501.
COHEN, R., and BOWLES, K.L., 1967, J. Geophys. Res., 72, pp. 885-894.
CROCHET, M., 1977, J. Atmosph. Terrest. Phys., 39, pp. 1103-1117.
FARLEY, D.T., 1963, J. Geophys. Res., 68, pp. 6083-6097.
FARLEY, D.T., and BALSLEY, B.B., 1973, J. Geophys. Res., 78, pp.227-239.
FARLEY, D.T., and FEJER, B.G., 1975, J. Geophys. Res., 80, pp. 3087-3090.
FEJER, B.G., FARLEY, D.T., BALSLEY, B.B., and WOODMAN, R.F., 1975, J. Geophys. Res., 80, pp. 1313-1323.
HANUISE, C., and CROCHET, M., 1977, J. Atmosph. Terrest. Phys., 39, pp. 1097-1101.
HANUISE, C., and CROCHET, M.,1978, J. Atmosph. Terrest. Phys., 40, pp. 49-59.
KAW, P.K., 1972, J. Geophys. Res., 77, pp. 1323-1326.
LEE, K., and KENNEL, C.F., 1973, J. Geophys. Res., 78, pp. 4619-4629.
LEE, K., KENNEL, C.F., and CORONITI, F.V., 1974, J. Geophys. Res., 79, pp. 249-266.
ROGISTER, A., and D'ANGELO, 1970, J. Geophys. Res., 75, pp. 3879-3887.

SUDAN, R.N., AKINRIMISI, J.,and FARLEY, D.T., 1973, J. Geophys. Res., 78, pp. 240-248.
WALDTEUFEL, P., 1965, Ann. Geophys., 21, pp. 579-604.

Reference is also made to the following unpublished material :

GREENWALD, 1972, NOAA Technical Report ERL-241, AL. 7.
KAW, P.K., IVANOV, A.A., and CHATURVEDI, P.K., 1973, First Lloyd Berkner Symposium, AGU, Dallas, Texas.

PART III

NONLINEAR EFFECTS

NONLINEAR WHISTLER-MODE INTERACTION AND TRIGGERED EMISSIONS IN THE
MAGNETOSPHERE : A REVIEW

Hiroshi Matsumoto
Ionosphere Research Laboratory, Kyoto University, Kyoto, Japan

A review is given of both experimental and theoretical aspects of coherent nonlinear effects in the whistler-mode interactions in the magnetosphere. Among various nonlinear wave-wave and wave-particle interactions in the magnetosphere, the whistler-mode interaction appears to be one of the most exciting phenomena which have been observed in space plasmas but not in laboratory plasmas. The present review may be far from complete since much of the research, both theoretical and experimental, is still in progress. Given the limitations of time and space, the present article will attempt to cover the following subjects :
(1) A brief history of the study of VLF triggered emissions in the magnetosphere.
(2) Recent results of the artificial VLF wave injection experiments with stress on the observed emission characteristics which need to be explained by theoretical work including effects of power-line harmonic radiation on the observed triggered emissions.
(3) Nonlinear behavior of resonant particles in a monochromatic or quasi-monochromatic whistler wave ---- basic concepts of the nonlinear interaction in the whistler mode.
(4) Theory of nonlinear interaction of quasi-monochromatic whistler wave in homogeneous and inhomogeneous plasmas.
(5) Computer simulations relevant to the nonlinear whistler-mode interactions.
(6) Some suggestions for future work.

1. INTRODUCTION

Nonlinear whistler-mode interaction is undoubtedly one of the most exciting phenomena which have been observed in space plasmas but not in laboratory plasmas. The present review will attempt to cover both experimental and theoretical aspects of nonlinear whistler mode interaction related to VLF triggered emissions in the magnetosphere. Triggered emissions are characterized by their narrow band frequency spectra and are believed to be a manifestation of *coherent* nonlinear cyclotron interaction between a monochromatic or quasi-monochromatic whistler wave and resonant electrons. The interaction is therefore rather different from the usual nonlinear turbulent wave-particle interaction which is based on a random phase approximation. The latter including usual nonlinear whistler instabilities have been extensively investigated both theoretically [e.g., Kennel and Petschek, 1966; Sonnerup and Su, 1967; Kimura, 1974 and Gendrin, 1975 for reviews and references therein] and by computer simulations [Davidson et al., 1972; Ossakow et al., 1972a,b, 1973a,b; Cuperman and Salu, 1973; Cuperman et al., 1973]. Nonlinear turbulent wave-particle interactions are thought to be the generation mechanisms of wide-band VLF emissions, whislter amplification and damping and wave-induced particle diffusion and resultant precipitation phenomena in the magnetosphere. We will not, however, consider the nonlinear turbulent interaction but focus our attention only on the coherent nonlinear particle interaction with a monochromatic whistler wave.

Experimental studies of triggered emissions will be chronologically reviewed in section 2. Intrinsic features of the triggered emissions which need to be accounted for by theoretical works are discussed in section 3. The fundamental concepts of phase bunching, phase trapping, second order resonance, interaction region and nonlinear resonant current which are essential to the coherent nonlinear whistler interaction are discussed in section 4. Theories and computer simulations of the coherent nonlinear whistler interactions are reviewed in section 5. An attempt is made to compare and sort the existing theories and simulations related to triggered emissions. Finally in section 6, a summary as well as some suggestions for future work on nonlinear triggered emissions is given.

2. EXPERIMENTS OF TRIGGERED EMISSIONS

The most commonly known wave phenomena related to the coherent nonlinear whistler interaction are VLF triggered emissions. They are triggered both by natural whistler waves and by man-made signals propagating in the whistler mode in the magnetosphere. Though the latter, often called ASE (Artificially Stimulated Emissions) have similar characteristics to those naturally triggered, they provide us with more information of the coherent nonlinear whistler interaction, because the nature of the triggering wave itself is known. Since there exist excellent reviews on triggered emissions so for [Helliwell, 1965; Kimura, 1967; Gendrin, 1972,1975], a brief history and main experimental results of the *artificially* triggered emissions are given.

The first ASE's, triggered by man-made Morse code signals from NPG on 18.6 kHz and NAA on 14.7 kHz, were discovered accidentally during the course of examination of natural VLF events [Helliwell et al., 1964]. Dynamic frequency spectra of the ASE's were published in Helliwell's Atlas in 1965. One example of the dynamic spectra of the ASE's triggered by the Morse code pulse is shown in Fig.1. They all show a *narrow bandwidth* ($\sim$ 100 Hz) and have a variety of spectral forms with *a sizable frequency variation* including risers, fallers, hooks, inverted hooks and simultaneous risers and fallers.

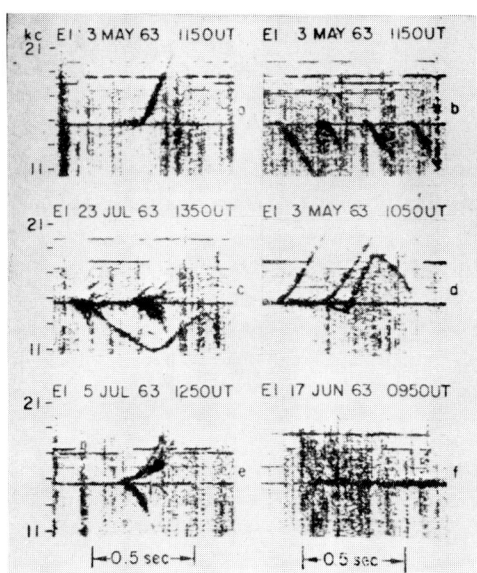

Fig.1 Examples of dynamic spectra of ASE's triggered by Morse code (after Kimura, 1967)

The striking feature of the ASE's triggered by Morse code signals is a *dash-dot anomaly*, i.e., a prevalence of triggering by dashes (150 msec) compared with dot (50 msec) [Helliwell et al., 1964; Helliwell, 1965; Lasch, 1968]. In connection with this anomaly, Helliwell [1969] showed that the triggered emission is significantly *delayed* with respect to the onset time of the triggering signal. The average delay time is about 100 msec which is shorter than the dash and longer than the dot. It was also pointed out [Kimura, 1967; Lee, 1968; Helliwell, 1969] that most of the NAA triggered ASE's apparently start at a slightly higher frequency above the triggering frequency. The *"offset"* frequency ranges from 0 to 450 Hz for the NAA triggering.

Kimura [1967, 1968] discovered that triggered emissions can also be triggered *by a 100 W transmitter*, Omega (10.2 kHz). This is several orders of magnitude smaller power than the 1MW NAA transmitter. The duration of the triggering pulse is, however, 1 sec, much longer than the dash of NAA. It should be noted that the Omega signal is intensified when triggering is observed, suggesting that amplification of the triggering signal may be associated with the triggering mechanism [Helliwell, 1966]. A similar amplification of the order of 10 db of NPG signal propagating in the whistler mode in the magnetosphere was shown by McNeil [1968] even though no triggered emissions were associated with the signal in this case.

During the monitoring of a slow inward drift of a whistler duct under the influence of the magnetospheric electric field, Carpenter [1968] found that the triggering activity of ASE's is substantially enhanced when the frequency of the triggering wave is equal to *one-half the equatorial cyclotron frequency* along the duct. This finding implies, in addition to the triggering preference at half the equatorial cyclotron frequency, that the triggered emissions are stimulated by a wave propagating *in a ducted mode*, as has often been assumed in many theories.

Interesting *pulsation phenomena* in amplitude and bandwidth were found [Bell and Helliwell, 1971] during the unmodulated (*"key-down"*) transmission from NAA (14.7 kHz and 17.8 kHz). The period of the pulsation was in the range from 300 to 600 msec.

(a)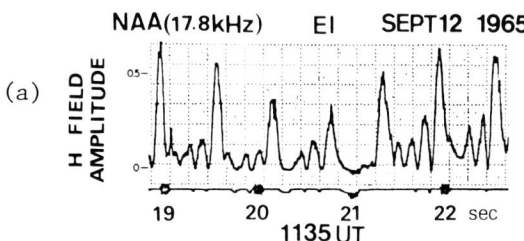

(b)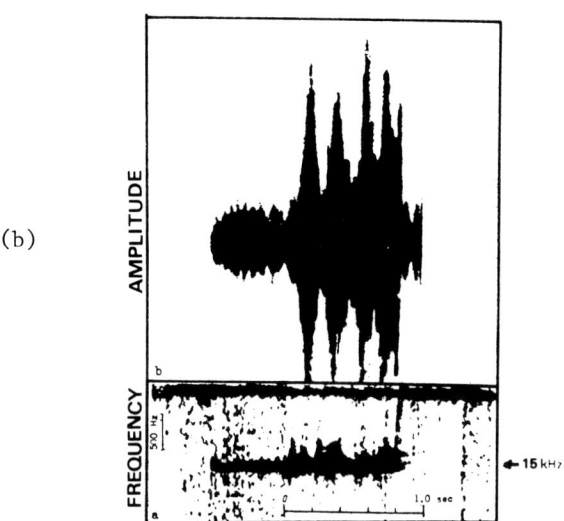

Fig.2 Pulsation phenomena. (a) Relative amplitude of key-down whistler mode signal transmitted from NAA (17.8 kHz) recorded at Eights Station, Antarctica (after Bell and Helliwell, 1971). (b) Pulsation in amplitude and corresponding spectral broadening of 800 msec whistler mode pulse (after Likhter et al., 1971).

Figure 2(a) shows one example of their result. The average spacing between the main peaks is shown to be equal to the one-hop whislter mode delay of the signal, while that of the subsidiary peaks is about 140 msec. Similar periodic modulation effects were reported by Likhter et al. [1971] for a pulse with a frequency of 15 kHz and duration of 800 msec. (see Fig.2(b)). They showed that the maxima of the spectrum broadening coincides with the maxima of the signal amplitude. The spectrum broadening is asymmetric with respect to the signal frequency with the higher frequencies predominating. Maximum spectrum broadening was about 100 to 400 Hz and the pulsation period was about 200 msec in their case. These pulsation phenomena may closely be related to the generation mechanism of triggered emission.

Controlled VLF wave injection experiments into the magnetosphere were begun in 1973 [see Helliwell and Katsufrakis, 1974; Helliwell, 1974 for a review].

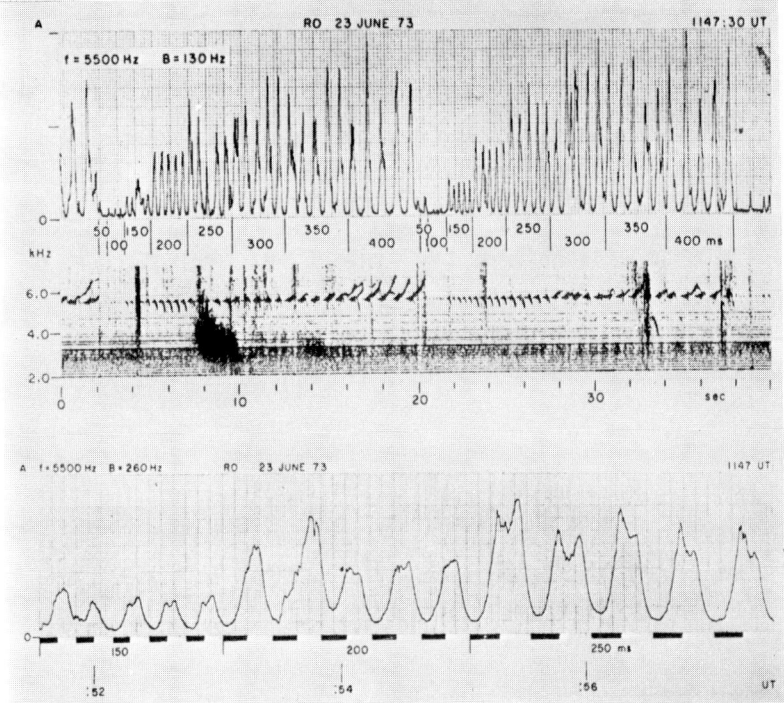

Fig.3 Result of the Siple controlled experiments.
(a) Varialble pulse length sequence received at Roberval. Lower panel shows the spectrum and upper panel the amplitude. Pulse lengths vary from 50 ms to 400 ms in 50 ms steps as indicated by the numbers between panels. (b) Corresponding pulse envelope of 150, 200, 250 ms in the second 18s sequence in (a). (after Helliwell and Katsufrakis, 1974).

A powerful (∼100 KW) VLF transmitter connected to a 21.3 km dipole antenna with variable frequency, pulse length and modulation was used to transmit a narrow band whistler mode signal from the Siple station, Antarctica (L=4). The first attempt to control whistler wave-particle interactions in the magnetosphere by artificial wave injection showed many interesting results.

Observations at the conjugate point showed triggered emissions as expected. Figure 3(a) shows test results of their variable pulse length experiment. The frequency was switched between 5.0 and 5.5 kHz with constant radiation power. The pulse length was varied from 50 msec to 400 msec in 50 msec steps. The emission activity is confined to 5.5 kHz in this case. The emission activity and intensities of the received signal and triggered emissions increase as the pulse length increases. Pulses shorter than 100 msec produce no triggered emissions, which is consistent with the dash-dot anomaly mentioned before. They commonly observed *a dependence of the spectral forms of emissions on the triggering pulse length*. As the pulse length increases, the triggered emissions change from weak fallers to strong risers. The lower panel of Fig.3(a) shows the amplitude in 130 Hz band centered on 5.5 kHz. The peak intensity shows a systematic increase as a function of the pulse length for pulses shorter than 300 msec. Corresponding individual pulse growth is illustrated in Fig.3(b). The solid bars at the bottom show the corrected timing of the corresponding pulse input by adding whistler mode time delay. It is seen that the amplitude of the received signal is minimum at the leading edge of the input signal and then *grows almost exponentially with time* until the end of the pulse, at which the intensity shows a flattening followed by a subsequent oscillation. The average growth rate in this case was 128 db/sec (this value ranges from 30 db/sec to 200 db/sec at other times [Helliwell, 1974]) and saturation was observed for pulses longer than 250 msec. The total growth is found to be about 30 db. They, thus, demonstrated that controlled VLF wave injection experiments can reproduce the triggered emissions and associated signal growth known at that time from the Morse code triggering.

Other features of interest revealed by the controlled experiments are summarized in Fig.4. Figure 4(a),(b) and (c) show *wave-wave coupling effects*. Figure 4(a) is an example of coupling between the triggered emissions by the Siple signal and *power line harmonic (PLH) radiation* which is seen as horizontal lines in the spectrum. Many of the triggered emissions are found either to terminate or to change their frequency-time slope at PLH radiation frequencies. This discovery of the coupling between the PLH radiation and triggered emissions stimulated studies of the PLH radiation and its effects on the magnetospheric wave-particle interactions [Helliwell, 1975; Helliwell et al., 1975; Park and Helliwell, 1977; Park, 1977]. These investigations revealed that the PLH radiation is actually amplified whistler wave with frequency shift of 20-30 Hz from the simple harmonics of the power line frequency. The PLH radiation has much larger bandwidth (∼30 Hz) than the original power line induction radiation and varies slowly in amplitude and frequency [Helliwell et al., 1975]. Park and Helliwell [1977] suggested that the PLH radiation plays a catalytic role in the triggereing of whistler precursors.

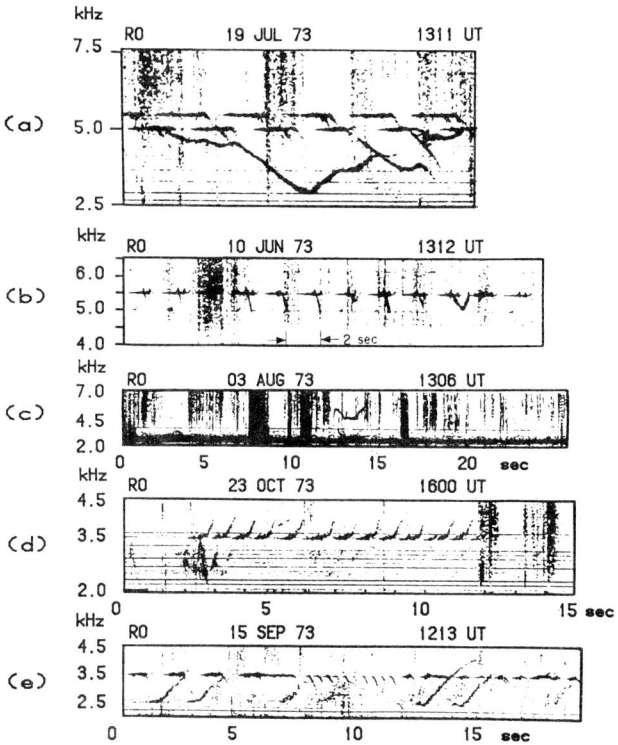

Fig.4 Interest features of triggered emissions discovered in the Siple experiment. (a) Coupling with power line harmonic radiation. (b) Cutoff at other weaker signal. (c) Entrainment of whistler triggered emissions by Siple signal. (d) Triggering of a series of repeated risers during a 10s key-down transmission. (e) The lowest frequency triggered emissions so far observed and their whistler echoes. (after Helliwell and Katsufrakis, 1974).

Figure 4(b) shows an example of another wave-wave coupling effect. The emissions triggered at one frequency (5.5 kHz) are frequently *cutoff at the other frequency* (5.0 kHz) even though the latter signal is weak [Helliwell and Katsufrakis, 1974; Helliwell, 1974]. It has also been found that the transmitted Siple signal not only triggers emissions but also *captures (or entrains)* the other emissions such as whistler triggered emissions as shown in Fig.4(c).

Figure 4(d) illustrates a new result for a 10 sec key-down transmission. During the key-down period, 13 successive risers with spacings averaging slightly less than 1 sec were triggered. This result appears to differ from the pulsation phenomena mentioned before [Bell and Helliwell, 1971], but implies that a relaxation mechanism is involved in the triggering process. The lowest panel (e) in Fig.4 is an example of the lowest frequency triggered emissions so far observed in the Siple experiment and their whistler mode echoes. The highest frequency that could trigger emissions is reported to be 7.6 kHz for the Siple experiments.

Another controlled VLF wave transmission experiment using a baloon-lofted transportable VLF (TVLF) transmitter has been performed between Alaska and New Zealand [McPherson et al., 1974; Koons et al., 1976; Dowden et al., 1978]. McPherson et al., 1974] found an amplification effect along the L=3 path which shows a weak dependence on pulse length with maximum intensity occurring around 200 to 400 msec. To determine the effect of a phase relation between a whistler mode wave and resonant electrons, waves (f=6.6 kHz, 100 W) with a controlled phase reversal were injected from Alaska (L=4) [Koons et al., 1976]. At times of the phase reversal, the amplitude of the signal received at the conjugate region went to zero and recovered to a saturation level of about 3.5 mγ with a characteristic time constant of 33 msec, showing a dephasing effect of phase trapped resonant electrons. Around the saturation level, the signal showed an amplitude oscillation with an upper sideband at about 40 Hz above the carrier frequency.

Dowden et al.[1978] conducted a controlled VLF injection experiment similar to the Siple experiment by their TVLF transmitter (93W) from Anchorage (L=4), Alaska. The transmitted signal with a frequency of 6.6 kHz is found to be nonlinearly amplified occasionally. During the non-linear amplification (NLA) events, the response to the transmitted pulse is similar to that in the Siple experiment. The wave grows exponentially for the first 100-200 msec at around 100 db/sec followed by a rather periodic amplitude oscillation with a period of 100 to 200 msec with no well-defined saturation lebel. At this stage, the received wave amplitude is around 100 μv/m which they interpret as corresponding to the peak wave magnetic field at the equator of 15 to 25 mγ. After the growth, the frequency shows a positive offset of 20-150 Hz above the transmitted frequency. They call this self-sustaining wave *"embryo emission (EE)"*. At or even before the pulse end, the EE may change continuously to a gliding frequency free emission.

The frequency-time behavior [Stiles and Helliwell, 1975] and the amplitude behavior [Stiles and Helliwell, 1977] of large number of ASE's were examined in great detail. All features but the offset phenomenon mentioned above were confirmed. They found that all emissions start at the triggering frequency. It was shown that the apparent frequency offset in the NAA triggering is caused by a frequent coincidence of the maximum amplitude of ASE's with a short (~40 msec) *plateau* in the dynamic spectrum roughly 200 Hz above the triggering signal. They also found that all emissions regardless of their final slope, initially rise on leaving the triggering frequency (see also Kimura, 1967).

In situ measurements during the VLF wave injection experiments are important for the study of the coherent nonlinear whistler interactions. However, few such experiments have been performed to this date. Inan et al.[1977] reported their first results of a Siple-satellites joint experiment. The observations by the Explorer 45 and Imp 6 spacecrafts showed that the equatorial area of the magnetosphere illuminated by the Siple transmitter is between 20°E and 15°W of the Siple longitude and between the L shells of 2.9 and 5. Signal amplitude measurements, which are of special importance in making theoretical models of the whistler interaction, were available on only one pass of the Imp 6. The result showed that the amplitude of the wave magnetic intensity varied from 0.01 mγ to 0.2-0.3 mγ on the off-equatorial orbit passing from L=4.8 to 4.0. The inferred equatorial wave magnetic intensity is then as large as 6-9 mγ if an amplification of the order of 30 db is available near the equator. Triggered emissions stimulated by the Siple transmitter pulse were observed on only three passes. The most surprising result is that on one pass of the Imp 6 emission triggering and/or wave entrainment (wave-wave interaction) effects appeared to be taking place far from the equator. If this is the case, the finding is very significant as most of the theories of the triggered emissions postulate that the nonlinear whistler interaction takes place near the equatorial region along the magnetic field.

A summary table of the experimental studies of the triggered emissions is given in Table 1.

DISCOVERY :	Helliwell et al.[1964]
FREQUENCY SPECTRA :	Helliwell [1965]
DASH-DOT ANOMALY :	Helliwell et al.[1964]
	Helliwell [1965]
	Lasch [1968]
LOW POWER OMEGA TRANSMITTER :	Kimura [1967]
	Kimura [1968]
AMPLIFICATION OF WHISTLER MODE :	McNeil [1968]
PULSATION PHENOMENA :	Bell and Helliwell [1971]
	Likhter et al.[1971]
SIPLE CONTROLLED EXPERIMENT :	Helliwell and Katsufrakis [1974]
	Helliwell [1974]
CONJUGATE TRANSMISSION EXPERIMENT :	McPherson et al.[1974]
PHASE REVERSAL EXPERIMENT :	Koons et al.[1976]
NONLINEAR AMPLIFICATION :	Dowden et al.[1978]
DETAILED ANALYSES :	Stiles and Helliwell [1975]
	Stiles and Helliwell [1977]
SATELLITE OBSERVATION :	Inan et al.[1977]

Table 1. Summary of experiments on triggered emissions

3. SALIENT FEATURES OF TRIGGERED EMISSIONS

In summary, the features of ASE's which have to be explained by theoretical works are;
(1) Narrow frequency bandwidth (∼ 100 Hz).
(2) Variable frequency giving a variety of spectral forms.
(3) Dash-dot anomaly and triggering delay (∼ 100 msec).
(4) Triggering by both high (1 MW) and low (100 W) power transmitters.
(5) Repeatability.
(6) Exponential growth of the triggering wave with time at rates ranging 25 db/sec to 250 db/sec giving a maximum final growth of 20 to 35 db.
(7) Beginning of emissions at the triggering wave frequency and initial rise in frequency.
(8) Frequency lock of emissions to or within 200 Hz above the triggering wave frequency during the growing phase.
(9) Release of emissions at times of growth stop or of the triggering wave termination.
(10) Predominance of risers for longer triggering pulse and fallers for shorter pulses.

Some other features that might have to be taken into account in theoretical works of the nonlinear whistler interactions are;
(11) Pulsation phenomena and/or associated triggering of successive series of emissions during the key-down transmission.
(12) Wave-wave coupling effects such as entrainment, cutoff and df/dt change by other emissions, signals and PLH radiation.
(13) Frequency shift of the PLH radiation of amount of 20 to 30 Hz from simple harmonic frequencies of the power line system.

4. BASIC PHYSICS IN COHERENT NONLINEAR WHISTLER INTERACTION

Particle motion in a coherent wave is of great importance in theories of the coherent nonlinear wave-particle interaction. Nonlinear behavior of resonant particles in a monochromatic or quasi-monochromatic whistler wave has been extensively investigated [Dungey, 1963; Roberts and Buchsbam, 1964; Laird and Knox, 1965; Bell, 1965; Lutomirski and Sudan, 1966; Dungey, 1969; Knox, 1969; Ashour-Abdalla, 1970; Dysthe, 1971; Nunn, 1971; Matsumoto, 1972; Laird, 1972; Gendrin, 1974; Matsumoto et al.,1974; Nunn, 1974; Crystal, 1975; Brinca, 1975]. In this section, we discuss fundamental concepts of the coherent nonlinear whistler interaction based on the electron motion in a monochromatic whistler wave which are of basic importance in the theory of triggered emissions.

4.1 Basic Equations of Electron Motion

We consider an electron of mass m and charge -e in a monochromatic, plane whistler wave propagating along the inhomogeneous geomagnetic field B_0. A curvature of the geomagnetic field is neglected and the direction of B_0 is taken to be parallel to the z-axis. Provided that the electron Larmor radius is much less than the scale length of inhomogeneity of B_o, the equations of motion are [see e.g., Dysthe,1971].

$$\frac{dv_\perp}{dt} = \Omega_w(v_\| - \frac{\omega}{k})\sin\psi + \frac{v_\| v_\perp}{2\Omega_e}\frac{\partial \Omega_e}{\partial z} \tag{1}$$

$$\frac{dv_\|}{dt} = -\Omega_w v_\perp \sin\psi - \frac{v_\perp^2}{2\Omega_e}\frac{\partial \Omega_e}{\partial z} \tag{2}$$

$$\frac{d\psi}{dt} = k(v_\| - V_R) - \frac{\Omega_w}{kv_\perp}(\omega - kv_\|)\cos\psi \tag{3}$$

where the velocity $\underline{v} = (v_\perp \cos\phi, v_\perp \sin\phi, v_\|)$ and $\psi = \phi + \int k dz - \pi$ is a relative phase angle between $\underline{v}_\perp$ and $-\underline{B}_w$, in which ϕ is the Larmor phase angle ($\phi = \int \Omega_e dt$) and $\underline{B}_w$ is the wave magnetic field. The quantities ω, k, Ω_w and Ω_e are the wave frequency, wave number, wave cyclotron frequency $\Omega_w = eB_w/m$ and cyclotron frequenct $\Omega_e = eB_o/m$, respectively. The quantity V_R is a resonance velocity defined by $V_R^o = (\omega-\Omega_e)/k = -(\Omega_e-\omega)^{3/2} c/\Pi_e \omega^{1/2}$, where use is made of the whistler dispersion of

$$k = (\Pi_e/c)[\omega/(\Omega_e-\omega)]^{1/2} \tag{4}$$

in which c and Π_e are the light speed and plasma frequency. The last terms in Eqs.(1) and (2) represent the inhomogeneity effect. The second term in the r.h.s. of Eq.(3), which comes from the wave electric and transverse Lorentz forces, is smaller than the first term coming from the longitudinal Lorentz force $-ev_\perp \times \underline{B}_w$ by the order of $(B_w/B_o)^{1/2}$ × cot α for resonant and nearly resonant electrons where α is a pitch angle of the electron. As B_w/B_o ranges from 10^{-5} to 10^{-6} for B_w = 1 mγ to 10mγ in the equatorial region along L = 4 magnetic field, we can well neglect the second term. From Eqs.(1) - (3) we obtain the following equations. Energy change:

$$\frac{dW}{dt} = -\frac{\omega}{k} m v_\perp \Omega_w \sin\psi \ ; \qquad W = \frac{1}{2} m(v_\perp^2 + v_\|^2) \tag{5}$$

Magnetic moment change:

$$\frac{d\mu}{dt} = ev_\perp(v_\| - \frac{\omega}{k})\frac{\Omega_w}{\Omega_e}\sin\psi \ ; \qquad \mu = \frac{1}{2} m v_\perp^2/B_o \tag{6}$$

Energy change in the wave frame:

$$\frac{d}{dt}\{ \frac{1}{2} m v_\perp^2 + \frac{1}{2} m(v_\parallel - \frac{\omega}{k})^2 \} = \frac{\omega}{k} \frac{1}{2\Omega_e} \frac{\partial \Omega_e}{\partial z} + (\frac{\omega}{k} - v_\parallel) \frac{d}{dt}(\frac{\omega}{k}) \tag{7}$$

Phase change:

$$\frac{d^2\psi}{dt^2} + \omega_t^2 \sin\psi = \frac{dk}{dt}(v_\parallel - V_R) - k(\frac{dV_R}{dt} + \frac{v_\perp^2}{2\Omega_e} \frac{\partial \Omega_e}{\partial z}) \tag{8}$$

where $\omega_t = (k v_\perp \Omega_w)^{1/2}$ is a *trapping frequency*. For resonant electrons, which satisfy $v_\parallel = V_R$, the following constant of motion [Karpman et al., 1974] is obtained from Eqs. (1) and (5).

$$v_\parallel^2 + v_\perp^2 (1 - \frac{\omega}{\Omega_e}) \equiv u^2 = \text{constant} \tag{9}$$

It is noted that the expressions for energy and magnetic moment change do not include the inhomogeneity effect explicitly. In the case of a homogeneous magnetic field, Eq.(8) becomes a simple pendulum equation of non-harmonic oscillation.

4.2 Phase Bunching and Phase Trapping

Trajectories of resonant electrons in a whistler wave are characterized by their phase behavior. The phase behavior has been described by two different terms, "phase trapping" and "phase bunching" [e.g., see Matsumoto,1972; Gendrin,1974; Crystal,1975]. "Phase trapping", which means a bounded phase oscillation of an *individual* resonant electron in a whistler wave, has been emphasized in analytic instability treatments [Sudan and Ott, 1971; Palmadesso and Schmidt,1971; Dysthe, 1971; Brinca, 1972a,b,; Istomin and Karpman,1972,1973a,b; Karpman et al.,1974a,b; Denavit and Sudan,1972,1975; Yamamoto,1976,1977; Roux and Pellat,1976,1978] "Phase bunching", on the other hand, is a concept emerging from visualization of temporal behavior of phases of *many* resonant electrons under the effect of a whistler wave. "Phase bunching" describes a state in which phases of resonant electrons are bunched around a certain phase angle. The phase bunching concept has been used for a calculation of nonlinear resonant current in many works [Helliwell,1967; Nunn,1971,1974; Matsumoto,1972; Helliwell and Crystal,1973; Matsumoto et al.,1974; Gendrin,1974; Crystal,1975].

Neglecting the inhomogeneous terms in Eqs(1) to (3) for the moment, we obtain the following constant of motion

$$\frac{1}{2} k^2 (v_{\|} - V_R)^2 - \omega_t^2 \cos\psi = \chi \qquad (10)$$

Provided the pitch angle α is not extremely small, we can well neglect a variation in $v_\perp$. In this case, solutions to Eq.(8) are given by Jacobi's elliptic function. Figure 5(a) is a phase diagram calculated from Eq.(10) in which the quantity V_t is a trapping velocity defined by $V_t = 2\omega_t/\kappa$ in which $\kappa^2 = 2\omega_t^2/(\chi+\omega_t^2)$. Electrons moving in the dotted region are phase-trapped while those outside are untrapped.

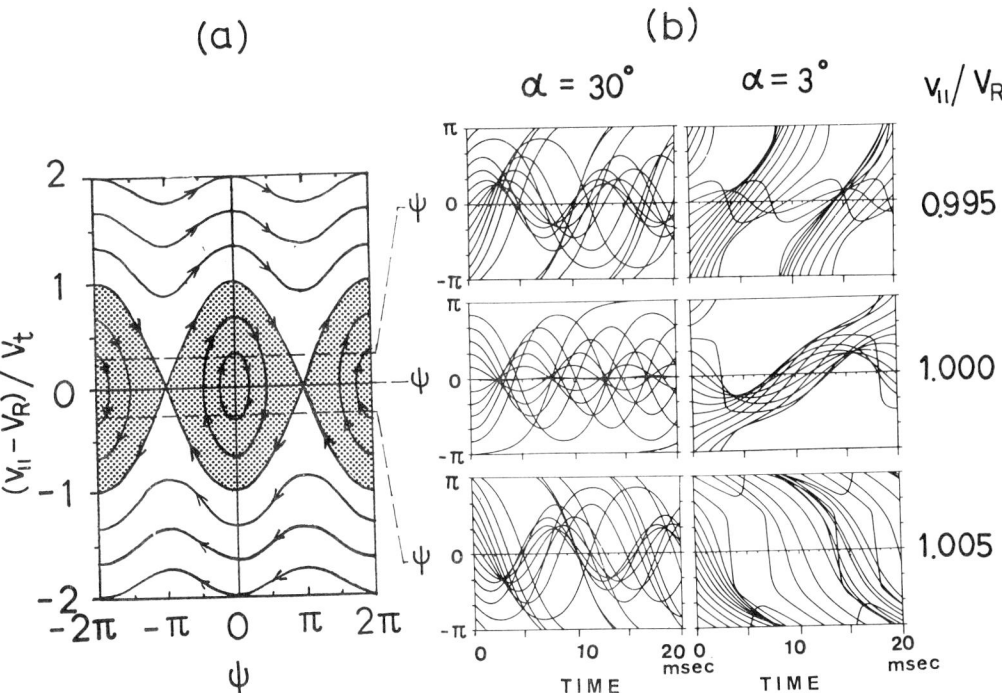

Fig. 5. (a) Phase diagram for phase motion of electrons in a whistler wave (Eq (10)). Dotted area shows a phase-trapping region. (b) Behavior of phases of electrons showing a phase bunching. A large B_w/B_o (= 6.8 x 10^{-4}) was taken to save computer time (after Matsumoto, 1972).

If the initial distribution in the $v_{\|}$-Ψ phase space is uniform (i.e., $\partial f/\partial v = \partial f/\partial \Psi = 0$) by Liouville's theorem there is no phase bunching thereafter [Dysthe, 1971; Crystal, 1975]. Therefore there can be a phase trapping without accompanying phase bunching.

Figure 5(b) is an example of time-behavior of set of electrons with different initial phase angles and with same initial $v_\parallel$ and α (or equivalently $v_\perp$). These trajectories are computed by integrating (1) to (3) dropping the inhomogeneity terms. Three values of $v_\parallel$, i.e., $v_\parallel > V_R$, $v_\parallel = V_R$ and $v_\parallel < V_R$ and two values of α, i.e., $\alpha = 30°$ and $\alpha = 0°$ are chosen. Electrons with initial $\alpha = 30°$ are phase-bunched initially around $\Psi \simeq 0$ for $v_\parallel = V_R$, $\Psi \simeq \pi/2$ for $v_\parallel > V_R$, and $\Psi \simeq -\pi/2$ for $v_\parallel < V_R$, while those with $\alpha = 3°$ are phase-bunched initially around $\Psi \simeq -\pi/2$ regardless of $v_\parallel$. The example of small α is shown to demonstrate that there can be a phase bunching even by untrapped electrons. It is noted that the latter bunching around $\Psi \simeq -\pi/2$ is mainly by the second term in the r.h.s. of Eq.(3), i.e., by the electric field ($\underset{\sim}{E}_w + \underset{\sim}{v}_\parallel \times \underset{\sim}{B}_w$) of the wave.

4.3 Second Order Resonance

In an inhomogeneous magnetic field, electrons generally cannot stay in resonance with the wave because both of the parallel velocity of electrons and the resonance velocity change along the geomagnetic field. Helliwell [1967] first pointed out that only those electrons which are able to keep resonance with a whistler wave play important role in generating triggered emissions. The condition necessary for the long-lasting resonance was called "consistent wave condition" and defined by

$$\omega_D \equiv \omega - kv_\parallel - \Omega_e = 0 \quad \text{and} \quad d\omega_D \equiv d(\omega_D - kv_\parallel - \Omega_e) = 0 \quad (11)$$

where ω_D is a Doppler shifted wave frequency seen by the resonant electron. Nunn [1971,1974] defined "second order resonance" by the following equations.

$$v_\parallel = V_R \quad \text{and} \quad \frac{dv_\parallel}{dt} = \frac{dV_R}{dt} \quad (12)$$

It can be shown that Helliwell's consistent wave condition and Nunn's second order resonance condition are quite equivalent. Only electrons which satisfy Eq.(12) can stay in resonance in an inhomogeneous medium and contribute to the formation of the resonant current necessary for the generation of triggered emissions. The second order resonance condition Eq.(12) is equivalent to

$$\frac{d\psi}{dt} = \frac{d^2\psi}{dt^2} = 0 \quad (13)$$

provided the second term in the r.h.s. of Eq.(3) can be neglected. The condition (12) combined with Eq.(2) requires

$$\frac{dV_R}{dt} = - \Omega_w v_\perp \sin\psi - \frac{v_\perp^2}{2\Omega_e} \frac{\partial \Omega_e}{\partial z} \qquad (14)$$

Therefore, there can be a phase angle $\Psi = \Psi_R$ which satisfies Eq.(14) *if*

$$\begin{aligned}\omega_t^2 \geq\; & k \left| \frac{dV_R}{dt} - \frac{v_\perp^2}{2\Omega_e} \frac{\partial \Omega_e}{\partial z} \right| \\ =\; & k \left| \frac{2\omega+\Omega_e}{2\omega} \left[\frac{d\omega}{dt} - \frac{3\omega}{2\omega+\Omega_e} V_R \left\{ 1 + \frac{\Omega_e - \omega}{3\Omega_e} \tan^2\alpha \right\} \frac{\partial \Omega_e}{\partial z} \right] \right|\end{aligned} \qquad (15)$$

where $d\omega/dt$ is a frequency change seen by the electron and use is made of Eq.(4) in eliminating dk/dt with an assumption of constant electron density [Helliwell, 1967]. The phase angle is then given by

$$\sin\psi_R = - \frac{k}{\omega_t^2} \left(\frac{dV_R}{dt} + \frac{v_\perp^2}{2\Omega_e} \frac{\partial \Omega_e}{\partial z} \right) \equiv S(t) \qquad (16)$$

Nunn [1971] called this angle Ψ_R a "second order resonance phase angle" and called the function $S(t)$ a "collective inhomogeneity factor". The condition (15) was found also by Dysthe [1971] and Sudan and Ott [1971] as a condition for phase trapping in an inhomogeneous medium.

If Eq.(15) is satisfied, the solution to Eq.(8) for nearly resonant electrons (i.e., $\Psi \simeq \Psi_R$, $v_\parallel \simeq V_R$) is given by

$$\psi = \psi_R + \widehat{\psi} \sin(\omega_t \sqrt{\cos\psi_R}\, t + \theta_\psi) \qquad (17)$$

$$v_\parallel = V_R + \frac{\omega_t}{k} \sqrt{\cos\psi_R}\, \widehat{\psi} \cos(\omega_t \sqrt{\cos\psi_R}\, t + \theta_\psi) \qquad (18)$$

for $|\Psi_R| \leq \pi/2$ and an aperiodic solution for $|\Psi_R| > \pi/2$, where $\widehat{\Psi}$ and θ_ψ are an amplitude and phase of the Ψ-oscillation which are determined by the initial condition of the electron. Solutions (17) and (18) mean that the phase trapping is possible only for $|\Psi_R| \leq \pi/2$ in an inhomogeneous plasma. The trapping frequency of phase oscillation around Ψ_R is reduced by a factor of $\sqrt{\cos\Psi_R}$ compared with the homogeneous case.

Figure 6 is an example of variation of the second order phase angle Ψ_R, $v_\perp$ and V_R as a function of distance z(km) from the equator along the L = 4 geomagnetic field line.

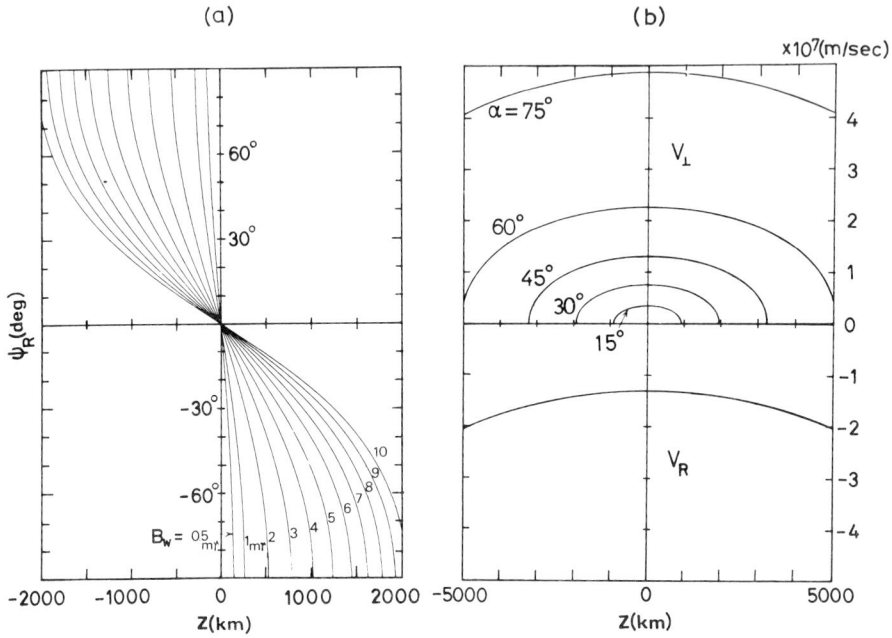

Fig. 6. (a) Variation of second order resonance angle Ψ_R versus distance from the equator. The interacting wave is propagating into a positive z direction. (b) Variation of resonant velocity V_R and $v_\perp$. Parameters used in the computation are; L = 4, f_{HEQ} = 13.75 kHz, f_{PEQ} = 157.5 kHz, and f = 0.5 f_{HEQ}.

The geomagnetic field is approximated by a parabolic dependence on z. The variation of $v_\perp$ is computed by Eq.(9) for various equatorial pitch angles. Note that the dependence of $v_\perp$ on z is quite different from that for *nonresonant* electrons which obey the first adiabatic invariant μ = constant instead of Eq.(9). As seen in the figure, the length of the region where the second order resonance is satisfied is quite dependent on the wave intensity B_w. The second order resonance breaks down either when $v_\perp$ becomes zero or $|\Psi_R|$ reaches $\pi/2$ whichever comes first. In Fig. 6, the wave frequency is assumed *constant* ($d\omega/dt$ = 0) and its intensity is varied from 0.5 mγ to 10 mγ.

The assumption of constant ω corresponds to a second order resonance with a triggering wave. For emissions with variable frequency, the condition (15) (or $|S(t)| \leq 1$) is less restrictive because the inhomogeneity effect can be compensated by the frequency variation. Therefore resonant electrons in a frequency-varying wave can stay in resonance much longer than in the constant frequency wave, if the frequency variation of the wave is such that

$$\frac{d\omega}{dt} = \frac{3\omega}{2\omega+\Omega_e} V_R \left(1 + \frac{\Omega_e - \omega}{3\Omega_e} \tan^2\alpha\right) \frac{\partial \Omega_e}{\partial z} + \Delta_n \; ; \; -\frac{3\omega}{2\omega+\Omega_e} \omega_t^2 \leq \Delta_n \leq \frac{3\omega}{2\omega+\Omega_e} \omega_t^2 \quad (19)$$

The first term is due to the inhomogeneity, while the second term is due to the nonlinear effect. Therefore the additional nonlinear effect Δ_n might be a possible cause of the complicated frequency spectra of triggered emissions because the possible maximum value of Δ_n is the same order of magnitude as the inhomogeneity term for $B_w \simeq 10$ mγ. We see that Helliwell's expression [1967, Eq.(12)] for frequency variation of emissions is a special case of Eq.(19), though he assumes the first adiabatic invariant for resonant electrons, while in Eq.(19) $v_\perp$ (or equivalently α) is computed by Eq.(9). This is equivalent to assuming that the second order resonance angle Ψ_R is zero, as seen from Eq.(6). In general, Ψ_R can be an arbitrary value between $-\pi/2$ and $\pi/2$.

4.4 Interaction Region

In the magnetosphere, the most effective whistler interaction is believed to take place in the vicinity of the equator where the resonant velocity is the lowest (hence the number of resonant particles is the highest) and the inhomogeneity is the lowest. Helliwell [1967] calculated the length L_i of the interaction region by equating a phase shift $\int \omega_D dz$ along z between the equator and $L_i/2$ to π and concluded that the length is of the order of 1000 to 2000 km along the L = 3 or 4 field line. However, as his estimation is based on the variation of $v_\parallel$ due to the first adiabatic invariant, the interaction length is independent of the wave intensity. Another possible estimation of the effective interaction region may come from the second order resonance condition. For constant frequency wave, we obtain the following expression for L_i from Eq.(15).

$$L_i = \frac{4}{27} k \tan\alpha \; (B_w/B_o)/\{1 + \frac{\Omega_e - \omega}{3\Omega_e} \tan^2\alpha\} \quad (20)$$

where the geomagnetic field is assumed to have a parabolic dependence on z, i.e., $\Omega_e(z) = \Omega_{eEQ} (1 + 4.5z^2/(LR_E)^2)$ in which Ω_{eEQ} and R_E are the equatorial cyclotron frequency and the earth's radius. The length of the second order resonance, thus, linearly increases with B_w giving L_i = 500 km to 4000 km for B_w = 1 mγ to 10 mγ along the L = 4 field line.

4.5 Nonlinear Resonant Current

As a result of phase bunching due to a nonlinear interaction, a resonant current is stimulated which is believed to generate triggered emissions. The resonant current is computed by integrating $f_R(v_\|, v_\perp, \Psi, z, t)\underset{\sim}{v}_\perp$ in the velocity phase space, where f_R is a perturbed velocity distribution function for resonant electrons after the interaction. The range in the integration over $v_\|$ should be taken enough wide to cover all resonant electrons, both trapped and untrapped, which exchange energy effectively with the whislter wave. By Liouville's theorem, $f_R(v_\|, v_\perp, \Psi, z, t) = f_{Ro}(v_{\|o}, v_{\perp o})$ where f_{Ro} is an unperturbed distribution function before the interaction and $v_{\|o}$ and $v_{\perp o}$ are the initial values of $v_\|$ and $v_\perp$ which are governed by the equations of motion (1) to (3). The unperturbed velocity distribution is assumed to be uniform over Ψ_o and z_o and may be expanded near $v_{\|o} = v_{Ro}$ as follows.

$$f_{Ro}(v_{\|o}, v_{\perp o}) \equiv f_{Ro}(v_{\|o}, C) \approx f_{Ro}(V_{Ro}, C) + \left.\frac{\partial f_{Ro}}{\partial v_{\|o}}\right|_{V_{Ro}} \cdot (v_{\|o} - V_{Ro}) \quad (21)$$

where C is a constant of motion containing $v_{\|o}$ and $v_{\perp o}$. In the homogeneous case C can be replaced by $(v_{\|o} - \omega/k)^2 + v_{\perp o}^2$ as expected from Eq. (7) and in the inhomogeneous case by u in Eq.(9) assuming a second order resonance. Thus

$$\underset{\sim}{J}_R(z,t) = -e\int_0^\infty v_\perp dv_\perp \int_0^{2\pi} d\psi \{v_{\|o}(v_\|, v_\perp, \psi, z, t) - V_{Ro}\}\underset{\sim}{v}_\perp dv_\| \quad (22)$$

Dysthe[1971], Brince[1972] and Newman[1977] treated the problem by the method mentioned above but the result of the analytic integration in Eq.(22) is a very complicated function even for the homogeneous case. In an inhomogeneous case, Nunn[1971,1974] and Dowden et al.,[1978] experss f_R in terms of W and μ (see Eq.(5) and (6)) and use the expression of

$$\underset{\sim}{J}_R(z,t) = e\int_0^\infty v_\perp dv_\perp \int_0^{2\pi} d\psi \int dv_\| \left[\frac{\partial f_{Ro}}{\partial W}\Delta W(v_\|, v_\perp, \psi, z, t) + \frac{\partial f_{Ro}}{\partial \mu}\Delta\mu(v_\|, v_\perp, \psi, z, t)\right]\underset{\sim}{v}_\perp \quad (23)$$

where ΔW and Δμ are computed by integration of Eqs.(5) and (6). However, further simplified analytic expression has not yet been obtained in this case. Extensive numerical integration of Eqs.(5) and (6) were performed to estimate W and μ by Nunn[1971,1974] and Das and Kulkarni[1975].

As for the direction of the resonant current relative to the wave magnetic field, there has not been a general agreement among inverstigators [Nunn, 1975; Helliwell and Crystal, 1975; Newman, 1977].

Analytic treatment like Eq.(22) [Dysthe,1971; Newman, 1977; and comment by Nunn,1975] claims that the resonant current is in the direction antiparallel to the wave electric field $\underline{E}_w$, i.e., $\Psi \simeq \pi/2$. On the other hand, numerically computed trajectories of resonant and nearly resonant electrons [Ashour-Abdalla,1970; Matsumoto,1972; Helliwell and Crystal, 1973; Matsumoto et al.,1974; Gendrin,1974] show that the phase bunching is observed mainly in the direction antiparallel to the wave magnetic field $\underline{B}_w$, i.e., $\Psi \simeq 0$. As seen in the phase trajetories for $\alpha = 30°$ in Fig.6(b), the phase bunching is almost symmetrical around $\Psi = 0$; a positive phase angle for electrons with $v_{\parallel o} > V_R$ and a negative phase angle for those with $v_{\parallel o} < V_R$. Therefore if $\partial f_{Ro}/\partial v_{\parallel o} > 0$ (i.e., $\partial f_{Ro}/\partial |v_{\parallel o}| < 0$), the resonant current which is obtained by integration over $v_{\parallel o}$ has a component antiparallel to $\underline{E}_w$. However, the main component is still parallel to $\underline{B}_w$ because the phase bunching around $\Psi \simeq \pm \pi$, which cancels the bunching around $\Psi \simeq 0$, does not take place at a bunching time $T_t/4$ $(=\pi/(2\omega_t))$ due to a slow phase change. Therefore, it would seem necessary to perform a further simulation of resonant current formation with a higher density in $v_{\parallel o}$ to resolve the above discrepancy. On the other hand, it also seems to be necessary to check the validity of expansion (21) in the analytic treatment.

In the inhomogeneous case, the direction of the resonant current is governed by the second order phase angle Ψ_R [Nunn,1971,1974]. Therefore electrons moving towards the equator bunch around a negative phase angle Ψ_R $(-\pi/2 \leq \Psi_R \leq 0)$ and those away from the equator around a positive Ψ_R $(0 \leq \Psi_R \leq \pi/2)$ (see Fig.6(a)).

4.6 Basic Equations of Wave Fields

Basic equations of whistler wave fields (sum of the triggering and triggered waves) are given by Maxwell's equations. Eliminating E_w and current due to the background cold plasma which supports the wave, we have

$$\left(\frac{\partial}{\partial t} - j\Omega_e\right) \frac{\partial^2 \dot{B}_w}{\partial z^2} - \frac{\Pi_e^2}{c^2} \frac{\partial \dot{B}_w}{\partial t} = -j\mu_o \left(\frac{\partial}{\partial t} - j\Omega_e\right) \frac{\partial \dot{J}_R}{\partial z} \tag{24}$$

where a dot on the top of B_w and J_R denotes a complex quantity representing a vetor. For $\dot{B}_w$ and $\dot{J}_R$ of the form of

$$\dot{B}_w = \widehat{B}_w \exp\{j\int \omega dt - j\int k dz\} \tag{25}$$

$$\dot{J}_R = \widehat{J}_R \exp\{j\int \omega dt - j\int k dz\} \exp(j\psi_B) \tag{26}$$

where Ψ_B is a bunching phase angle and $\widehat{B}_w$ and $\widehat{J}_R$ are amplitudes, we get the following two equations from real and imaginary parts of (24), assuming $\partial \Psi_B/\partial t = \partial \Psi_B/\partial z = 0$, $\partial k/\partial z = \partial \omega/\partial z = \partial J_R/\partial z = 0$, and $\partial^2 B_w/\partial t \partial z = 0$.

$$(\frac{\partial}{\partial t} - \frac{\partial \omega}{\partial k}\frac{\partial}{\partial z})\widehat{B}_w = \frac{1}{2}\mu_o \frac{\partial \omega}{\partial k} \{(J_R \sin\psi_B) + \frac{1}{\Omega_e - \omega}\frac{\partial}{\partial t}(J_R \cos\psi_B)\} \quad (27)$$

$$\Delta\omega = -\frac{1}{2}\mu_o \frac{\partial \omega}{\partial k} \{(J_R \cos\psi_B) - \frac{1}{\Omega_e - \omega}\frac{\partial}{\partial t}(J_R \sin\psi_B)\} \quad (28)$$

where $\Delta\omega$ is a frequency shift from $\bar{\omega}$ which satisfies the dispersion equation (4), i.e., $\Delta\omega = \omega - \bar{\omega}$.

Though Eqs.(27) and (28) are valid only in the homogeneous case in which spatial variations of k, ω and $\widehat{J}_R$ can be neglected, they are instructive to show relations among $\underset{\sim}{J}_R$, $\partial \underset{\sim}{J}_R/\partial t$, $\underset{\sim}{B}_w$ and $\Delta\omega$. The quantities $\widehat{J}_R \sin\Psi_B$ and $\widehat{J}_R \cos\Psi_B$ are the magnitude of current components antiparallel to $\underset{\sim}{E}_w$ and parallel to $\underset{\sim}{B}_w$, respectively. Resonant current antiparallel to $\underset{\sim}{E}_w$ contributes the wave growth or damping according to whether $\Psi_B > 0$ or $\Psi_B < 0$, while its time derivative changes wave frequency in such a way $\Delta\omega > 0$ if $\Psi_B > 0$ and $\Delta\omega < 0$ if $\Psi_B < 0$. Resonant current parallel to $\underset{\sim}{B}_w$, on the other hand, decreases the wave frequency. However, it contributes to the wave growth or damping if it increases or decreases with time. These fundamental relations can be understood by a simpler physical consideration of vector relations among $\underset{\sim}{B}_w^t$, $\underset{\sim}{E}_w^t$, $\underset{\sim}{J}_c$, $\underset{\sim}{J}_R$, $\underset{\sim}{B}_w^e$ and $\underset{\sim}{E}_w^e$ where $\underset{\sim}{J}_c$ is a current due to the cold plasma and the superscripts t and e denote triggering and emission fields, respectively. Fundamental physics of the whislter interaction in the inhomogeneous case may be inferred by a similar consideration as above if we replace Ψ_B with the second order resonance angle Ψ_R by assuming a slow change of Ψ_R.

5. THEORIES AND COMPUTER SIMULATIONS OF TRIGGERED EMISSIONS

There have been many theoretical and computational works which attempt to explain triggered emissions. In this section, theories and computer simulations are reviewed according to their different approaches.

Brice[1963] and Dungey[1963] first pointed out the importance of cyclotron resonance interaction in the generation process of VLF emissions. Whistler mode cyclotron instabilities and amplifications were then studied within the framework of linear and quasilinear approach [Kennel and Petschek,1966; Lutomirski and Sudan,1966; Kimura,1967; Matsumoto and Kimura,1971; Liemohn,1974 and references therein]. However, these attempts could not successfully explain triggered emissions.

A pioneering phenomenological theory of triggered emissions was proposed by Helliwell[1967]. The importance of a coherent nonlinear interacion in an inhomogeneous geomagnetic field is pointed out and many basic concepts inherent to the interaction are shown. His idea is as follows: Resonant electrons are phase-bunched by a triggering wave and give rise to a nonlinear resonant current which, acting like an endfire array antenna, in turn causes emissions. Dysthe[1971] extended Helliwell's theory and gave mathematical explanation of the basic concepts and physical effects. An independent mathematical formulation of Helliwell's idea was also made by Sudan and Ott[1971] who, however, investigated the possibility of a trapped particle instability caused by a phase-bunched electrons beam, instead of computing a nonlinear resonant current.

Nunn[1971,1974] stressed the importance of nonlinear current formed by second order resonant electrons. He developed a study of triggered emissions which is based partially on analytic equations and partially on numerical computations. By integrating a set of self-consistent equations for electron motion (Eqs.(2),(3)), wave amplitude (similar to Eq.(27)), wave frequency (similar to Eq.(28)), and second-order resonance phase angle (Eq.(16)), he could reproduce emissions like risers, fallers and waves with amplitude oscillation. However, he was forced to make many assumptions and simplifications in his numerical simulation. For example, he presumes that only stably trapped electrons which satisfy the second order resonance contribute to the resonant current and therefore follows the trajectories of only a few particles representing stably trapped beams with constant $v_\perp$ (This seems to be inconsistent with the constant of motion u in Eq.(10)). In the equations for fields, the inhomogeneity and time-varying effects such as $\partial \widehat{J}_R/\partial z$ and $\partial \widehat{J}_R/\partial t$ seem to be dropped. In addition, for the sake of numerical stability, he artificially supressed a sideband instability and introduced a heavy smoothing of all the fields. It would be necessary to justify these points. Similar numerical study has been done by Das and Kulkarni[1975] based on Nunn's basic set of equations.

Relating to the nonlinear resonant current, proposed by Helliwell [1967], many authors have investigated phase bunching phenomena based on a test particle approach [Matsumoto,1972; Gendrin,1974; Matsumoto et al.,1974; Crystal,1975], while exact particle orbits in a monochromatic whistler wave had been investigated with a stress on the phase-trapping characteristics [Roberts and Buchsbaum,1964; Laird and Knox,1965; Dungey, 1969]. Ashour-Abdalla[1970] examined particle trajectories in a whistler wave packet in a homogeneous medium and found that the packet grows at the front and decays at the rear with a net result of growth due to a nonlinear current. However, the nonlinear current in the packet is not sufficiently large to explain triggered emissions.

In order to explain a self-sustaining exponential growth of triggered emissions, Helliwell and Crystal[1973] performed a self-consistent simulation of the whistler interaction in a homogeneous plasma in which a feedback between the stimulated emissions and the incoming resonant electrons is included. Motion of resonant sheets, each consisting of 12 representative electrons, is computed to give a resonant current due to the phase bunching, and then the emission field is added to the incoming wave field. They could show a self-sustaining oscillation grow exponentially even in a stable plasma when the system is triggered by a short pulse. It is, however, noted that this result differs from Nunn's simulation result in which the plasma is required to be unstable for the emission growth and may be related to a method of the feedback in Helliwell and Crystal. They took the direction of the emission magnetic field normal to that of the incoming wave magnetic field, $\underline{B}_w^{in}$, even though their phase-bunched current is almost parallel to $\underline{B}_w^{in}$. This should be, however, parallel to $\underline{B}_w^{in}$ in order to satisfy the Maxwell equation. If $\underline{J}_R$ is parallel to $\underline{B}_w^{in}$ and constant, there should not be a wave growth as seen from Eq.(27).

Another approach to a study of the coherent nonlinear whistler interaction is to investigate an instability caused by a distorted velocity distribution function of resonant electrons after the interaction with a monochromatic whistler wave. Das[1968] first pointed out the importance of the distortion effect. Ashour-Abdalla[1972] computed an evolution of the distribution function by solving a Fokker Planck equation and found that a slot develops in the distribution function at a velocity corresponding to the triggering wave frequency yielding an enhanced growth rate near the wave frequency. The distortion caused by a whistler wave is also examined by Brinca[1972a,b] and Denavit and Sudan[1972]. They found that the resultant distribution function gives rise to a whistler side-band instability. This mechanism was used as a generation mechanism of emissions triggered near the triggering frequency. However, as the triggered emissions is generated at the triggering frequency [Stiles and Helliwell, 1975], the sideband instability may rather be related to the "embryo emissions" or "plateau" in the spectrogram of triggered emission. Roux and Pellat[1976,1978] have recently shown that a hole or a beam in the spatially averaged velocity distribution function is created at the triggering wave termination due to different behavior of trapped and untrapped resonant electrons in an inhomogeneous magnetic field as previously discussed in section 4.3. They showed that the distored distribution of these suddenly detrapped electrons give rise to an instability even in a linearly stable plasma.

Behavior of a whistler wave packet is studied by Knox[1969]. He found that when the amplitude of a packet becomes large enough, the packet grows in length rather than the amplitude. Istomin and Karpman [1972,1973a,b] examined a nonlinear evolution of a quasi-monochromatic whistler packet in a homogeneous plasma by solving the kinetic equations. They found that the packet steepens in a stable plasma and stretches out in an unstable plasma and that a periodic amplitude oscillation is observed for a packet with a sufficiently steep leading front. Karpman et al.[1974] analytically studied a coherent nonlinear whistler interaction in an inhomogeneous plasma by the kinetic equation in the adiabatic approximation. They showed that the nonlinear growth (or damping) rate at large distances from the front edge of the packet is determined by a difference between the average distribution functions of trapped and untrapped resonant electrons.

In order to explain the pulsation phenomenon associated with key-down transmission, Karpman et al.[1974] and Newman[1977] proposed kinetic theories of the coherent nonlinear interaction. Karpman et al. showed that a modulational instability is caused by a nonlinear frequency shift arising from the nonlinear interacion in an inhomogeneous plamsa. It is shown that an inhomogeneity factor should be large, for the effective amplitude modulation which is possible for sufficiently long wave train such as in the key-down case. Newman derived an integral equation which governs modulation of a continuous whistler wave by resonant electrons in a homogeneous plasma. In contrast to the result by Karpman et al., it is shown that an amplitude oscillation can be attributed to the nonlinear whistler interaction which is restricted to a small equatorial homogeneous portion of the signal ray path.

Brinca[1975] examined a possibility of triggering of emissions by an obliquely propagating monochromatic whistler wave. He showed that, in analogy with the parallel propagation case, similar phase bunching current is organized. Brinca[1977] also examined a modification of dispersion equation for turbulent whistler modes by a monochromatic whistler wave and successfully explained a noise-free bands which appear in the VLF spectrum obtained in the Siple VLF injection experiment.

Other theoretical treatments of monochromatic whistler interaction, though not directly connected to theoretical explanation of triggered emissions, have been given by Palmadesso and Schmidt[1971] in which nonlinear quenching of cyclotron damping and subsequent amplitude oscillation are shown. Similar arguments are also given by Yamamoto[1976,1977] by a different approach.

Full self-consistent computer simulations, in which simultaneous equations of motion for a large number of resonant and nonresonant electrons and Maxwell's equations are solved, have been performed in connection to the coherent nonlinear whistler interactions. Ossakow et al.[1972] carried out a computer simulation of a monochromatic whistler wave and showed an initial linear damping and subsequent amplitude oscillation. The result agreed well with the theory by Palmadesso and Schmidt [1971]. Denavit and Sudan[1972] performed a computer simulation of whistler sideband instability and observed that after an amplitude oscillation stops, both upper and lower sidebands grow but only the upper sideband survives later. Denavit and Sudan[1975] and Sudan and Denavit[1973] examined an evolution of large-amplitude whistler wavepackets in a homogeneous plasma by a computer simulation. They could show that along with an elongation of the wavepacket, an emission with variable frequency emerges at the rear of the wavepacket. Matsumoto and Yasuda[1976] carried out a computer simulation of coherent nonlinear interaction between a monochromatic whistler wave and a resonant electron beam. They observed an initial growth of the wave as well as a subsequent amplitude oscillation. They demonstrated an associated phase bunching and evolution of the distribution function of resonant electrons. Recently Rathmann et al.[1978] developed a new simulation code which enables us to treat long-time-scale phenomenon by resonant electrons, and showed similar results to those by Matsumoto and Yasuda [1976].

A summary of classification of these theoretical and simulational works of triggered emissions is given in Table 2.

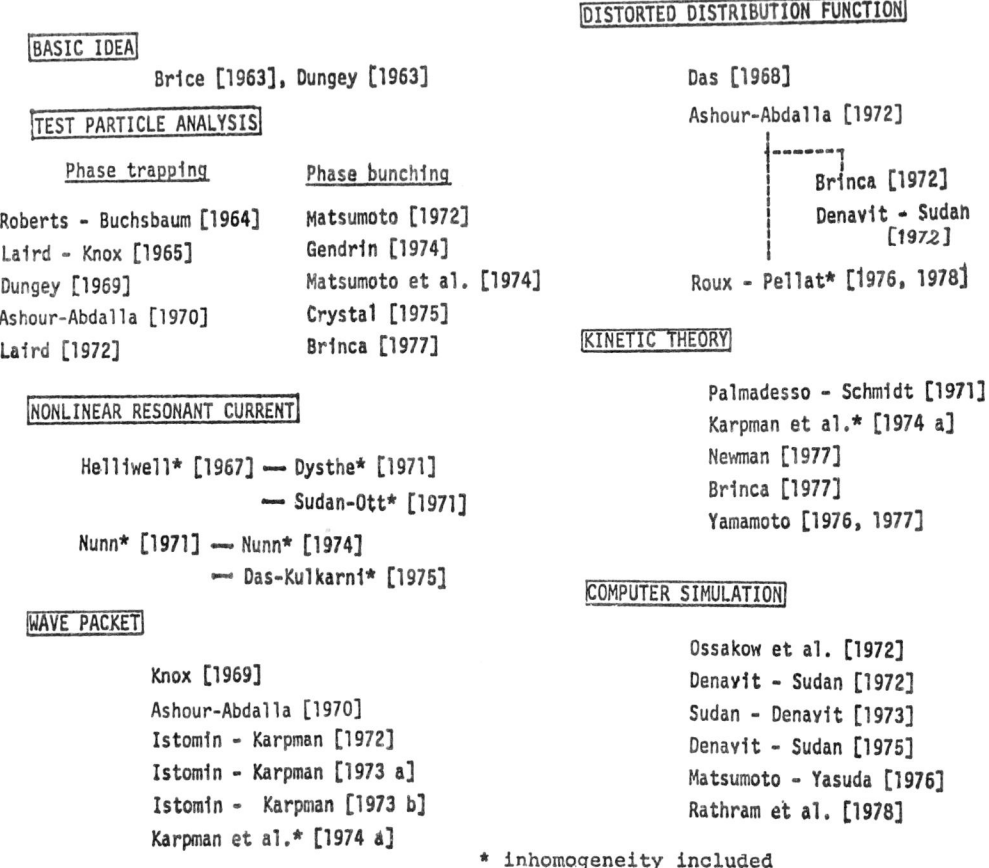

Table 2 Summary of theories and computer simulations on triggered emissions

6. CONCLUSION

Coherent nonlinear effects in the whistler interaction associated with triggered emissions in the magnetosphere are quite interesting in two senses; One is that such nonlinear effects have never been found in laboratory plasmas. Triggered emissions are, thus, an example of the value of the "space plasma laboratory". The other is that their complicated features, which could not be explained by existing theories of plasma instabilities, "triggered" many theoretical works.

In spite of many experimental and theoretical (including computer-simulational) evdeavors, a full understanding of triggered emissions and associated nonlinear effects have not yet been reached. There exist, however, a general agreement that the triggered emissions are generated as a result of whistler interaction in which both nonlinearity and inhomogeneity are involved. An outline of the generation mechanism may be as follows. In the triggering phase, a wave with a constant frequency interacts effectively with second-order resonant electrons in the inhomogeneous magnetic field and grows exponentially with time. During the triggering phase, the frequency of the wave field is locked to the triggering frequency (or within a very small, almost constant frequency shift). When the triggering wave terminates or reaches a saturation in amplitude, the frequency lock is released. In the emission phase, a self-sustained emission is generated by a nonlinear resonant current and/or by an instability under a distorted velocity distribution function of resonant electrons. The emission frequency variation may be that necessary to keep the second order resonance in the self-sustaining mechanism in the presence of an inhomogeneity. The emission frequency seems to be easily effected by other propagating on the same path like power line harmonic radiation.

However, the details of the generation mechanism are not yet clear and much work remains to be done both experimentally and theoretically. Work in the following areas is especially needed :
(1) In-situ measurements by satellites on both waves and particles in the midst of the interaction region.
(2) Detailed analyses of wave-wave coupling, especially on the effect of PLH radiation on the emissions, and on wave entrainment effects.
(3) A unified theory which could explain both triggered emissions and pulsation phenomenon.
(4) Nonlinear current aspect analysis and distorted distribution aspect analysis may be the same in essense but a clearer understanding of the relation between the two is necessary.
(5) The release of emissions and the associated frequency unlocking should be explained quantitatively, as should the subsequent frequency variation of the released emission.
(6) A theory for wave-wave couplings related to the wave entrainment, and abrupt df/dt change of emissions should be developed.
(7) Computer simulations of the coherent nonlinear whistler interaction in the inhomogeneous medium would be helpful for further ing of the basic physics underlying.

ACKNOWLEDGEMENT

I would like to offer my thanks to Prof. I. Kimura for his encouragement and useful discussions. Thanks are due to Dr. R.M. Harper for his careful reading of the manuscript and to Miss Mitsuko Tatsukawa for her typing of the manuscript.

REFERENCES

Ashour-Abdalla,M.:1970, Planet.Space Sci., 18, pp.1799-1812.
Ashour-Abdalla,M.:1972, Planet.Space Sci., 20, pp.639-662.
Bell,T.:1965, Phys. Fluids, 8, pp.1829-1939.
Bell,T. and Helliwell,R.A.:1971, J.Geophys. Res., 76, pp.8414-8419.
Brice,N.M.:1963, J. Geophys. Res., 68, pp.4626-4628.
Brinca,A.L.:1972a, Ph. D. thesis, Stanford Univ., Inst. Plasma Phys., Rept., #489.
Brinca,A.L.:1972b, J. Geophys. Res., 77, pp.3508-3523.
Brinca,A.L.:1975, J. Geophys. Res., 80, pp.203-206.
Brinca,A.L.:1977, Planet. Space Sci., 25, pp.879-885.
Cuperman,S. and Salu,Y.:1973, J. Geophys. Res., 78, pp.4792-4796.
Cuperman,S., Salu,Y., Bernstein,Y. and Williams,D.J.:1973, J. Geophys. Res., 78, pp.7372-7387.
Crystal,T.L.:1975, Stanford Univ., Tech. Rept. #3465-4.
Das.A.C.:1968, J. Geophys. Res., 73, pp.7457-7471.
Das.A.D. and Kulkarni,V.H.:1975, Planet. Space Sci., 23, pp.41-52.
Davidson,R.C., Harmmer,D.A., Haber,I. and Wagner,C.E.:1917, Phys. Fluids., 15, pp.317-333.
Denavit,J. and Sudan,R.N.: 1972, Phys. Rev. Lett., 28, pp.404-407.
Denavit,J. and Sudan,R.N.: 1975, Phys. Fluids, 18, pp.575-584.
Denavit,J. and Sudan,R.N.: 1975, Phys. Fluids, 18, pp.1533-1541.
Dowden,R.L., McKay,A.D. and Amon,L.E.S.: 1978, J. Geophys. Res., 83, pp. 169-181.
Dungey,J.W.: 1963, J. Fluid Mech., 15, pp.74.
Dungey,J.W.: 1969, Plasma Waves in Space and in the Laboratory (Ed. by Thomas and Landmark), Edinburgh Univ. Press, 1, 407.
Dysthe,K.B.: 1971, J. Geophys. Res., 76, pp.6915-6931.
Gendrin,R.: 1972, Solar-Terrestorial Physics/1970, part 2(Ed. by E.R.Dyer), D.Reidel,Dordrecht,Holland,236.
Gendrin,R.: 1974, Astrophys. Space Sci., 28, pp.245-266.
Gendrin,R.: 1975, Space Sci. Rev., 18, pp.145-200.
Helliwell,R.A.: 1967, J. Geophys. Res., 72, pp.4773-4790.
Helliwell,R.A.: 1969, Particles and Fields in the Magnetsophere (Ed. by B.M.McCormac), D.Reidel Publishing Co., Dordrecht,Holland, pp.292-301.
Helliwell,R.A.: 1969, Plasma Waves in space and in the Laboratory(Ed. by Thomas and Landmark), Edinburgh Univ. Press, 1, pp.335-360.
Helliwell,R.A.: 1974, Space Sci. Rev., 15, pp.781- .
Helliwell,R.A.: 1975, Phys. Today, Dec. pp.17-19
Helliwell,R.A. and Crystal,T.L.: 1973, J. Geophys. Res., 78, pp.7357-7371.
Helliwell,R.A. and Katsufrakis,J.P.: 1974, J. Geophys. Res., 79, pp.2511-2518.
Helliwell,R.A. and Crystal,T.L.: 1975, J. Geophys. Res., 80, pp.4399-4400.
Helliwell,R.A.,Katsufrakis,J.P.,Trimpi,M. and Brice,N.: 1964, J. Geophys. Res., 69, pp.2391-2394.
Helliwell,R.A., Katsufrakis,J.P., Bell,T.F. and Raghuram,R.: 1975, J. Geophys. Res., 80, pp.4249-4258.
Inan,U.S., Bell,T.F., Carpenter,D.L. and Anderson,R.R.:1977, J. Geophys. Res., 82, pp.1177-1187.
Istomin,Ya.N. and Karpman,V.I.:1972, Planet. Spce Sci., 20, pp.1790-1793.

Istomin,Ya.N. and Karpman,V.I.:1973a, Soviet Phys. JETP., 36, pp.69-74.
Istomin,Ya.N. and Karpman,V.I.:1973b, Soviet Phys. JETP., 36, pp.897-900.
Karpman,V.I., Istomin,Ya,N. and Shklyar,D.R.:1974a, Plasma Phys., 16, pp.685-703.
Karpman,V.I., Istomin,Ya.N. and Shklyar,D.R.:1974b, Planet.Space Sci.,22, pp.859-871.
Karpman,V.I. and Shklyar,D.R.:1977, Planet.Space Sci., 25, pp.395-403.
Kennel,C.F. and Petschek,H.E.:1966, J. Geophys.Res., 71, pp.1-28.
Kimura,I.:1967, Planet.Space Sci. 15, pp.1427-1462.
Kimura,I.:1968, J.Geophys.Res., 73, pp.445-447.
Kimura,I.:1974, Space Sci. Rev.,16,pp.389-411.
Knox,F.B.:1969, Planet.Space Sci., 17, pp.13-30.
Koons,H.C., Dazey,M.H., Dowden,R.L. and Amon,L.E.S.:1976, J.Geophys.Res., 81, pp.5536-5540.
Laird,M.J. and Knox,R.B.:1965, Phys.Fluids,8, pp.755-756.
Laird,M.J.:1972, J. Plasma Phys.,8, pp.255-260.
Lee,B.G.:1968, Tech.Rep. SU-SEL-68-01, Radiosci. Lab. Stanford Univ.
Liemohn,H.B:1974, Space Sci. Rev., 15, pp.861-889.
Likhter,Ya.I., Molchanov,O.A. and Chmyrev,V.M.:1971, Sov. Phys. JETP Lett., 14, pp.325-327.
Lutomirski,R.F. and Sudan,R.N.:1966,Phys. Rev.,147, pp.156-165.
Matsumoto,H.:1972, Ph. D. Thesis, Kyoto Univ.
Matsumoto,H. and Kimura,I.:1971, Planet. Space Sci., 19, pp.567-608.
Matsumoto,H., Hashimoto,K. and Kimura,I.:1974, J. Geomag.Geoelectr., 26, pp.365-383
Matsumoto,H. and Yasuda,Y.:1976, Phys.Fluids, 19, pp.1513-1522.
McPherson,D.A., Koons,H.C., Dasey,M.H., Dowden,R.L., Amon,L.E.S. and Thomson,N.R.:1974, J. Geophys. Res., 79, pp.1555-1557.
McNeil,F.A.:1968, J. Geophys. Res., Space Phys., 73, pp.6860-6862.
Newman,C.E.:1977, J. Geophys. Res.,82, pp.105-115.
Nunn,D.:1974,Planet. Space Sci.,22, pp.349-378.
Nunn,D.:1971,Planet. Space Sci.,19, pp.1141-1167.
Nunn,D.:1975,J. Geophys.Res.,80, pp.4397-4398.
Ossakow,S.L., Haber,I. and Sudan,R.N.:1972, Phys. Fluids, 15, pp.935-937.
Ossakow,S.L., Haber,I. and Ott,E.:1972, Phys. Fluids, 15, pp.1538-1540.
Ossakow,S.L., Ott,E.and Haber,I.:1972, Phys. Fluids, 15, pp.2314-2326.
Ossakow,S.L., Ott,E.and Haber,I.:1973, J. Geophys. Res.,78, pp.2945-2958.
Ossakow,S.L., Ott,E.and Haber,I.:1973, J. Geophys. Res.,78, pp.3970-3975.
Palmadesso,R. and Schmidt,G.:1971, Phys. Fluids, 14, pp.1411-1418.
Park,C.G.:1977, J. Geophys.Res., 82, pp.3251-3260.
Rathmann,C.E., Vomvoridis,J.L. and Denavit,J.:1978,Tech. Rept., Northwestern Univ.,Evanston, Illinois.
Roberts,C.S. and Buchsbaum:1964, Phys. Rev., 135, pp.A381-A389.
Roux,A. and Pellat,R.,1976, Magnetospheric Particles(Ed. by B.M.McCormac), D. Reidel Publishing Co., Dordrecht, Holland, pp.209-221.
Roux,A. and Pellat,R.:1978, J. Geophys.Res., 83, pp.1433-1441.
Sonnerup,B.U.O. and Su,S.Y.:1967, Phys. Fluids, 10, pp.462-462.
Stiles,G.S. and Helliwell,R.A.:1975, J. Geophys. Res., 80, pp.608-618.
Stiles,G.S. and Helliwell,R.A.;1977, J. Geophys. Res., 82, pp.523-530.
Sudan,R.N. and Ott.E.:1971, J. Geophys. Res., 76, pp.4463-4476.
Sudan,R.N. and Denavit,J.:1973,Phys. Today, 26, pp.32-39.

Yamamoto,T.:1976, J. Plasma Phys., 15, pp.357-370.
Yamamoto,T.:1977, J. Plasma Phys., 27, pp.27-38.

SIPLE STATION EXPERIMENTS ON WAVE-PARTICLE INTERACTIONS IN THE MAGNETOSPHERE

R. A. Helliwell
Radioscience Laboratory
Stanford University
Stanford, CA 94305

ABSTRACT

Natural and controlled whistler-mode signals have been used at Siple Station (and its conjugate, Roberval, Quebec) to study nonlinear mechanisms of wave growth and wave-wave interactions (WWI) in the magnetosphere. Three general classes of WWI (triggering, suppression, and entrainment) are identified and interpreted in terms of a model based on cyclotron resonance interaction. A new type of triggered emission, the 'band-limited impulse' (BLI), often appears at the end of an amplified signal. It covers a frequency range of about 150 Hz above the carrier. It is interpreted in terms of the switching of phase-bunched currents from their driven mode at the carrier frequency f_o to their natural modes at $f_o + \Delta f$ where Δf depends on the change in medium parameters over the interaction region. In a related type of BLI, which appears before the termination of the amplified pulse, the frequencies are symmetrically distributed about the carrier. Assuming that this BLI is caused by wave-induced spread in the $v_{\shortparallel}$ of the interacting electrons, the frequency range ($\pm$ 100 Hz) of the BLI gives an estimate of the local values (~20 pT) of the wave field. Triggering and entrainment of self-sustained emissions are interpreted using the same model and their properties lead to independent estimates of signal strength (~1 pT) at the input to the interaction region (IR). Measured temporal growth typically ranges from 20-30 dB, giving output fields of 10-100 pT in the equatorial region at L = 4. It is estimated that a step function wave of this value can produce an initial pitch precipitation flux of ~0.1 ergs/cm^2-sec for $E \geq 1$ keV through pitch angle scattering, sufficient to explain observed transient bursts of X-ray fluxes and E-region ionization enhancements.

When two coherent signals are transmitted simultaneously with a small frequency spacing, they may interact with one another. The critical separation, called the coherence bandwidth (CB), is typically near 50 Hz and leads to estimated values for the wave intensity of 2.5-10 pT, comparable with those estimated from the BLI.

When the transmitter output power is reduced below some critical level (varies with time), temporal growth ceases. This initial power level depends on the individual ducted path and may be related to the signal attenuation from transmitter to IR on the path.

I. INTRODUCTION

This paper reports some new results of recent controlled VLF wave-particle interaction (WPI) experiments on the magnetosphere from Siple Station, Antarctica (L = 4.2). Coherent signals from a 100 kW variable frequency transmitter travel along one or more field-aligned ducts to the conjugate point at Roberval, Quebec, as shown in Figure 1. Signal amplification of 30 dB or more, triggered emissions, and various wave-wave interactions (WWI) are often observed. Waves of this type precipitate energetic electrons that create X-rays, enhanced ionization, ligh and heat in the ionosphere.

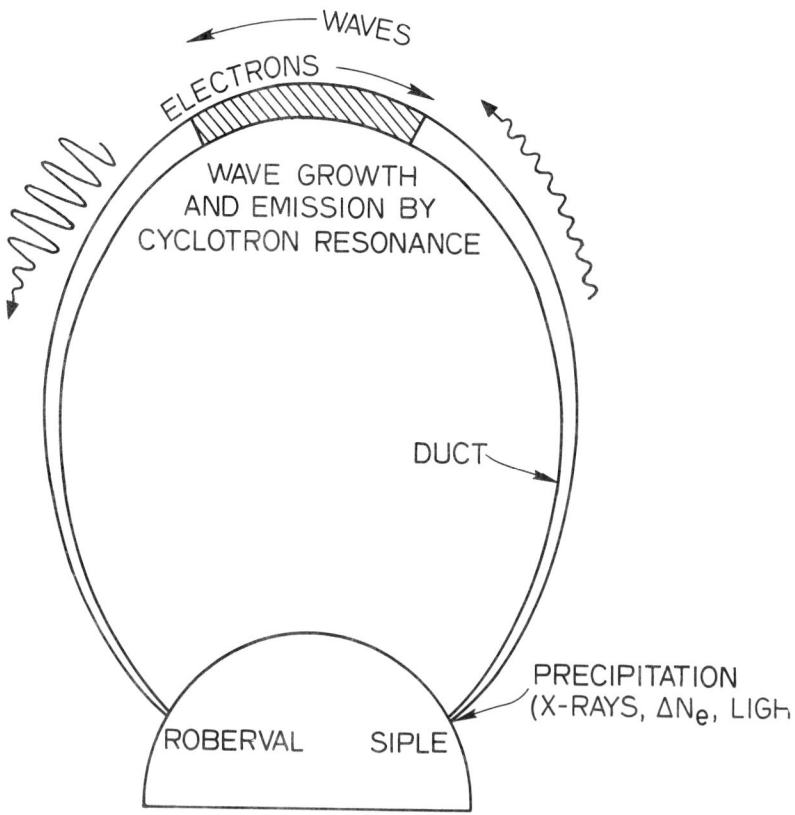

Figure 1. Schematic showing ducted VLF waves from the VLF transmitter at Siple Station, Antarctica, being amplified by counter-streaming electrons. Wave-induced precipitation of energetic electrons into the ionosphere generates X-rays, light, heat, and enhanced ionization.

SIPLE STATION EXPERIMENTS

The object of these experiments is to understand, and learn how to control, the interaction between the waves and particles. Some possible applications of these results include (1) studies of nonlinear phenomena of plasma physics, (2) diagnostics of the radiation belts, (3) controlled precipitation experiments on the ionosphere, (4) communications at VLF, ELF and ULF, and (5) depletion of the radiation belts.

Types of WWI, including triggering (T), suppression (S), and entrainment (E), and combinations of T, S and E, are classified according to their spectral forms in Figure 2. Signals entering the interaction region from downstream sources are designated by the term 'driven' (D) while those originating within the interaction region are designated by the term 'natural' (N). The titles of the remaining sections of this paper are: II. Model of WWI; III. Case Studies of WWI; IV. Wave Intensities; V. Power Threshold; VI. Wave-Induced Precipitation; and VII. Conclusions. A detailed discussion of WWI is given elsewhere [Helliwell, 1979].

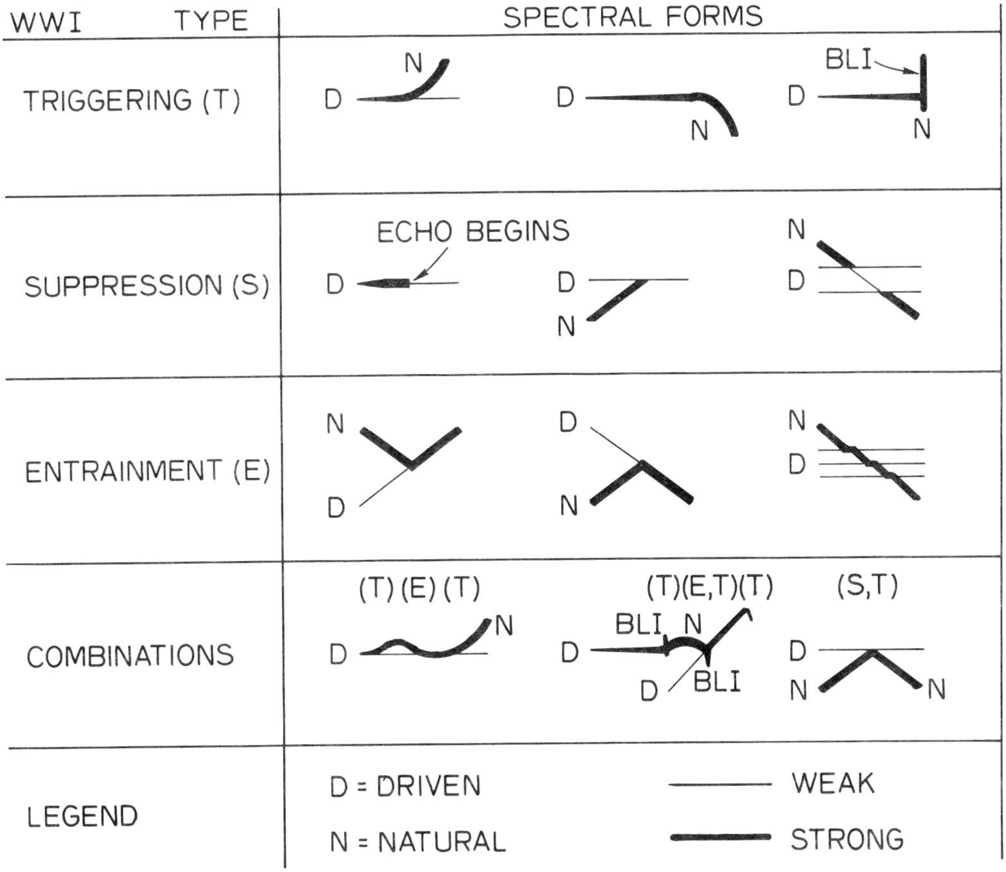

WWI TYPE	SPECTRAL FORMS
TRIGGERING (T)	
SUPPRESSION (S)	
ENTRAINMENT (E)	
COMBINATIONS	
LEGEND	D = DRIVEN, N = NATURAL, ——— WEAK, ▬▬▬ STRONG

Figure 2. Spectral forms of typical VLF wave-wave interactions (WWI), where one of the interacting waves is a natural oscillation, labeled 'N' and the other is a driven wave, labeled 'D.'

II. MODEL OF WAVE-WAVE INTERACTIONS (WWI)

Extended interaction between the right-hand circularly polarized whistler-mode waves and counter-streaming energetic electrons occurs through cyclotron resonance near the equatorial plane (see Figure 1). When an electron spiralling along the static magnetic field $\bar{B}_0$ sees a doppler-shifted wave frequency equal to its own gyrofrequency f_H, it is said to be in cyclotron resonance with the wave, as given by (1) of Figure 3, where f = wave frequency, k = wave number, and $v_{\parallel}$ = electron parallel velocity. The vicinity of resonance is called the particle interaction region (PIR) and is defined by the condition that the change in the unperturbed value of θ from its resonance value is no more than $\pm \pi/2$. For f = constant, the PIR has its greatest length (~100 km) near the equatorial plane where the parameters in (1) change slowly with latitude. Within its PIR an electron exchanges energy with the wave as a result of work done on it by $\bar{E}$, the electric field of the wave. For strong waves PIRs well off the equator become significant, causing the length of the wave interaction region (WIR) to exceed the length of any single PIR.

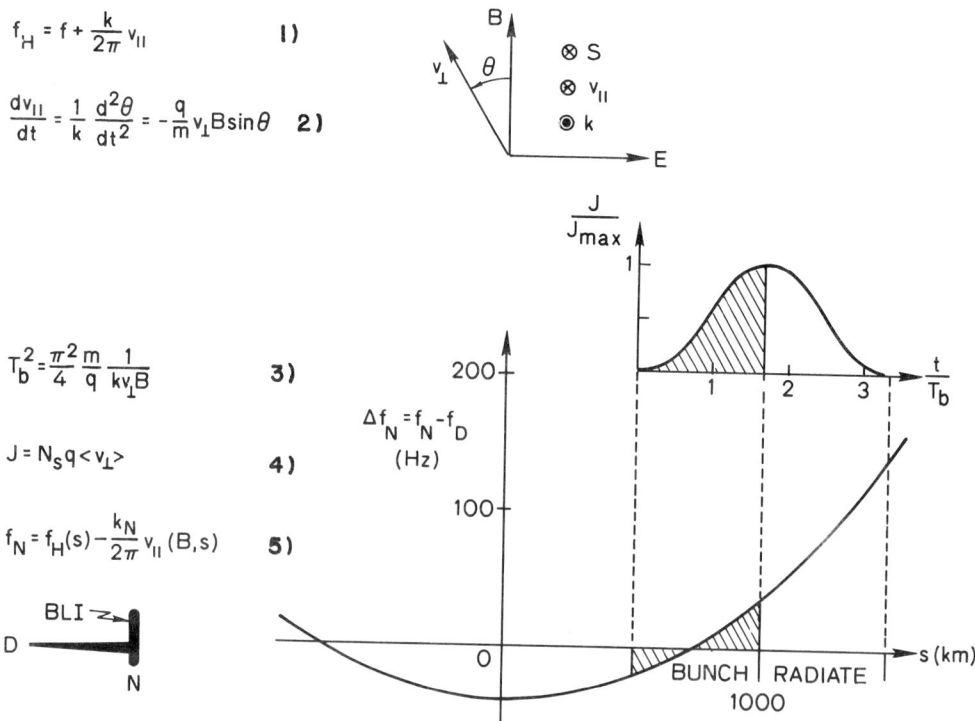

Figure 3. Equations for cyclotron interaction, phase-bunching and natural radiation (see text for definitions). Sketch on right shows the relation of the phase-bunched transverse current to the inhomogeneity ($\Delta f_N(S)$) that is hypothesized as the source of the unsymmetrical BLI. $S = v_{\parallel} t$.

Although all energy exchange is produced by $\overline{E}$, the wave-induced perturbation of the electron's orbit depends mainly on the magnetic field $\overline{B}$ of the wave, and is given by the pendulum equation (2), where θ = angle between the perpendicular velocity $\overline{v}_\perp$ and $\overline{B}$. Solution of (2) for small θ gives the classical pendulum period, one-fourth of which is taken to be the so-called 'bunching time' T_b, given by (3).

As individual electrons are forced by the wave from their unperturbed positions, a transverse current J builds up. Depending on the electron distribution function (usually unknown), this current, given by (4), where N_s = effective energetic electron concentration and $<v_\perp>$ = average transverse vector velocity, may have components both parallel to $\overline{B}$ and parallel to $\overline{E}$. We shall consider only the latter here. An idealized spatial distribution of the normalized current J/J_{max}, for electrons resonant with a fixed-frequency wave at S = 1000 km, is sketched as a function of t/T_b in Figure 3.

Under steady-state conditions (spatial variations of $\overline{B}$ permitted), $\overline{J}$ is fixed in time and the system is said to be undergoing a forced or driven response. Effectively, any transverse current at a fixed location rotates at the frequency f_D of the driving wave. Hence the spectrum of the output signal is monochromatic (zero bandwidth). Changes in amplitude or frequency of the wave may excite the transient or natural (N) responses of the system. Each component current having a particular $v_\parallel$ at the instant of the change switches to its natural mode given by (5), a version of (1) rewritten to emphasize the dependence of the natural frequency on $\overline{B}$ and the medium parameters. (The dispersion relation connects k_N to f_N and f_o, the plasma frequency.) Thus we expect to see combinations of natural and forced responses in most WWIs.

A key to interpreting several WWI phenomena is the 'band-limited impulse' (BLI), illustrated in Section III, and shown in idealized form in the sketch below (5). The BLI can be interpreted in terms of the sketch on the right of Figure 3, in which $\Delta f_N = f_N - f_D$ is plotted vs distance S from the equator. In the shaded region around resonance, at S = 1000 km, the wave force (for a particular wave intensity) controls the perturbation and a phase-bunched current is formed as shown. At larger S the inhomogeneity force prevails, causing significant phase shift as the current dies away (unshaded region). In the steady-state (driven response), this region simply causes phase shift but generates no new frequencies.

Now suppose that the forcing wave train is suddenly terminated. Then, as the tail end of the wave travels in the -S direction, the existing transverse currents are progressively switched to their natural modes whose frequencies are given by $\Delta f_N(S)$. Each associated wave packet travels at nearly the group velocity of the forcing wave, causing all frequency components to reach the receiver at about the same time, giving rise to the BLI. If none of these components triggers a self-sustaining emission, the result is an 'unsymmetrical' BLI whose frequencies are predominantly above the carrier and whose duration $\Delta t = 1/\Delta f$, where Δf = bandwidth of the BLI. As can be seen from (2) and (5), BLI components will also arise from the wave-induced symmetrical spread in $v_\parallel$, giving rise to a 'symmetrical' BLI.

III. CASE STUDIES

The unsymmetrical BLI, explained in Section II, is illustrated in Figure 4. In part (a) the amplitude of a one-second pulse and its associated BLI is shown. Parts (b), (c) and (d) show three analyses of the BLI spectrum using different analyzer bandwidths. In these spectra the BLI appears as a nearly vertical 'hammer-head' trace at the end of the pulse and extends roughly 120 Hz above and 100 Hz below f_D. Its duration at any of the filter settings used is roughly the reciprocal of the filter bandwidth, a fact that first suggested the term 'BLI.' In parts (c) and (d) the BLI spectrum shows striations, the upper one of which coincides with the 55th harmonic of the local 60-Hz power system at Roberval, Quebec. The lowest striation occurs at the frequency of the 52nd harmonic. Closer to the stimulating pulse are two sidebands symmetrically spaced from the carrier by ~30 Hz. They are associated with the amplitude fluctuation in (a), whose period is about 30 ms.

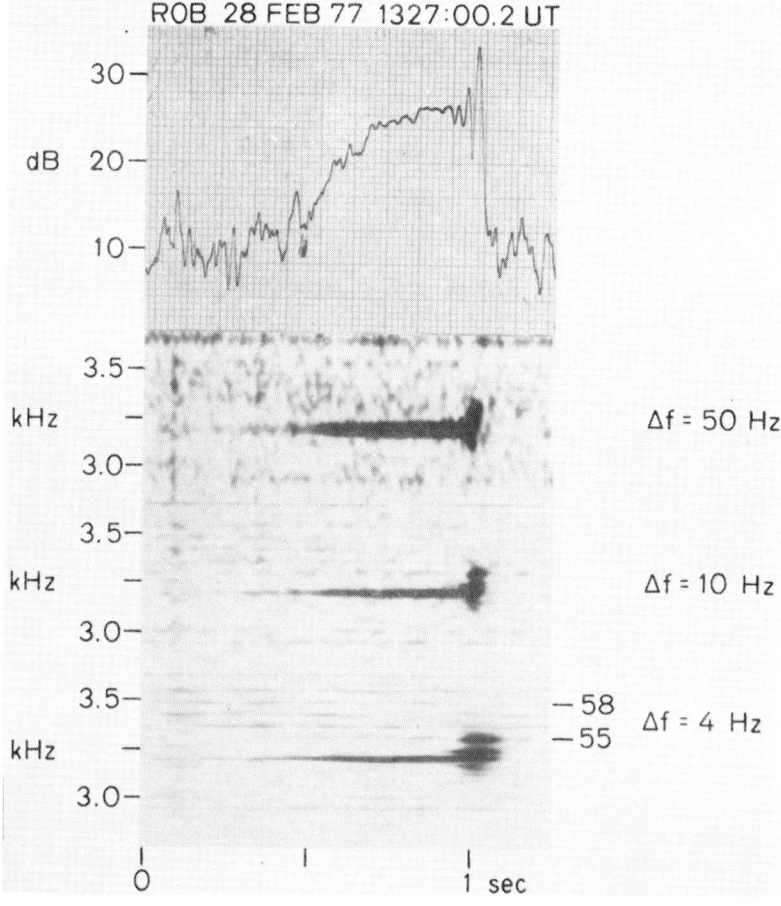

Figure 4. BLI observed at end of 1 s pulse on 3.2 kHz at Roberval on 28 Feb 1977, at 1327:00 UT. (a) Amplitude in 300 Hz bandwidth centered on 3.2 kHz. (b), (c) and (d) Dynamic spectra with filter bandwidths of 50, 10 and 4 Hz, respectively.

Another basic type of WWI is called <u>entrainment</u> and is said to occur when an external signal captures a narrowband emission (e.g., an ASE) causing the frequency of the latter to follow the frequency of the external signal. The external or driven signal may be as much as 20 dB weaker than the natural emission. A striking aspect of entrainment is that the df/dt of the emission can switch rapidly to that of the locking signal with no appreciable change in average emission amplitude.

Entrainment is illustrated in Figure 5. Two Siple one-second pulses at 3.7 kHz trigger falling ASEs, each of which is entrained by a Siple rising ramp as its trace crosses that of the ASE. The ramp rises at 2 kHz/s on the left and 4 kHz/s on the right. Both ramps show curvature caused by dispersion over the one-hop path from Siple to Roberval. The monitoring frequency of the tracker is set at 3.7 kHz so that it follows the triggering pulse from its beginning. Periodic fading is produced by the combination of the Siple pulse and the 61st harmonic of the Canadian power grid.

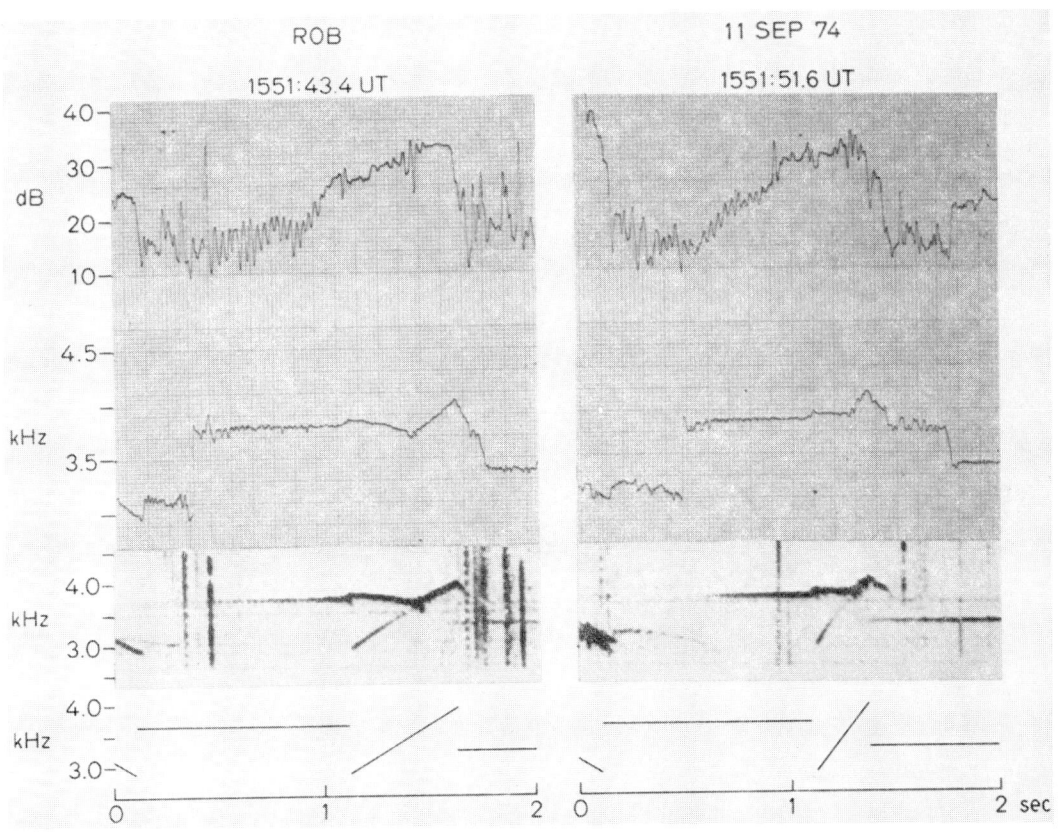

Figure 5. Entrainment of ASEs by rising ramps, 11 Sep 74. (s) Tracker output, (b) tracker frequency, (c) signal spectrum, (d) transmitted format.

Both examples show perturbations in amplitude and frequency in the vicinity of the entrainment point as well as the triggering point (at 1 s). In the left-hand example the amplitude at encounter shows a few cycles of regular fading whose period is about 40 ms. The spectrum also shows a 100-Hz negative BLI at about the time of the largest dip in the amplitude fluctuations. This dip can be interpreted as a beat between components of the frequency-broadened signal. The amplitude of the emission after entrainment is about 3 dB greater than before.

The right-hand record behaves similarly except that there is a 2 dB decrease in intensity after entrainment. Impulsive spectral broadening is observed both before and after encounter, suggesting that a natural instability may be present, related to the BLI. It is remarkable that the average intensity is not much affected by this turbulence or fluctuation in bandwidth.

Invoking the condition $T_r \geq T_b$ [Helliwell, 1979], a crude criterion for entrainment of a free-running oscillation is that the input $B_n > 5.6 \times 10^{-4}$ df/dt, where df/dt is the difference in the slopes of the emission and the capturing signal. (T_r = interaction time.)

A BLI of the symmetrical type is shown in Figure 6. Here a free-running or natural oscillation is triggered at the top of an amplified nose whistler. The tracker locks on to the whistler at 4.5 kHz and follows the ASE through its entrainment at 4.5 kHz by a 1.0 sec Siple pulse that is just below the threshold of detection. Other nearby records reveal a faint trace of the Siple pulse at roughly 20 dB below the ASE. About 100 ms after entrainment by the Siple pulse a second whistler crosses the entrained emission and appears to capture some of its energy from the Siple pulse for about 50 ms. Then the Siple-entrained emission rises about 100 Hz above the Siple carrier frequency at 4.5 kHz and abruptly returns with a sharp drop of about 6 dB in intensity, accompanied by a negative BLI. The amplitude recovers in about 100 ms. Before the end of the Siple pulse the ASE separates again, rising irregularly in frequency until it terminates at 6.0 kHz. Various BLIs with $\Delta f \approx \pm 100$ Hz are seen on both the entrained and free-running portions of the emission. The spread in frequency can be related to B through (2) and (5) [Helliwell, 1979] and for $\Delta f = \pm 100$ Hz the corresponding B is 17 pT.

IV. WAVE INTENSITIES

Using the model outlined in Section II, it is possible to estimate the wave field intensity associated with different types of WWI. Selected cases thought to be typical of the WWI phenomena near L = 4 were analyzed by the methods discussed here and elsewhere [Helliwell, 1979]. The results are shown in the first three lines of Figure 7, in which the predicted field quantity is shown boxed. In each case the temporal growth rate was measured so that the other value of B (B_{in} or B_{out}) could be calculated.

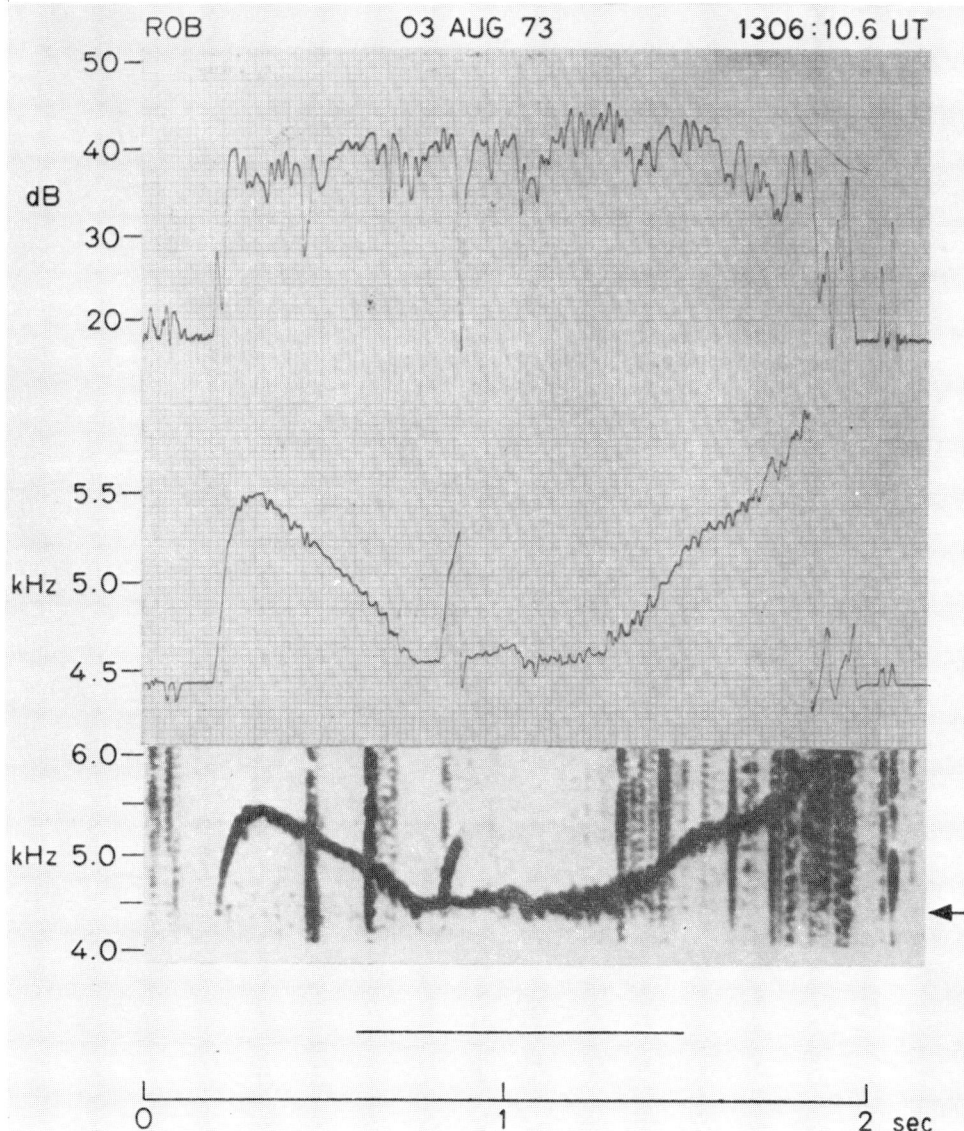

Figure 6. Symmetrical BLIs with $\Delta f \approx \pm 100$ Hz. Format same as Figure 5. Whistler-triggered falling emission is partially entrained by weak (undetectable on this record) 1-sec Siple pulse on 4.5 kHz. A second whistler entrains a portion of the Siple-entrained emission.

Also shown in Figure 7 are two representative *in situ* measurements of VLF wave fields on satellites, both made at high altitudes. The IMP6 measurement of a Siple pulse gives an unamplified field intensity of 0.2 mγ [Inan *et al.*, 1977], within an order of magnitude of the values of the input fields predicted using the model. This measured value is probably lower than those commonly encountered in the Siple-Roberval experiments because the foot of the field line passing through IMP6 was no closer to Siple than about 1000 km. The peak level (28 pT) of a natural chorus emission measured on OGO-1 at L = 4 is close to the predicted B_{out}, giving support to these ideas.

WAVE-WAVE INTERACTION INTENSITIES

EVENTS	SPECTRUM	B_{IN}	AMPLITUDES GROWTH	B_{OUT}
TRIGGERING (TYPICAL, SI-RO)	f_i = 5.0 kHz df/dt = 1 kHz/s	0.5 pT	30 dB	15 pT
ENTRAINMENT (11 SEP 74, SI-RO)	f_i = 3.7 kHz df/dt = 4 kHz/s	≥ 2.2	20	22
SYMMETRICAL BLI (3 AUG 73, SI-RO)	f_i = 4.5 kHz Δf = ± 100 kHz	>1.7	20	>17
UNAMPLIFIED PULSE (28 JUN 73, SI-IMP 6, L = 4)	f = 5.62 kHz	0.2		
CHORUS EMISSION (3 NOV 66, OGO-1, L = 4)	f_i = 6.0 kHz	<0.14	>46	28

Figure 7. Wave-wave interaction intensities. First three entries give B values deduced from the model. Third line gives a satellite-measured value of the unamplified Siple pulse near the equator. Fourth line gives a typical satellite-measured peak of intensity of a chorus emission.

V. POWER THRESHOLD EXPERIMENT

At times of no detectable noise from the magnetosphere, strong emissions can frequently be excited by the Siple signals [Helliwell and Katsufrakis, 1974], suggesting that there is a threshold for the excitation of the coherent wave instability. An experiment to find this threshold was performed by varying the transmitter output in steps of 4 dB while transmitting a standard diagnostic program. The results are shown in Figure 8. In part (c) with 0.6 kW into the antenna, the power

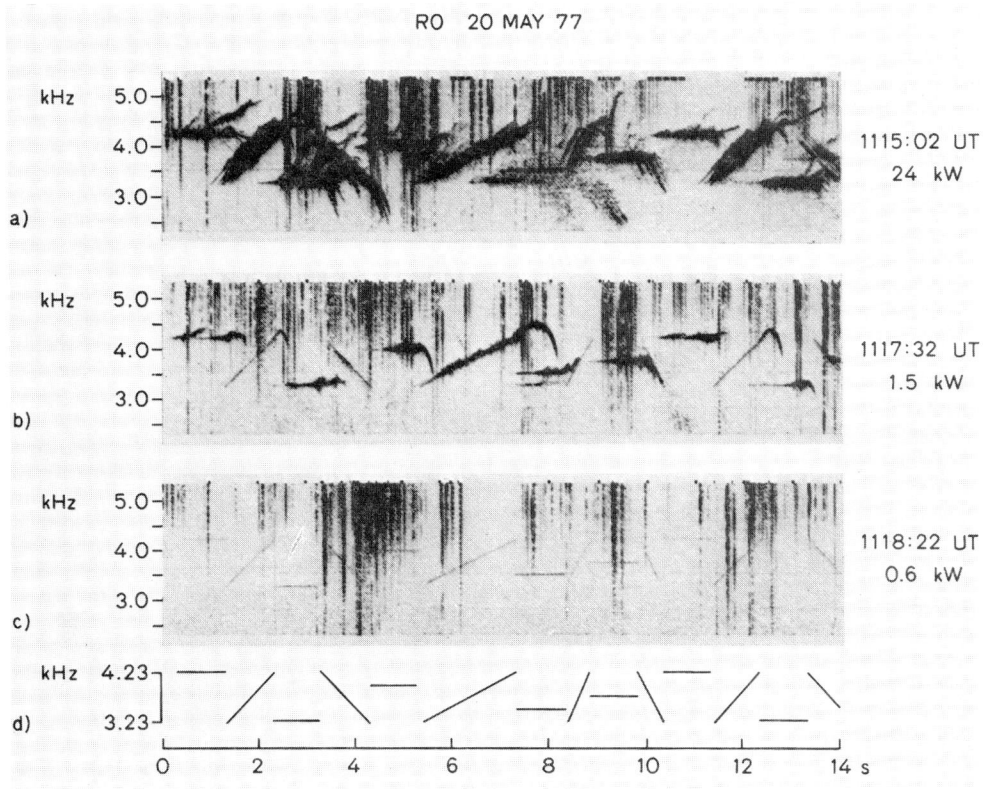

Figure 8. Power step experiment. Format at bottom was transmitted at three different powers, as shown in (a), (b) and (c). Threshold for growth on the most active path lies between 0.6 and 1.5 kW into the antenna (efficiency $\approx 0.5\%$). At maximum power (24 kW) at least four identifiable paths exhibit signal growth and emission triggering. Whistler-mode echoes (2-hop period ≈ 4.2 s) causes some suppression of growth of later signals.

radiated is estimated to be 3 watts and there is no detectable growth or triggering. In part (b), 1.5 kW, temporal growth of about 20 dB and triggering occur on one path, while in part (a), 24 kW, growth and triggering occur simultaneously on four paths. Other records show a time variation in these thresholds. Thus we conclude that the coherent wave instability may have a definite input power threshold (in this case about 7 watts radiated) that varies with time and path. Above the threshold on a given path the peak output intensity usually increases no faster than the input, and sometimes appears to level off, suggesting the existence of an ultimate saturation level in the magnetosphere. However, much stronger input signals are required to adequately probe this feature.

The results help to explain the high degree of variability in the appearance of PLR effects. The radiated power in PLR is typically about 1 watt per harmonic line [Park and Chang, 1979] and hence is well below the normal Siple radiated power. When moderately high particle fluxes are present, the threshold power is lowered and PLR begins to play a controlling role. At even higher fluxes the effective threshold may go to zero, giving rise to spontaneous emissions, such as hiss.

VI. WAVE-INDUCED PRECIPITATION

Using the full equations of interaction between a coherent whistler-mode wave and electrons, the wave-induced scattering has been calculated [Inan et al., 1979]. For example, a 10 pT wave can easily scatter an electron near the loss cone at L = 4 as much as $2\frac{1}{2}$ degrees. A precipitated flux can be estimated by allowing a step function wave to interact with all resonant electrons in a given energy range, including those that return more than once from their mirror paths. A typical flux vs time is shown in Figure 9, starting at 0.2 erg/cm^2-sec and dropping rapidly in the first few seconds and then more slowly. However, even at 100 sec the precipitated flux is still more than one-third of its initial value. These values are thought to be sufficient to explain the observed VLF wave-induced transient bursts of X-rays, light and E-region ionization enhancements.

VII. CONCLUSIONS

Controlled VLF experiments work well, giving reproducible results that cannot be obtained with natural signals. New phenomena (e.g., BLI) have been found, leading to new ideas (e.g., forced and natural response) that help to unify the observations and theories. A wide range of applications is suggested, including controlled experiments on the ionosphere as well as the magnetosphere. Modification of both magnetosphere and ionosphere using wave-induced particle precipitation is now possible. For example, the radiation belts might be temporarily depleted using VLF transmitters on satellites. These experiments can also aid in the design and use of VLF/ELF/ULF communications links through the magnetosphere.

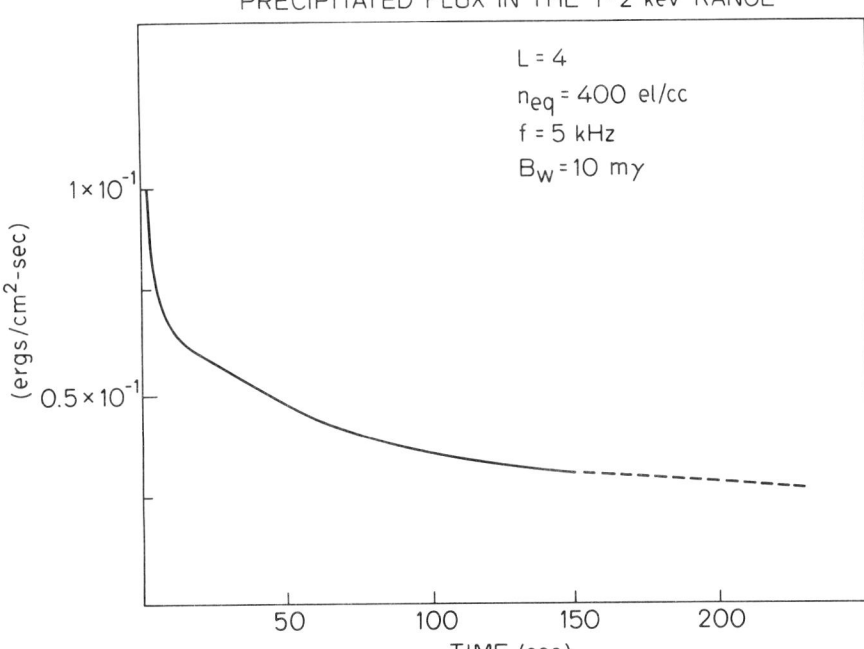

Figure 9. Precipitated electron flux in the 1-2 keV range for L = 4, n_{eq} = 400 el/cc, f = 5 kHz, B_w = 10 mγ.

Acknowledgements: I thank T. R. Miller for aid with the text and figures. The research was supported in part by the Division of Polar Programs of the National Science Foundation under grant DPP76-82646, and the Atmospheric Sciences Section of the National Science Foundation under grants ATM75-07707 and ATM78-05746, and by the National Aeronautics and Space Administration under grant NGL-05-020-008.

REFERENCES

Helliwell, R. A.: 1979, Radioscience Lab., Stanford Univ., Stanford CA 94305, in preparation.
Helliwell, R. A., and Katsufrakis, J. P.: 1974, J. Geophys. Res., 79(16), p. 2511.
Inan, U. S., Bell, T. F., Carpenter, D. L., and Anderson, R. R.: 1977, J. Geophys. Res., 82(7), p. 1177.
Inan, U. S., Bell, T. F., and Helliwell, R. A.: 1979, Radioscience Lab., Stanford Univ., Stanford, CA 94305, in preparation.
Park, C. G., and Chang, D. C. D.: 1979, Radioscience Lab., Stanford Univ., Stanford, CA 94305, in preparation.

NON LINEAR WAVE PARTICLE INTERACTION THEORY APPLIED TO SIPLE TRIGGERED EMISSIONS

David Nunn
Royal Aircraft Establishment, Farnborough, UK.

ABSTRACT

The work considers the basic theory of nonlinear wave pacticle interactions applied to the case of triggering of VLF emissions by Siple transmissions. Important aspects of the theory are underlined. Parameters for Siple triggered emissions are used and "quasi-static" generation regions are constructed. Siple results are found to fit the theory very well.

1. INTRODUCTION

This paper is intended as a sequel to Nunn (1974), and will elaborate and extend the theory of triggered VLF emissions laid out in that paper. The theory will also be applied to recent observations of Siple triggered emissions (Carpenter et al 1976), and appropriate quasi-static generation regions will be constructed.

2. EXPERIMENTAL WORK

A great deal of experimental work involving active VLF experiments in the magnetosphere has been done in recent years. Most notable has been the construction of Stanford's Siple antenna in Antarctica at L=4.2 and the subsequent program of active VLF transmissions. (see Inan et al 1977) Also extensive experimental research into the artificial triggering of VLF emissions has been undertaken by Dowden et al (1977) and also by the Sheffield group working in Antarctica.

The main conclusion from this work is that there is overwhelming evidence that triggered VLF emissions result from fully non linear wave particle interaction effects. Furthermore, during some Siple triggering events in situ satellite measurements of wave amplitudes are available thus enabling the non linear theory to be checked.

3. GENERAL COMMENTS UPON NON LINEAR WPI THEORY

The key to a correct theory of triggered emissions is the calculation of the non linear resonant particle current as a function of the wave field. It has to be calculated by integrating the resonant particle distribution function over 3 dimensional velocity space. Values of F_{res} ($\underline{V}$,z,t) are found by integrating non linear resonant particle trajectories backwards in time and by applying Liouville's theorem. The so called particle bunching approach to the problem gives incorrect values for resonant particle current.

The second point is that one has to consider explicitly the parabolic variation in background magnetic field strength, and integrate resonant particle trajectories in this inhomogeneous medium. The non linear WPI problem in inhomogeneous media is totally different from the homogeneous case. It is never correct to consider ' locally homogeneous' regions in the wave field, since particles entering such a region have already undergone a substantial interaction in the inhomogeneous region outside.

The third point is that in a fully nonlinear problem it is usually unnecessary, confusing and irrelevant to insert 'test waves' into the field, or to try to subdivide the wave field into component parts. The real problem is the self consistent nonlinear interaction between the total wave field and the total resonant particle current, as a function of both space and time. In non linear sideband theory (Nunn, 1973) it can be useful to insert a small amplitude sideband wave, but then one must consider the resulting linear perturbation of <u>nonlinear</u> current. In these circumstances one must never adopt the quasi linear approach and use normal linear expressions for the sideband growth rate i.e. proportional to the time average gradients of the distribution function.

4. RESONANT PARTICLE EQUATIONS OF MOTION

We assume a narrow band, parallel propagating (ducted) whistler wave field, and a parabolic variation of geomagnetic field strength along the field line

$$B/BEQ = \beta = 1 + \tfrac{1}{2} \mathcal{X} z^2 \qquad (1)$$

where z is the distance from the equator, and all units are dedimensionalised as in Nunn (1974). The basic resonant particle equation of motion is

$$\ddot{\psi} + \omega_{tr}^2 (\cos \psi - S(t)) = 0 \qquad (2)$$

where $\omega_{tr}^2 = \{RK_0|V_\perp|/\omega_0\}$; ψ is the angle between $\underline{V}_\perp$ and $\underline{E}$ as as shown in fig 1. The quantity R(z,t) is the dimensionless wave amplitude. S(T) is the total inhomogeneity as seen by the particle in question and is given by an expression of the form

$$S = -\frac{\ddot{x}Z}{R}\left\{\frac{A}{|V_\perp|} + \frac{|V_\perp|}{A_1}\right\} - \frac{\ddot{\phi}}{R|V_\perp|}B + 0\frac{\partial K_0}{\partial Z} \qquad (3)$$

Here $\ddot{\phi}$ is the 'acceleration' of additional phase of the wave field in the particle's frame of reference, and is closely related to the rate of change of frequency in a generation region

$$\ddot{\phi} = \left(\frac{\partial}{\partial t} + V_{res}\frac{\partial}{\partial z}\right)^2 \phi = \frac{\partial^2\phi}{\partial t^2} + 2V_{res}\frac{\partial^2\phi}{\partial z \partial t} + V_{res}^2\frac{\partial^2\phi}{\partial z^2}$$

$$\simeq 4\left(\frac{d\omega}{dt}\right)_{gr} \qquad (4)$$

If $|S|>1$ no trapping occurs and resonant particle motion is approx. linear. If $|S|<1$ particles become stably phase trapped about an angle

$$\psi = P_0 = \cos^{-1}S \; ; \; 0 > P_0 > -\pi \qquad (5)$$

Such trapped particles undergo relatively large changes of energy and magnetic moment, since

$$\hat{W} = -R|V_\perp|\cos\psi \sim -R|V_\perp|S(t) \qquad (6)$$

The expression for resonant particle current may be written in the form (Nunn, 1974)

$$J_{RES} = J_R + iJ_I = -\text{const}\iiint\left\{\frac{\partial F_0}{\partial W} + \frac{2}{\omega_0}\frac{\partial F_0}{\partial \mu}\right\}|V_\perp|^2 \Delta W e^{i\psi}d|V_\perp|$$

$$d\psi dV_Z \qquad (7)$$

Stably trapped particles tend to dominate this integral, provided they have been trapped for at least 1-2 trapping periods. Note that they are important in the sense that they have large ΔW and correspond to the dominant distortion in F_{res}. Untrapped resonant particles have a small ΔW, but they exchange significant amounts of energy with the wave 'in toto' because they are far more numerous than trapped particles.

Note that the sign of J_r is determined by the gradients in F_0. In general a linearly unstable system will also be nonlinearly unstable and vv. Nonlinearity will not destablize a stable plasma!

It might be thought that nonlinear growth rates in inhomogeneous media could be made very large, since ΔW could be of order W. However in practice in the parabolically varying medium this does not seem to happen, and nonlinear growth rates do not exceed linear growth rates by more than a factor of 2-3. Note however that in the nonlinear inhomogeneous case power transfer from particles to wave field will be consistent in space/time. This is not so for homogeneous media when nonlinearity occurs.

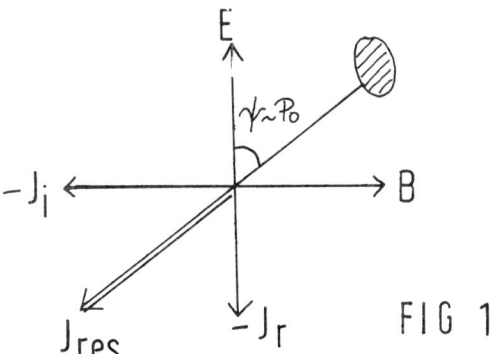

FIG 1

It will be noticed that the inhomogeneity S and thus phase trapping angle P_0, are functions of $|V_\perp|$. At low and high pitch angles $|S|>1$ and particle motion will be linear in any case. The dominant non linear contribution to J_{res} will probably come from a narrow range of pitch angles, corresponding to the maximum of the quantity $|V_\perp|^2 \{\frac{\partial F_0}{\partial W} + \frac{2}{\omega_0}\frac{\partial F_0}{\partial \mu}\}_{V_z=V_{res}}$ as a function of $|V_\perp|$. Since S increases rapidly at pitch angles below about 30 degrees, it is often convenient in theoretical developments to assume that the entire nonlinear contribution to J_{res} comes from particles with a pitch angle of 45 degrees ($|V_\perp| \sim 1$).

5. SIPLE WAVE AMPLITUDES

Carpenter et al have made 'in situ' estimates of wave amplitudes of Siple emissions using satellite data. They consider that magnetic field amplitudes are in the range 2-10mγ. It is instructive to see if these values concur with our requirement for non linear wave particle interaction. Following the notation in Nunn (1974) consider a long constant frequency pulse ($\phi \sim 0$), with $\omega_0 = k_0 = 1$. Consider a 45 degree pitch angle, or $|V_\perp| = 1$. The point $z = z_t$ where $|S| = 1$ is at a distance $R/(2.5\chi)$ from the equator in this instance. If we require that a resonant particle experience 2 full trapping periods in a distance Z_t, then we get a rough value for R, (R_{nl}), above which nonlinear trapping dominates.

$$R_{nl} \cong (10\pi\chi)^{2/3} \qquad (8)$$

Let us now take a typical Siple event. Let f_{eq} = 9.1kHz, f = 3.6 kHz, N_{eq} = 500 el/cc. We thus have ω_0 = 0.8, K_0 = 0.6 and $\chi = 7.5\ 10^{-10}$ for L = 4.1. With these numbers we get that $R_{nl} \sim 8\ 10^{-6}$. This corresponds to a magnetic amplitude of 1.3mγ. This figure agrees very well with Carpenter's lower limit for Siple triggered emissions, and reinforce our notion that triggered emission result from non linear trapping of cyclotron resonant electrons. The claim in Carpenter et al

INTERACTION THEORY

(1977) that the theory of Nunn (1974) requires wave amplitudes ~ 50 mY for Siple emissions is not correct, and results from a misunderstanding.

6. DEVELOPMENT OF THE FIELD EQUATIONS

The basic field equations are derived from the linearised equation of motion of the cold plasma electrons, and from Maxwell's equations. All quantities are assumed to vary only in the z direction (i.e. we assume parallel propagation). We make a narrow band assumption that the wave field and current spectra are confined to a trapping frequency either side of a 'base' frequency. In dimensionless units (Nunn, 1974) the field equations become

$$\left(\frac{\partial}{\partial t} + V_g \frac{\partial}{\partial z}\right) R = -AJ_R - \zeta R \qquad (9)$$

$$\left(\frac{\partial}{\partial t} + V_g \frac{\partial}{\partial z}\right) \phi = -\frac{AJ_I}{R} \qquad (10)$$

where $R(z,t)$ is the wave profile and $\phi(z,t)$ is the additional phase over and above that due to the base frequency. In deriving equations 9 and 10 a number of small terms in $\frac{\partial^2 \phi}{\partial z^2}, \frac{\partial^2 R}{\partial z^2}, \frac{\partial J_I}{\partial t}, \frac{\partial J_R}{\partial t}, \frac{\partial^2 \phi}{\partial z \partial t}, \frac{\partial^2 R}{\partial z \partial t}$ have been dropped. This is fully in accord with the narrow band approximation. As shown in figure 1, Jr is the in phase component of resonant particle current, and Ji the out of phase component. This method of representing the current is obviously useful when the current results from trapped particle bunches phase locked to the wave field. In equation 9 the term $-\zeta R$ is largely unknown, and represents Landau damping and loss of wave energy due to the unducting induced by the non linear wpi.

The rate of change of frequency at a given point in the wave field is readily obtained from equations 9 and 10 and is given by

$$\frac{\partial \omega}{\partial t} = \frac{\partial^2 \phi}{\partial t^2} = -A \frac{\partial}{\partial t}\left(\frac{J_I}{R}\right) + V_g \frac{\partial}{\partial z}\left(\frac{J_I}{R}\right) + V_g^2 \frac{\partial^2 \phi}{\partial z^2} \qquad (11)$$

The first term is obviously small and can never change the wave frequency by more than a small amount. The second term is a rate of change of frequency due to the steady winding up of wave phase caused by the z variation in Ji. We see at once that herein lies a possible mechanism for producing the sweeping frequency of triggered emissions since we know that in an inhomogeneous medium, stably trapped particle bunches can produce a consistent Ji with a marked z dependence. However to continuously wind up the phase $\phi(z,t)$ over the whole wave field requires that one end be 'free', but this is indeed the case at the low amplitude end of a generation region, where trapped particle bunches spiral independently from the wave field.

The third term in equation 11 is simple the advection of different frequencies from upstream of the point in question. One could of course argue directly that the $\partial/\partial z\, (^JI/R)$ term will cause the sweeping frequency in an emission, but it is not obvious that, in a self consistent integration starting from the initial square pulse, this term will not be cancelled out by the ϕ_{zz} term. It was the aim of the self consistent model described in Nunn (1974) to clarify this point.

The nonlinearity of Ji (and Jr) are manifested in their z and t dependence and in how this effects the self consistent time integration. An out-of-phase current Ji could always be produced by injecting extra cold plasma, but in this case Ji/R would not be a function of t. Rewriting equation 11 in the form

$$\frac{\partial^2 \phi}{\partial t^2} = -\frac{\partial}{\partial t}\left(^{J}I/R\right) - V_g \frac{\partial^2 \phi}{\partial z \partial t} \qquad (12)$$

we see that if $\partial \phi/\partial t$ is initially everywhere zero, then $\frac{\partial^2 \phi}{\partial z \partial t} \equiv 0$. This is reasonable, since we do not expect a linearly produced Ji to cause a sweeping frequency!

7. THE SELF CONSISTENT APPROACH

The self consistent model of triggered VLF emissions described in Nunn (1974) uses, out of necessity, a model for the nonlinear current. This assumes that the wave field remains narrow band during the integration with a bandwidth less than a trapping frequency, and that the wave amplitudes are sufficiently large for nonlinear trapping to take place. The model assumes that single pitch angle, stably trapped bunches dominate the current. These bunches have a phase relative to the wave field that is close to the local value of Po. The model current may be roughly represented by

$$J_{res} = J_R + iJ_I \simeq -Ce^{iPo}\left\{\int_{-\infty}^{t} R[t']S(t')dt'\right\} R(t)^{\frac{1}{2}} \cdot K \qquad (13)$$
$$K = P_o(t)^2 \quad \text{or} \quad (Po + \pi)^2 \text{ if } s<0.$$

The actual model employed is a little more sophisticated in that the particle bunch is allowed to respond dynamically to variations in R, ϕ and S, but diffuses back towards a phase Po with a time constant of a trapping period. Detrapped particle bunches are allowed to spiral freely according to the usual equations of motion, and phase mix away in a few trapping periods. The model remains valid for the particular triggering events presented in Nunn (1974). It breaks down when sidebands appear or when a significant reversal of sign of S is experienced by trapped particles.

The main aim of the simulation was to follow the phase behaviour, since the lossterm $-\zeta R$ is largely unknown. A loss term that was an increasing function of R was used. This served to stabilize the simulation and models the unducting effects of non linear wpi in a cylindrical duct.

Great difficulty was caused by numerical instability during the runs, but this was overcome by numerical smoothing. Criticisms that this somehow invalidates the results are not valid, as this is a standard technique. The main requirement is that the runs should make good physical sence and not carry the current model beyond its range of validity. Unsatisfactory runs were discarded.

Another problem is sideband instability which is 'real'. This was partially suppressed by making ζ a function of R. In reality discrete sidebands do not seem to occur with TE's, although spectral broadening is very much in evidence. Again a unifield theory of nonlinear wpi in a cylindrical duct should account for the suppression effect.

The chief result of the self consistent program was to show that indeed the $\partial/\partial z$ (J_T/R) term could cause a steadily changing wave frequency. The model also elucidated many details of generation region structure and position that were in fact far from obvious a priori. Simulations of fallers were not too stable, but nevertheless something was learned from them.

8. THE QUASI STATIC GENERATION REGION APPROACH

The generation regions of triggered emissions appear in reality to be remarkably stable structures, in contrast to the difficulty of simulating them in the time domain. In this respect they resemble phenomena such as vortices in non linear fluid mechanics. This suggests the technique of constructing static or quasi-static generation regions that approximately satisfy the relevant equations. This promises to be a more useful approach than time domain simulation when it comes to investigating the details of triggered emission behaviour. In this paper we will approximately construct generation regions for a typical Siple riser and faller.

Generation regions are principally characterised by a number of external parameters, namely exit amplitude, frequency sweep rate, frequency, the dimensionless constant χ which depends upon L shell and plasma density, the perpendicular velocity of 'active' particles $|V_\perp|$, and the degree of plasma instability or constant 'C' in equation 13. The static profile equation becomes

$$\frac{\partial R}{\partial z} = - \frac{A J_R}{V_g} - \frac{\zeta}{V_g} R \qquad (14)$$

where unfortunately ζ is to a large extent unknown. It very probably increases sharply as J_{res} becomes nonlinear. The frequency sweep rate is a property of the whole emission g.r. and may be regarded as a constant, so

$$\left(\frac{d\omega}{dt}\right)_{gr} = \text{const} = V_g \frac{\partial}{\partial z}\left(\frac{JI}{R}\right) + V_g^2 \frac{\partial^2 \phi}{\partial z^2} \qquad (15)$$

The phase and current in the g.r. 'adjust' themselves so that the sum of the two terms on the rh side of equation 15 is everywhere constant. The current J_i must of course be self consistently calculated, and may be obtained from the inhomogeneity factor which is a function of z

$$S(z) = -\frac{\chi z}{R}\left(\frac{A}{|V_\perp|} + \frac{|V_\perp|}{A_1}\right) - \frac{B}{R|V_\perp|}\left[\left(\frac{d\omega}{dt}\right)_{gr}\left\{1 - 2\frac{V_{res}}{V_g} + \frac{V_{res}^2}{V_g^2}\right\}\right.$$
$$\left. - \frac{V_{res}^2}{V_g}\frac{\partial}{\partial z}\left(\frac{J_I}{R}\right)\right] \qquad (16)$$

Note how, in the quasistatic case, there is a z dependent term in S derived from the resonant particle current itself. In the stable trapping regime $|S| < 1$, and provided that $S(z)$ varies reasonable slowly with z, the current phase will be given approximately by $P_0 \pm \pi = \cos^{-1}S \pm \pi$. (equation 13)

Figure 2 shows the structure of a typical Siple rising freqency emission. We take the following parameters exit amplitude = 8m γ or $R = 5 \cdot 10^{-5}$, frequency sweep rate= 1kHz/sec or $d\omega/dt = 7.7 \cdot 10^{-6}$ in dim. units, $|V_\perp| = 0.9$, $f = 3.6kHz = 0.4f_{eq}$, $\chi = 7.5 \cdot 10^{-10}$. The kind of structure illustrated in figure 2 is almost exactly the kind of thing produced, without prompting, by the self consistent model. These structural details are thus more than just idle speculation.

The g.r. divides into 3 zones. Zone A on the rh side has $S < -1$ and the current is nearly linear owing to the strong inhomogeneity (i.e. $J_i \sim 0$). In zone B non linear trapping occurs, and resonant particles may undergo several complete trapping oscillations in traversing this region. Zone B is about 2000km long and lies on the transmitter side of the equator. The phase of J_{res} rotates steadily as S varies from -1 on the rh side to 0 on the lh side. This gives a consistent positive value for $\partial/\partial z$ (J_I/R) in zone B and causes the frequency rise. At $z=z_h = -3380km$ trapped particle bunches become detrapped and in zone C spiral freely, more or less independent of the wave field. Zone C is fairly difficult to deal with and has been looked at in some detail by French workers (Roux & Pellat, this vol). In this theory it is maintained that zone B is the important one. In order to maintain a constant wave profile we require synchronism between the current and the field vectors for as far as possible into zone C. This is achieved by having zero net inhomogeneity, or $RS \sim 0$, at the point at which particles are detrapped ($z=z_h$). Otherwise $-J_{res} \cdot E \sim 0$ in zone C and the whole g.r. would slip downstream.

Another less obvious boundary condition at the B/C interface is that $\frac{\partial^2\phi}{\partial z^2} \sim 0$, or that higher frequencies are not advected from the free particle zone. This boundary condition was suggested by the self consistent model but it is difficult to prove that it must be satisfied. The frequency shift mechanism does not work by advecting higher frequencies from regions of the field where wave and current amplitudes are very

INTERACTION THEORY

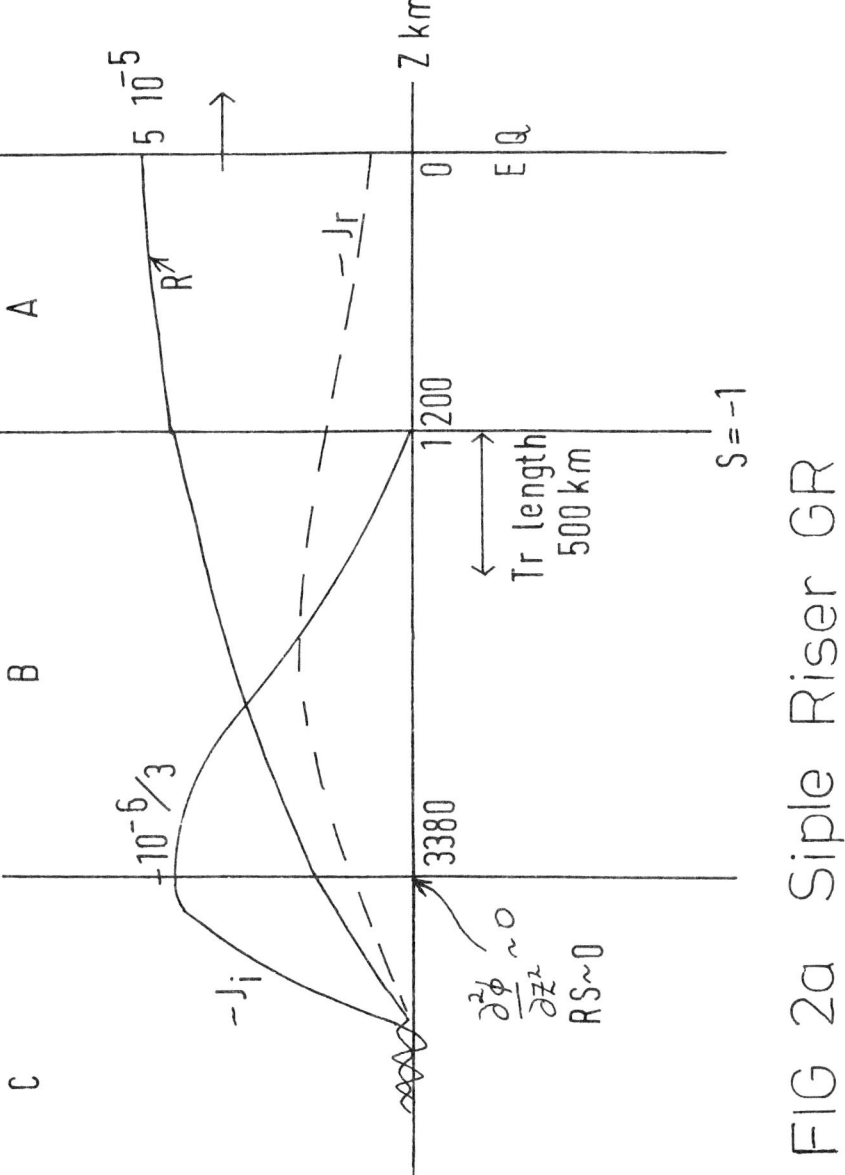

FIG 2a Siple Riser GR

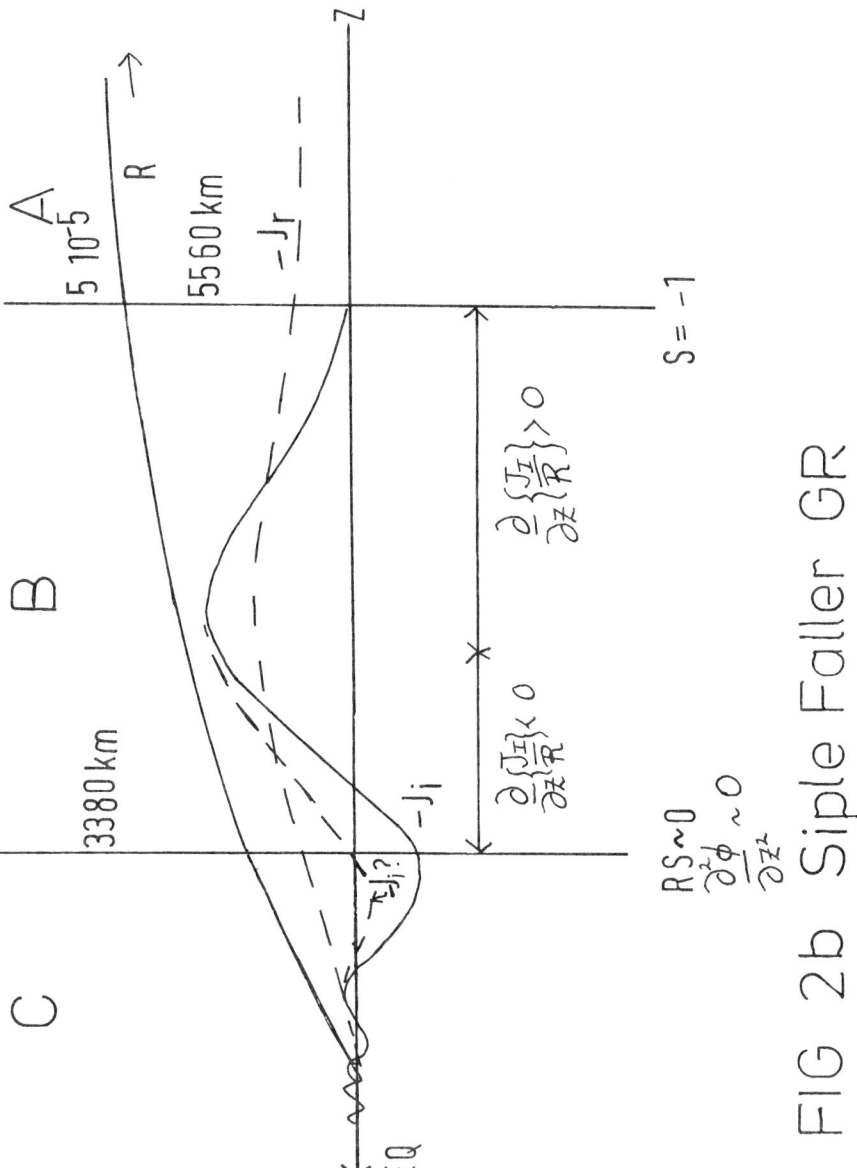

FIG 2b Siple Faller GR

small. Such fields as are generated in zone C are overwritten by the much larger fields produced in zone B, where the current is locked to the wave field. The resonant particle current in zone C is in fact entirely dictated by what is happening in zone B.

Figure 3 shows the projected structure for a Siple faller g.r. Because of the RS $\sim$ 0 requirement at $Z=Zh$ the whole g.r. is now well downstream from the equator, in fact 3-5000kms away. Note the fundamental asymmetry between the riser and faller. To get a faller we require that $\partial/\partial z$ (J_I/R) be negative at the point where $\partial^2\phi/\partial z^2 = 0$ ($Z=Zh$). This is achieved by swinging the current vector round anticlockwise in zone B. The resulting marked z dependence in $\partial/\partial z$ (J_I/R) gives a negative kick to S which is what is required to cause Ji to change sign (see equation 16). The self consistent model did produce fallers, but although the g.r.'s were correctly positioned, the structure was messy and the emissions were short lived. The change in sign of Ji was however apparent.

One feature of both riser and faller g.r.'s is that particles are stably trapped at negative S, and thus undergo increases in energy and pitch angle. It would seem then that particles precipitated by triggered emissions are the untrapped 'passing particles'.

An important aspect of self consistent gr's is the necessity of satisfying the profile equation (equation 14). In constructing gr's, the observed frequency sweep rates fix the order of magnitude of Ji, and thus of Jr. In general it will be found that the resulting Jr is substantially greater than that required to maintain the profile i.e. the term $\zeta R \gg V_g \partial R/\partial z$. For example in the NAA riser simulated in Nunn (1974) the power excess amounted to a factor of about 6. The surpluss is lost due to unducting effects induced by the nonlinear wpi itself. It is perhaps fortunate that the numbers work out this way. If the available power was insufficient, doubt would be cast upon the whole theory.

9. A UNIFIED THEORY OF NONLINEAR WPI IN A DUCT

A more advanced theory of triggered VLF emissions will consider self consistent non linear wpi taking place in a cylindrical duct. The problem is obviously extremely complex but certain aspects will be apparent at once.

As explained in Nunn & Laird (1975) the VLF wave energy in a cylindrical duct is distributed amongst a large number of modes. With non linear wpi the out of phase current Ji will become a function of r and will thus rotate the k vector away from the duct axis-in other words there will be a VERY strong intermodal coupling into escaping modes. Also the lowest order mode, with fields going as $J_o(\lambda r)$ will tend to become dominant with nonlinear wpi, since the higher order modes will be rapidly unducted.

Thus, before non linear effects set in the wave fields will tend to grow very rapidly with minimal unducting. When non linear wpi commences rapid mode conversion and energy loss will take place. This kind of situation in which the damping increases drastically above a certain amplitude ceiling will considerably stabilize an emission gr, and in fact this was the only way in which the self consistent model could be persuaded to work.

The whole problem might be tackled by doing a 2 dimensional simulation in a slab geometry, but even then the most powerful computers will be required.

REFERENCES

 NUNN, D. (1974) : 'A self consistent theory of triggered VLF emissions'. Planetary & Space Science, vol 22, pp.349-378.

 NUNN, D. (1973) : 'The sideband instability of electrostatic waves in an inhomogeneous medium'. Plan. & Space Sc, 21,67.

 DOWDEN, R.L., McKAY, A.D., and AMON, L.E.S. (1978) : 'Linear and nonlinear amplification in the magnetosphere during a 6.6kHz transmission'. JGR vol 83 no A1, pp.169-181.

 INAN, U.S., BELL, T.F., and CARPENTER, D.L. (1977) : 'Explorer 45 and Imp 6 observations in the magnetosphere of injected waves from the Siple station VLF transmitter'. JGR vol 82, no 7 pp.1177-1187.

 CARPENTER & MILLER (1976) : 'Ducted magnetospheric propagation of signals from the Siple, Antarctica VLF transmitter'.'. JGR, vol 18, no 16 pp.2692-2699.

 LAIRD, M.J., and NUNN, D. (1975) : 'Full wave VLF modes in a cylindrically symmetric enhancement of plasma density'. Plan. & Space Sc, vol 23, pp.1649-1657.

QUENCHING OF NATURAL CYCLOTRON INSTABILITY BY LARGE AMPLITUDE MONO-
CHROMATIC WAVES PROPAGATING IN AN INHOMOGENEOUS MEDIUM

Nicole Cornilleau-Wehrlin and Roger Gendrin
Centre de Recherches en Physique de l'Environnement Terrestre
et Planétaire
CNET/ETE, 92131 Issy-les-Moulineaux (France)

ABSTRACT

Recent experiments have shown that a long duration monochromatic wave generated by a high power VLF transmitter may quench natural magnetospheric hiss emissions over a frequency range Δf of the order of 50-150 Hz below the transmitter frequency. We show that this effect can be interpreted by trapping, inside the monochromatic wave, of a certain amount of particles which were contributing to the generation of hiss in the equatorial region before the transmitter was switched on. The competition between the trapping and detrapping forces (the last one being due to the inhomogeneity of the medium) is studied. We show that the frequency range Δf which is concerned by this effect is proportional to b^2, where b is the field intensity of the monochromatic wave. Δf is also an increasing function of the transmitter frequency.

1. INTRODUCTION

Recently, Raghuram et al. (1977) have shown that strong VLF monochromatic waves ($f_o \sim 5$ kHz), injected from Siple, Antarctica, into the magnetosphere suppressed mid-latitude hiss in a band up to 150 Hz wide, located just below the transmitter frequency. Often, a weak side band line appears just at the lower border of the quiet band. The width Δf of the quiet band increases with the transmitter power (from ~ 40 to ~ 150 Hz when the power is increased by a factor of ~ 4), and with the transmitter frequency (from ~ 55 Hz to ~ 80 Hz when f_o varies from 5.35 kHz to 5.65 kHz). The attenuation of the natural noise with respect to its previous level (when the transmitter was not switched on) also increases with the transmitter power, but there a saturation phenomenon seems to appear. The onset and decay times for the quiet band development are very different and vary in opposite directions when the transmitter frequency is changed. Similar results have also been reported by Helliwell (1977).

We will try to interpret this phenomenon in the framework of non-linear trapping and detrapping of particles in a monochromatic wave (Roux et al., 1973; Roux and Pellat, 1978).

2. NONLINEAR THEORY

Trapping and detrapping of particles inside a monochromatic wave propagating in an inhomogeneous medium is a well known phenomenon (see Matsumoto (1979) for a review). To demonstrate how such a phenomenon can be invoked to explain the quenching of natural emissions, let us consider a monochromatic wave of angular frequency ω_0 propagating along a magnetic field line. z is the distance along the field line and z = 0 corresponds to the equator (figure 1). Particles propagating in the opposite direction interact with this wave in the electron cyclotron mode when their parallel velocity is equal to the resonant velocity V_r. V_r is an increasing function of z. But in the absence of wave, particles would follow an adiabatic trajectory such that their parallel velocity is a decreasing function of z. If $v_{//0}$ is the equatorial parallel velocity that a given particle would have had in the absence of wave, there is a distance $z_a(v_{//0})$ at which the local parallel velocity of the particle and the resonant velocity are equal. When the wave is present, the particle becomes trapped at this distance z_a and then it continues its trajectory towards the equator at the resonant velocity V_r. Such a particle disappears therefore from the equatorial distribution function. If, before the transmitter was switched on - this particle was unstable with respect to cyclotron waves for angular frequencies around $\omega = \Omega_0 - k_0 v_{//0}$ (where Ω_0 is the equatorial gyro-

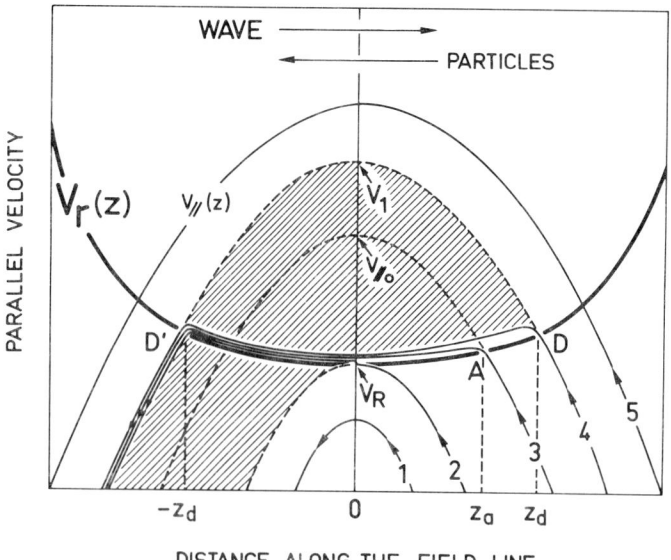

Figure 1. Schematic representation of the nonlinear trapping and detrapping mechanism. The hatched area corresponds to particles which disappear from the initial distribution function and which are released by the wave, all together, at point D'. In this drawing, pitch angle effects have been neglected.

pulsation and k_o the equatorial wave number), there is now no more particle which can resonate at the frequency ω and the emission around this frequency is quenched. This process works for particles such that $v_{//o} > V_R$, where V_R is the equatorial minimum resonant velocity. Because V_R verifies $\omega_o = \Omega_o - k_o V_R$, $v_o > V_R$ implies $\omega < \omega_o$. In other words, the quenching of natural emissions must occur below the transmitter frequency, as it is observed.

The frequency range $\Delta f = f_o - f_1$ which is involved is defined by the maximum equatorial parallel velocity V_1 for which particles can be trapped (particle 4 of figure 1). V_1 is determined by the maximum distance z_d for which the trapping force $m\Gamma_1 = m\omega_T^2/k$ (due to nonlinear effects) exceeds the detrapping force $m\Gamma_2 = \mu \partial B/\partial z + mV_r \partial V_r/\partial z$ (due to the inhomogeneity). In these equations $\omega_T = (eb/m)^{1/2} (kv_\perp)^{1/2}$ is the trapping angular frequency and $\mu = mv_\perp^2/2B$ is the first adiabatic invariant (b is the wave field intensity, B is the DC magnetic field intensity and $v_\perp$ is the perpendicular velocity of the particle). At the equator, $\Gamma_2 = 0$, whereas Γ_1 is $\neq 0$. Trapping can always occur at the equator. By introducing the laws of variation of k, $v_\perp$, B and V_r with z, it is easy to show that there is a distance z_d above which $\Gamma_2 > \Gamma_1$. Therefore, a particle like particle 5 cannot be trapped and it still participates to the cyclotron instability at the equator. There is a limited frequency range over which quenching can occur.

Clearly, Γ_1 increases when b increases, whereas Γ_2 does not depend on b. Therefore, the larger is the wave field intensity, the larger will be the distance for which the equality $\Gamma_1 = \Gamma_2$ is satisfied. So will be the velocity range $\Delta v = V_1 - V_R$ involved by the trapping mechanism and so will be the frequency range $\Delta f = (\omega_o - \omega_1)/2\pi$ over which quenching is efficient. For the values of the magnetospheric parameters which are given in table 1 and which are typical of the experiment, it can be shown that the relation between Δf and Δv is $\Delta f \sim 5.2 \Delta v/V_R$ where f is expressed in kHz. The principle of the interpretation is therefore clear, but in order to get a quantitative evaluation, we must take into account the pitch angle dependence of the trapping and detrapping forces.

TABLE 1 - NUMERICAL VALUES

L value	$L = 3.5$	
Electron density	$N_e = 700$	cm^{-3}
Electron gyrofrequency	$\Omega_o/2\pi = 20.25$	kHz
Wave frequency	$f_o = 5.5$	kHz
Wave number (at f_o)	$k_o = 3 \times 10^{-3}$	m^{-1}
Resonant velocity	$V_R = 3 \times 10^7$	$m.s^{-1}$
Inhomogeneity length	$Z_o = 1 \times 10^7$	m

3. QUANTITATIVE EVALUATION

For a given wave field intensity, the distance z_d at which detrapping occurs ($\Gamma_1 = \Gamma_2$) can be computed as a function of α_0, the value of the pitch angle which the particle would have had at the equator in the absence of the wave field (thick curves of figure 2). We see that z_d is maximum for $\alpha_0 = \alpha_m \simeq 60°$ (this fact will be used later on). Similarly, for a given $v_{//0}$, we can compute the distance z at which a particle will resonate with the wave, for any value of α_0 (dashed curves of figure 2). If, for a given wave field b and a given pitch angle α_0, z is smaller than z_d, the particle can be trapped, because at the distance z it resonates with the wave and Γ_2 is smaller than Γ_1. On the contrary, if z is larger than z_d, the particle cannot be trapped. We see that only particles with large pitch angles are trapped. Fortunately for our theory, those are the particles which contribute the most to the growth rate of the natural emission. For instance, it can be shown that for typical values prevailing in the magnetosphere, 90% of the growth rate at frequencies around 5.5 kHz is due to particles with pitch angles larger than 60° (Cornilleau-Wehrlin and Gendrin, 1979). Examination of figure 2 reveals that, for b = 4 mγ, the growth rate will be reduced by at least 90% for all waves which resonate at the equator with particles having parallel velocities between V_R and $\sim V_R (1 + 0.015)$. Applying equation (1) we see that quenching will occur over a frequency range of the order of 80 Hz. Figure 2 shows also that for a given pitch angle range, doubling the wave field intensity is equivalent to multiplying the parallel velocity range by 4. In other words, the frequency range Δf over which the quenching can occur varies like b^2, in agreement with the experiment.

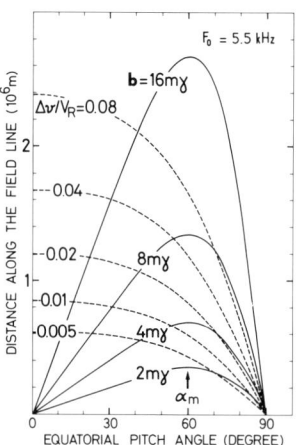

Figure 2. Variation of the detrapping distance z_d with the equatorial pitch angle, for different wave intensities (full curves). The dashed curves represent the distance at which a particle with a given pitch angle and a given equatorial parallel velocity ($v_{//0} = V_R + \Delta v$) resonate with the wave. To compare with Raghuram et al's (1977) experiment, the curves have been plotted for f_0 = 5.5 kHz and L = 3.5.

An analytic expression can be obtained (Cornilleau-Wehrlin and Gendrin, 1979) if we consider the point z_d (α_m) as defining the limiting value $V_\perp$ (this is reasonable because $\alpha_m \sim 60°$ and 90% of the growth rate is due to particles with $\alpha > 60°$. This expression is :

$$\Delta f/f_o = C(eb/m)^2(Z_o/V_R)^2$$

where Z_o is the characteristic length of the inhomogeneity and C a coefficient which depends on the ratio ω_o/Ω_o and which, in the present situation, is of the order of 0.25. A numerical application gives $\Delta f = 50$ Hz for $b \sim 3.2$ mγ and $f = 150$ Hz for $b \sim 5.6$ mγ. Such values for the equatorial wave field intensities are currently expected (Helliwell, 1977; Inan et al., 1978). Consequently, the theory also agrees in order of magnitude with the experiment.

4. VALIDITY CONDITIONS

Trapping is efficient only if the length z_d over which particles remain trapped is at least of the order of one trapping length $\mathcal{L}_t = 2\pi v_\parallel/\omega_t$ (Helliwell and Crystal, 1973; Gendrin, 1974). The trapping length $\mathcal{L}_t$ is easily computed as a function of the wave intensity and of the pitch angle (thick lines of figure 3). We see that for low wave field intensities (~ 2mγ), the trapping length is much larger than the maximum trapping distance. For these low field intensities, an efficient trapping cannot occur and the quasi-linear diffusion theory (Ashour-Abdalla, 1972) must be applied. For large wave field intensities (~ 8mγ), trapping occurs over distances larger than $\mathcal{L}_t$ for a wide range of pitch angle and the present theory is fully applicable.

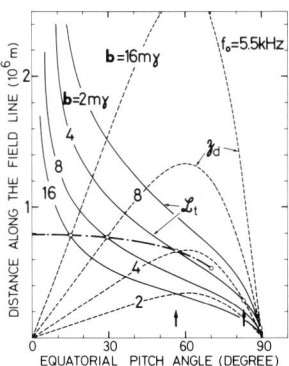

Figure 3. The trapping length $\mathcal{L}_t$ as a function of the equatorial pitch angle for different wave intensities (full curves). The detrapping length z_d has been plotted (dashed curves) for purpose of comparison. Arrows indicate the two pitch angles between which particles are trapped over more than one trapping length, for $b = 4$ mγ. As for figure 2, $f_o = 5.5$ kHz and $L = 3.5$.

By taking into account the condition $z_d > \mathcal{L}_t$, it is also possible to study the efficiency of the proposed mechanism as a function of the transmitter frequency (Cornilleau-Wehrlin and Gendrin, 1979). One finds that there is a minimum frequency below which no quiet bands can be generated ($f_o \sim 4.8$ kHz for the assumed values of the plasma parameters). Above that frequency, the frequency range Δf increases when f_o is increased, as experimentally observed.

Another condition which must be fulfilled is $\Delta f > \omega_t/2\pi$, otherwise we are in the trapped particle instability regime (Budko et al., 1972). For a pitch angle of $\sim 45°$ and a field intensity of $\sim 2m\gamma$ we have $\Delta f \sim \omega_t/2\pi \sim 30$ Hz. But ω_t varies like $b^{1/2}$ whereas Δf varies like b^2. Consequently as soon as $b > 2m\gamma$, $\Delta f > \omega_t/2\pi$ and again the present theory is applicable.

5. CONCLUSION

Trapping and detrapping of particles inside a monochromatic wave propagating in an inhomogeneous medium has been invoked to interpret magnetospheric emissions triggered by VLF pulses (Roux and Pellat, 1978). This mechanism may also explain the generation of "quiet bands" below the monochromatic frequency when the transmitter is operating continuously, as suggested by the Stanford group (Raghuram, 1977; Helliwell, et al., 1978). The theory is able to interpret quantitatively the value of Δf which has been observed as well as its variation with the transmitter power or frequency. The time scales which are involved in this mechanism (~ 10-50s) may also be qualitatively interpreted (Cornilleau-Wehrlin and Gendrin, 1979).

REFERENCES

Ashour-Abdalla, M. : 1972, Planet. Space Sci., 20, pp. 639-662.
Budko, N.I., Karpman, V.I., and Pokhotelov, O.A. : 1972, Cosmic Electrodynamics, 3, pp. 147-164.
Cornilleau-Wehrlin, N., and Gendrin, R. : 1979, J. Geophys. Res., to be published.
Gendrin, R. : 1974, Astrophys. Space Sci., 28, pp. 245-266.
Helliwell, R.A. : 1977, Phil. Trans. Roy. Soc. London, 279, pp.213-224.
Helliwell, R.A., and Crystal, T.L. : 1973, J. Geophys. Res., 78, pp. 7357-7371.
Helliwell, R.A., Katsufrakis, J.P., and Bernhardt, P.A. : 1977, in "Agard Conference Proceedings Nr.192 on Artificial Modification of Propagation Media", ed. by H.J. Albrecht, Agard-Nato, Neuilly, France, pp. 11-1 to 11-8.
Inan, U.S., Bell, T.F., and Helliwell, R.A. : 1978, J. Geophys. Res., 83, pp. 3235-3253.
Matsumoto, H. : 1979, this issue.
Raghuram, R. : 1977, Techn. Re. Nr. 3456-3, Stanford University, Stanford.
Raghuram, R., Bell, T.F., Helliwell, R.A., and Katsufrakis, J.P. : 1977, Geophys. Res. Lett., 4, pp. 199-202.
Roux, A., Gendrin, R., Wehrlin, N., Pellat, R., and Welti, R. : 1973, J. Geophys. Res., 78, pp. 3176-3181.
Roux, A., and Pellat, R. : 1978, J. Geophys. Res., 83, pp. 1433-1441.

NONLINEAR EFFECTS INVOLVED IN THE GENERATION OF TYPE III SOLAR RADIO BURSTS

Dean F. Smith
High Altitude Observatory, National Center for Atmospheric Research and Department of Astro-Geophysics, University of Colorado
Dwight R. Nicholson
Department of Astro-Geophysics, University of Colorado

ABSTRACT: Recent observations of the simultaneous measurement of plasma waves and radiation, and electrons and radiation during type III bursts are reviewed. One-dimensional quasilinear relaxation results are considered and found to be in reasonable agreement with observations, but often inconsistent when weak turbulence wave-wave and strong turbulence processes are included. Strong turbulence processes are considered and soliton collapse is found to be important, but the steady-state plasma wave distribution in a plasma of three-dimensionally collapsing solitons is presently unknown. Radiation processes are considered from both weak and strong turbulence. Subsidiary problems such as the origin of observed and inferred plasma wave clumping are briefly considered and directions for future research are suggested.

1. INTRODUCTION

In the four years since the last major review on the interpretation of type III solar radio bursts has been published (Smith 1974a) exciting observations and important theoretical results have been obtained. The most important observations are the direct measurement of the plasma waves and associated radiation in the solar wind (Gurnett and Anderson, 1976, 1977) and the simultaneous measurement of type III bursts and the associated electrons near the earth (Lin, 1974; Fitzenreiter, Evans and Lin, 1976; Kurt, Mishin and Pissarenko, 1977). The most important theoretical results are on possible processes in stream propagation (Papadopoulous, Goldstein and R. Smith, 1974; R. Smith, Goldstein and Papadopolous, 1976; R. Smith, Goldstein and Papadopolous 1978; Bardwell and Goldman, 1976; Nicholson et al., 1978; Zaitsev et al., 1974; Melrose, 1974; Takakura and Shibahashi, 1976; Takakura, 1977; Magelssen and Smith, 1977) and on possible processes in radiation (Melrose, 1975; Smith and Riddle, 1975; Smith, 1977; Reiter et al., 1978). Many other subsidiary observational and theoretical analyses have been made during this time, e.g. an explanation of Type IIIb stria in terms of modulational instability (Zaitsev, 1977), but spatial limitations demand that we restrict our review to the two "main-line" topics mentioned above. Even here no pretense is made toward completeness and the interested reader is encouraged to peruse recent workshop and symposium proceedings (Lin 1976; Radiofizika, Vol. 20, Issue 9, 1977). We shall endeavor to cover the major advances in the above two topics.

Our approach is to review recent observations to see what new insights they give us into stream propagation, plasma wave generation and radiation in Section 2. We then suppose that type III streams are sufficiently diffuse so that they can propagate undergoing quasilinear relaxation alone (see e.g. Fig. 3 of Smith (1974a)), but taking into account the inhomogeneous nature of the stream and corona and see to what extent this approach can explain the observations (Section 3). We consider the possibility that type III streams are sufficiently dense and/or plasma wave levels sufficiently high for strong turbulence processes such as collapse to occur and analyze the consequences in Section 4. In Section 5 the problem of radiation near the fundamental and second harmonic of the plasma frequency is considered, primarily in the weak turbulence approximation, but with some comments on what can be expected with strong turbulence. Finally we discuss the results and indicate areas in need of further research in Section 6.

2. OBSERVATIONS

We consider the direct measurement of plasma waves and associated radiation in the solar wind (Gurnett and Anderson, 1976, 1977; Gurnett et al., 1978) whose interpretation has been considered in some detail by Smith (1977). While the observation that the probability of such simultaneous observations increases dramatically with decreasing distance from the sun (Gurnett et al., 1978) is interesting and deserves consideration, besides noting the general trend of decreasing plasma wave levels with increasing radial distance for the eleven bursts thus far recorded, we limit our review to the few intense bursts inside 0.45 A.U. which provide the most important clues for a theory. The only burst thus far measured in the high time resolution mode (one measurement every 0.07s) belongs to this group. Whereas the radiation changes slowly with time, the plasma waves have a very bursty character consisting of a succession of spikes whose electric field amplitudes are on the order of 10^3 larger than the mean.

It might be thought that this bursty character could be taken as evidence of strong turbulence processes such as collapse (Section 4). However, any support for such processes in the observations must be qualified. Despite their spiky character, there is no indication in the data that each spike is not temporally resolved in the high time resolution mode although only the burst of April 1, 1976, 10:50 U.T. has been so observed (Gurnett, private communication). The time scale of the shortest spikes is 0.35 s. The spacecraft velocity is about 70 km s^{-1} which is almost perpendicular to the radial direction and the solar wind velocity is of order 300 km s^{-1} in the radial direction. Thus the radial and transverse spatial scales of the shortest spikes are $\sim$105 and $\sim$25 km, respectively. A typical space scale for the regions of intense plasma waves in the final stage of collapse (Section 4) is

$$d = 20\lambda_e = 140(T_e/n_o)^{1/2} \text{ cm} , \qquad (2.1)$$

where λ_e is the electron Debye length and T_e is the electron temperature and n_o is the average background density. For appropriate temperatures and densities $T_e = 1.2 \times 10^5$ K and $n_o = 42$ cm^{-3}, d = 74m. Thus the 105 or 25 km minimum scale sizes from observations cannot be taken as definitive support for the presence of collapse or other coherent parametric processes, but also in no way exclude their possible existence. The possible origin of these large scale sizes is considered in Section 6.

The one indication of a definite limiting value to the maximum intensity of plasma waves comes from the burst of April 18, 1977, 2:50 U.T. which occurred at 0.317 A.U. close to the perihelion of the Helios trajectory. The maximum electric field strength for this burst appears clamped at $\sim$4 mV m^{-1} for several minutes while the average electric field amplitude varies randomly, but also reaches values as high as 2mV m^{-1} which is atypical (Gurnett et al., 1978). While this type of clamping is very suggestive of a nonlinear process with a built-in feedback mechanism, there is still insufficient information in the observations to exclude other possibilities. For example, a stream of fairly constant density n_s undergoing quasilinear relaxation could mimic the observed behavior (Magelssen and Smith, 1977).

We turn now to the observations of type III bursts and their associated electrons (Lin, 1974; Fitzenreiter et al., 1976; Kurt et al., 1977). It has recently become apparent that interplanetary electrons consist of two components -- a more or less isotropic halo and a highly collimated core which propagates within a narrow cone of opening angle $\Delta\theta \lesssim 10°$ (Kurt et al., 1977). There is a one-one correspondence between the arrival of the core electrons and the start of a type III burst on April 27, 1977, 21:45 U.T. which lends credence to the suggestion of Kurt et al. (1977) that plasma oscillations are excited only by the core electrons. This result is also consistent with theory since an isotropic electron distribution is stable (Tsytovich, 1970) and cannot lead to highly nonthermal plasma wave levels. It is possible that previous electron observations with poor angular resolution represent some mixture of core and halo electrons with the relative percentages of each varying from observation to observation. As a result only qualitative features and general trends from previous observations can by taken as meaningful. Quantitative results must await more observations which clearly distinguish the core from the halo.

An example of the gross anisotropy of a type III stream is shown by Figure 39 of Lin (1974) which shows the angular distribution of >45 keV electrons associated with a type III burst on May 16, 1971 averaged over the time interval 13:05–13:45 U.T. This shows a 2:1 anisotropy in the antisolar direction. This should be compared with our Figure 1 which shows the deep spin modulation and the inferred instantaneous electron angular distribution for the type III burst of April 27, 1972, 20:45 U.T. The solar directed electrons in Figure 39 of Lin (1974) can be produced by pitch-angle scattering of the halo or A component of our Figure 1, at end.

Fitzenreiter et al. (1976) investigated the relation of burst intensity to electron flux at 1 A.U. and found two classes of behavior. At low electron fluxes, the radio emission (presumed to be second harmonic) is proportional to the first power of the electron flux. At fluxes greater than ~ 60 cm^{-2} s^{-1} ster^{-1}, the radio intensity becomes proportional to the ~ 2.4 power of the electron flux.

In summary, the electrons associated with type III bursts are found in two different configurations. The observations of Lin (1974) indicate electrons of a diffusive character, forming a halo in velocity space having only $\sim 2:1$ anisotropy. By contrast, the observations of Kurt et al. (1977) indicate a narrow (in velocity space) core of electrons which coexists with a halo. It is concievable that a typical type III burst has, at the earth's orbit, some spatial regions which contain only halo electrons, some spatial regions which contain only core electrons, and some spatial regions which contain both core and halo electrons.

3. STREAM PROPAGATION INCLUDING QUASILINEAR EFFECTS

We are left with no definitive clue from observations concerning how type III streams propagate, i.e. whether the levels and distributions of Langmuir plasma waves generated fall into the quasilinear or strong turbulence regimes. The distinction between these two regimes is the following. The quasilinear regime is a subclass of the weak turbulence regime. Both weak and strong turbulence regimes involve low frequency effects which are quadratic in the high frequency electric field and thus nonlinear. Weak turbulence refers to situations where the quadratic nonlinear correction to the Langmuir wave equation is much smaller than the linear dispersive contribution. Strong turbulence refers to cases where the quadratic nonlinear correction is of the same magnitude or exceeds the linear dispersive contribution. We first suppose that the exciting stream is sufficiently weak and sufficiently broad in velocity so that only the weak turbulence effect of quasilinear relaxation which is a wave-electron effect is important. Since the physical process and relevant equations have been reviewed in Smith (1974a, Sections 2.2 and 2.3) we shall not repeat them here. Rather we review recent results obtained using Eqs. (2.13)-(2.14) of Smith (1974a).

Zaitsev et al. (1974) extended the results of Zaitsev et al. (1972) to the case of long stream injection times. However, the more realistic numerical calculations of Magelssen and Smith (1977) which remove assumptions made by Zaitsev et al. (1974) such as instantaneous velocity diffusion in real space largely supersede their results. Where the results do not depend on such assumptions, the numerical results agree with Zaitsev et al. (1974). Melrose (1974) suggested that type III streams do not undergo significant quasilinear relaxation because he could not see how significant reabsorption by slow electrons of plasma waves produced by fast electrons could occur. Again, although it must be admitted that the physical process is not completely understood, both the numerical results of Takakura and Shibahashi (1976) and of Magelssen and Smith (1977) show that such reabsorption does occur. The set of quasilinear equations is simply too complex to use physical arguments to guess results. Of course physical intuition which explains numerical results <u>a posteriori</u> is invaluable in furthering our understanding. It should be noted that the major source of the complexity is the fact that the relaxation is inhomogeneous in both time and space. As a result streams can propagate for several A.U. in contrast to the results of homogeneous quasilinear theory which led several authors to consider wave-wave processes for stabilizing the stream (e.g., Smith, 1970).

The one-dimensional numerical work of Takakura and Shibahashi (1976) and Takakura (1977) showed clearly that almost all of the plasma wave energy was produced before the peak of the stream had passed and that in the absence of collisional damping, almost all of this energy was reabsorbed by electrons after the passage of this peak. The qualitative explanation for this recycling offered by Takakura and Shibahashi (1976) is that the velocity of electrons which emitted plasma waves with a phase velocity v_{ph} became slightly less than v_{ph} and that of electrons which absorbed the energy of these plasma waves became slightly greater than v_{ph}. This single particle explanation leaves much to be desired because it does not answer the criticism of Melrose (1974) which goes as follows. If we consider a single stream of electrons with some velocity spread Δv_s about some mean velocity, then electrons in the approximate interval $v_s - \Delta v_s$ to v_s always lose energy to the waves and those in the interval v_s to $v_s + \Delta v_s$ always gain energy from the waves. Thus it is difficult to see how recycling with a single stream can occur.

Magelssen and Smith (1977) who obtained the same recycling as Takakura and Shibahashi (1976) and Takakura (1977), but with realistic stream densities for the whole space from the sun to the earth, offered the following explanation for recycling. The above observation of Melrose is true, but the type III situation corresponds to an effectively multiple stream situation in the sense that the type III stream is divided into n sections. The slower electrons of the first section which is the head of the type III stream excite plasma waves. The fast electrons of the second section absorb these plasma waves while the slower electrons of the second section excite plasma waves which are absorbed by the fast electrons of the third section, etc. There is always some loss because the fast electrons of the first section have no plasma waves to absorb and there is no (n + 1)th section to absorb the plasma waves produced by the last (nth) section. However, the numerical calculations indicate that the amount of loss per unit beam length is small. Since the numerical calculations of Magelssen and Smith (1977) are the only ones performed with realistic stream densities and over the entire distance from the sun to the earth, thus allowing meaningful comparison with observations, we state briefly the assumptions and results.

The calculation is one-dimensional which is not justified for all plasma wave energy densities obtained (see Section 5), but is all that was within the capability of the large computer used at the time the calculation was performed. For weak streams associated with weak bursts the one-dimensionality can be justified on the basis of calculations with two-dimensional wave spectra which show that 90% of the wave energy is contained within a cone of $\sim 25°$ and the one-dimensional results are of interest in any case since they serve as a point of comparison for future more complete results. Realistic temperature and density distributions were used; collisional damping of plasma waves, collisional scattering of electrons, and the decrease of stream density due to radial expansion were included. Equations equivalent to (2.13)-(2.14) of Smith (1974a) were solved with injection of a power law velocity distribution which had a smooth temporal variation, rising to a maximum and then decreasing. With initial stream densities in the range 1.5-7.5×10^4 cm^{-3} adequate agreement with electron observations near the earth was obtained. The ratio $W = W_p/n_o KT_e$ was $\sim 10^{-4}$ at 0.5 A.U. for the case of an initial stream density 7.5×10^4 cm^{-3}; W_p is the electric field energy in plasma waves. This is already in the range where strong turbulence effects such as collapse should occur and Nicholson et al. (1978) have shown that collapse would occur using the Magelssen and Smith (1977) plasma wave distribution as an initial condition.

Thus while adequate agreement with electron observations near the earth can be obtained using one-dimensional quasilinear theory, it leads to wave levels which indicate that weak turbulence wave-wave processes (Section 5) and strong turbulence processes (Section 4) will be important. A completely self-consistent treatment would require a two-dimensional analysis (assuming axial symmetry) of all of these processes. In the absence of such a treatment, whose feasibility is discussed in Section 6, we consider the necessary pieces separately using the results of one-dimensional quasilinear theory as a guide to plasma wave levels. Since these levels are within an order of magnitude of those observed and the effect of the processes excluded will be to decrease the efficiency of the recycling process and thus lower the wave levels, it may well turn out that the order of magnitude results from one-dimensional quasilinear theory are not far off the actual values which would come from a more complete analysis.

4. STREAM PROPAGATION INCLUDING STRONG TURBULENCE EFFECTS

We now suppose that the plasma wave level is sufficiently high to make the quadratic nonlinear contribution of the same magnitude or larger than the linear dispersive contribution. The nonlinear contribution to Langmuir wave physics is the result of a single nonlinear effect, the so-called ponderomotive force. By combining the ponderomotive force with the physics of linear Langmuir and ion acoustic waves, we can understand all of the results which involve electrostatic waves that are called weakly or strongly turbulent. We begin with a short derivation of the fundamental equations, first deriving the expression for the ponderomotive force (see, e.g., Landau and Lifshitz (1960)).

A charged particle moving in a high frequency electric field $E(t) = E_o \cos \omega t$ satisfies the force equation

$$m\ddot{x}_1 = q E_o \cos \omega t \qquad (4.1)$$

and thus has the orbit

$$x_1 = \frac{-qE_o}{m\omega^2} \cos \omega t \quad . \qquad (4.2)$$

Consider the case where the electric field is a slowly varying function of position, $E(x,t) = E_o(x) \cos \omega t$. We still expect oscillatory motion as in (4.2), but new components of the orbit caused by the spatial variation of the electric field may also be present. We look for a new component of the orbit with very slow time variation, by writing the orbit $x(t) = x_o(t) + x_1(t)$ with $x_1(t)$ given by (4.2); the slowly varying component $x_o(t)$ is often called the oscillation center of the orbit. The new force equation reads

$$\begin{aligned} m(\ddot{x}_o + \ddot{x}_1) &= q\, E(x,t) \\ &= q\, E_o(x) \cos \omega t \\ &= q\left[E_o(x_o) + x_1 \left.\frac{\partial E_o}{\partial x}\right|_{x_o}\right] \cos \omega t \quad, \quad (4.3) \end{aligned}$$

where the electric field has been Taylor expanded about the oscillation center x_o. We obtain an equation for x_o by averaging each term in Eq. (4.3) over the fast time scale $2\pi/\omega$. One high frequency term on each side vanishes, and using (4.2) for $x_1(t)$ the last term on the right contributes the time average of $\cos^2 \omega t$ or $1/2$. We find

$$m\ddot{x}_o = \frac{-q^2}{4m\omega^2} \frac{d}{dx}(E_o^2) \quad . \qquad (4.4)$$

Thus, the charged particle feels a "ponderomotive" force which directs it away from regions of intense electric field. Since the mass appears in the denominator of (4.4), the ponderomotive force acts much more strongly on electrons than on ions.

We now combine the ponderomotive force with the physics of linear Langmuir and ion acoustic waves to obtain the nonlinear Schrödinger equation first discussed by Zakharov (1972). For the high frequency Langmuir waves, the Fourier transform of the well known linear dispersion relation

$$\omega^2 = \omega_e^2 + 3k^2 v_e^2 \qquad (4.5)$$

in space and time leads to

$$\partial_t^2 \tilde{E}(x,t) = \frac{-4\pi e^2}{m_e}(n_o + n)\tilde{E}(x,t) + 3v_e^2 \partial_x^2 \tilde{E}(x,t) \quad, \qquad (4.6)$$

where the plasma frequency ω_e in (4.5) has been taken as $[4\pi e^2(n_o+n)/m_e]^{1/2}$, $n(x,t)$ is a low frequency fluctuation in the background density associated with ion acoustic-like disturbances (the same for electrons and ions because of quasineutrality), and v_e is the electron thermal speed. The high frequency electric field $\tilde{E}(x,t)$ can be written as a slowly varying amplitude $E(x,t)$ times a rapidly varying phase whose frequency is given by the average background plasma frequency $\omega_{pe} = (4\pi n_o e^2/m_e)^{1/2}$. Thus the real high frequency electric field is

$$\tilde{E}(x,t) = E(x,t)\exp(-i\omega_{pe}t) + E^*(x,t)\exp(i\omega_{pe}t) . \quad (4.7)$$

Inserting this form into (4.6), multiplying by $\exp(i\omega_{pe}t)$ and averaging over the fast time scale $2\pi/\omega_{pe}$ to pick out the terms involving $E(x,t)$, and ignoring second time derivatives of the slowly varying amplitude $E(x,t)$, we find

$$i\,\partial_t E(x,t) + \frac{3v_e^2}{2\omega_{pe}}\partial_x^2 E(x,t) = \frac{\omega_{pe}}{2}\frac{n}{n_o}E(x,t) \quad (4.8)$$

which is the desired equation for the low frequency complex envelope $E(x,t)$ of the real high frequency Langmuir electric field $\tilde{E}(x,t)$.

We rederive the ion acoustic wave equation for the low frequency density perturbation $n(x,t)$ with the addition of the ponderomotive force due to the high frequency electric field envelope $E(x,t)$. Comparing the definition of the high frequency field $\tilde{E}$ in (4.7) to that used in the derivation of (4.4), we find that E_o in (4.4) should be replaced by $2|E|$, so that the low frequency electron force equation becomes

$$m_e \frac{\partial v_e^L}{\partial t} = -e\,E^L - \frac{T_e}{n_o}\partial_x n - \frac{-e^2}{m_e\omega_{pe}^2}\partial_x|E|^2 \quad (4.9)$$

where the right side includes forces due to the low frequency electric field, the low frequency pressure, and the ponderomotive force. Adding the ion force equation neglecting ion pressure,

$$m_i \frac{\partial v_i^L}{\partial t} = e\,E^L - m_i \nu_i v_i^L \quad (4.10)$$

to (4.9), we find, neglecting terms of order m_e/m_i,

$$m_i \frac{\partial v_i^L}{\partial t} + m_i \nu_i v_i^L = \frac{-T_e}{n_o}\partial_x n - \frac{e^2}{m_e\omega_{pe}^2}\partial_x|E|^2 , \quad (4.11)$$

where the term $-m_i \nu_i v_i^L$ on the right of (4.10) is a phenomenological damping term which can be used to model ion Landau damping. Taking the spatial derivative of Eq. (4.11) and eliminating v_i^L by the ion continuity equation

$$\frac{\partial n}{\partial t} + n_o \partial_x v_i^L = 0 \quad , \tag{4.12}$$

we find

$$\partial_t^2 n + \nu_i \partial_t n - c_s^2 \partial_x^2 n = \frac{1}{4\pi m_i} \partial_x^2 |E|^2 \tag{4.13}$$

where the ion-acoustic speed is $c_s = (KT_e/m_i)^{1/2}$.

Equations (4.8) and (4.13), generalized to two or three spatial dimensions, are the desired set of coupled equations, often called the nonlinear Schrödinger equation, because (4.8) resembles the time dependent Schrödinger equation of quantum mechanics, with the energy term $i\partial_t$, the momentum term ∂_x^2, and the nonlinear potential n whose evolution is described by (4.13). A rich variety of physical effects is contained in (4.8) and (4.13). If we choose the phenomenological low frequency damping ν_i in (4.13) equal to twice the ion acoustic frequency, we have a good fluid model of the process of induced scattering on the polarization clouds of ions which is the main weak turbulence process formerly used in attempts to stabilize the type III beam-plasma instability (Smith, 1970). This process is also called the parametric decay instability; it involves two Langmuir waves and one low frequency disturbance. Given a single large amplitude Langmuir wave, the set (4.8), (4.13) can also give rise to instabilities involving two other Langmuir waves and one low frequency disturbance; these are called the oscillating two-stream and the stimulated modulational instabilities. Finally, the generalization of (4.8) and (4.13) to two or three dimensions describes a completely nonlinear phenomenon known as Langmuir soliton collapse; this process, which turns out to be the principle one of interest, cannot be considered in terms of the interaction of a certain fixed number of almost linear waves.

The distinction between weak and strong turbulence can be seen easily from (4.8) and (4.13). Neglecting the first two terms on the left of (4.13) leads to $n \sim |E|^2/4\pi KT_e = W$. Thus the ratio of the nonlinear term in (4.8) to the linear dispersive term becomes $\sim W/(k\lambda_e)^2$, where k is a typical wavenumber or spread in wavenumbers. When $W \ll (k\lambda_e)^2$, we have weak turbulence; when $W \stackrel{\sim}{>} (k\lambda_e)^2$, we have strong turbulence.

We review the application of the ideas contained in Eqs. (4.8), (4.13) to the physics of type III bursts. In pioneering work, Papadopoulos et al. (1974) showed that the Langmuir wave intensities encountered in type III bursts could be large enough to become strongly turbulent, driving the oscillating two-stream instability which could remove wave energy from phase velocities in resonance with the beam. Thus, the beam plasma instability could be stabilized at wave energies lower than those predicted by quasilinear theory. This work was extended by Papadopoulos (1975), R. Smith et al. (1976), Hubbard and Joyce (1976), Rowland and Papadopoulos (1977), Galeev et al (1975), and R. Smith et al. (1978). All of these papers consider purely one-dimensional wave motion along the stream direction. The papers of Papadopoulos et al. (1974) Papadopoulos (1975), R. Smith et al. (1976), and R. Smith et al. (1978) replace (4.8), (4.13) with a set of coupled mode equations in which some modes grow with the growth rate they would have if all of the beam excited waves were concentrated in one monochromatic wave. By contrast, the work of Hubbard and Joyce (1976) and of Rowland and Papadopoulos (1977) involves a many particle computer simulation. In the work of R. Smith et al. (1976), some agreement was obtained between the theoretical threshold for strongly turbulent effects and the observed behavior of second harmonic emission near the earth. The importance of the second spatial dimension, perpendicular to the stream direction, was first pointed out by Bardwell and Goldman (1976). For type III parameters very close to the sun's surface, again assuming that all of the stream excited wave energy can be placed in one monochromatic wave, they find that the maximum parametric growth rates belong to waves which form a cone about the beam excited wavenumber. These waves remain in resonance with the stream, and their production does not help to stabilize the stream-plasma instability. However, the calculation also shows regions of substantial growth in two other locations of wavenumber space; one of these regions corresponds to the parametric decay instability or the fluid analog of the weak turbulence process of induced scattering. Waves produced in this region are indeed out of resonance with the beam. In a careful treatment of the effects of the finite bandwidth of the beam excited waves, they show that it is only marginally valid to treat all of the beam excited wave energy as belonging to only one monochromatic wave.

The two dimensional work of Bardwell and Goldman was extended by Nicholson et al. (1978), Nicholson and Goldman (1978), and Goldman and Nicholson (1978). In these paper, Eqs. (4.8), (4.13) were extended to tw spatial dimensions (the extension is straightforward, but must be done carefully) and solved analytically and numerically. The numerical calculations of Nicholson et al. (1978) are appropriate to type III parameters at 0.5 A.U. They found that for reasonable wave intensities and wavenumber distributions, the most important effect is an immediate soliton collapse in two spatial dimensions; this collapse is independent of and proceeds faster than any associated parametric instabilities. The bandwidth of the initial stream excited waves leads to regions of constructive interference in real space via the uncertainty principle $\Delta x \Delta k \sim 1$. The regions of constructive interference push plasma away via the ponderomotive force and the waves are nonlinearly refracted into the resulting regions of reduced plasma density, leading to even greater wav intensity and ponderomotive force, etc.

Eventually, the intense soliton collapses to a size of only $\sim 10 \lambda_e$ at which point linear Landau damping or other nonlinear effects presumab enter to deplete almost all of the wave energy of the collapsing soliton In spatial regions which do not directly participate in the soliton collapse, the wave energy remains undepleted. A sufficiently large fractio of all wave energy is affected by soliton collapses to take almost all of the wave energy out of resonance with the beam. We note that this scenario for the final stages of soliton collapse could not be carefully studied in the work of Nicholson et al. (1978) because of numerical limitations, but can be inferred from many studies of multi-dimensional soliton collapse (see Nicholson et al. (1978) for references on this subject).

This scenario is only valid when the inequality $k_o \lambda_e < (m_e/m_i)^{\frac{1}{2}}$ is satisfied; here, k_o is the typical stream excited Langmuir wavenumber. This inequality is very important because it determines when the group velocity of stream excited Langmuir waves is less than the ion sound speed. Thus, ponderomotive effects can resonate strongly with ion motions, facilitating direct soliton collapse. In this regime, Goldman and Nicholson (1978) and others have shown that the correct way to compare nonlinear effects in Eqs. (4.8), (4.13) with linear effects is through the inequality $W > (\Delta k_o \lambda_e)^2$ where Δk_o is the spread of stream excited wavenumbers about the central stream excited wavenumber k_o. This inequality gives the threshold for the soliton collapse effects discussed above.

At a given spatial location, the comparison of $k_o \lambda_e$ to $(m_e/m_i)^{\frac{1}{2}}$ may be different at different times in the same burst. The above scenario is usually valid at distances from the sun greater than roughly 10 $R_\odot$. At distances from the sun closer than roughly 10 $R_\odot$, we have the opposite inequality $k_o \lambda_e > (m_e/m_i)^{\frac{1}{2}}$. In this case the linear group velocity of the stream excited Langmuir waves is larger than the ion-acoustic speed, ponderomotive effects can not resonate strongly with ion motions, and direct soliton collapse is prohibited at early times. When the waves become intense, they perform a parametric decay instability (induced scattering on ions) producing wavenumbers of roughly equal magnitude, but opposite direction to the stream excited waves. The spread in wavenumbers is then of the same order as the wavenumber k_o of the initial stream excited waves, and the criterion for soliton collapse effects becomes $W > (k_o \lambda_e)^2$. The soliton collapse is now initiated by the stream excited waves acting together with the oppositely traveling waves produced by the parameteric decay instability. This subject has been studied in detail by Nicholson and Goldman (1978), but has not been specifically applied to type III paramaters.

In summary, Langmuir waves grow during type III bursts due to the linear stream plasma instability. The wave levels may be limited by inhomogeneous quasilinear effects to values low enough that strong turbulence effects do not enter. If the wave levels grow to satisfy $W > (k_o \lambda_e)^2$ when $k_o \lambda_e > (m_e/m_i)^{\frac{1}{2}}$ (at distances from the sun less than 10 $R_\odot$) or $W > (\Delta k_o \lambda_e)^2$ when $k_o \lambda_e < (m_e/m_i)^{\frac{1}{2}}$ (at distances from the sun greater than 10 $R_\odot$), then strongly turbulent effects, especially three-dimensional soliton collapse, become important. The resulting wave energy dissipation will tend to limit the wave levels to values reversing the above inequalities. As noted in Section 3 some order of magnitude agreement was obtained between the numerically predicted wave levels of Nicholson et al. (1978) and the observations of Gurnett and Anderson (1977) near 0.5 A.U.

We would like to know the steady-state wave distribution in the three-dimensional strongly turbulent regime in order, for example, to predict the resulting second-harmonic emission discussed in Section 5. Unfortunately, this problem is not at all well understood at the present time, and very little can be said quantitatively. Preliminary results have been obtained by Galeev et al. (1975) and Reiter, Goldman and Nicholson (1978).

We have seen that inhomogeneous quasilinear theory, in the absence of weak turbulence wave-wave and strong turbulence effects, yields results consistent with the propagation of electrons from the sun to the earth and beyond. Let us now show by means of very crude estimates that the strongly turbulent situation can also be consistent with the propagation of electrons to these distances; such estimates have also been made by R. Smith et al. (1978). Consider an oversimplified model of a type III burst consisting of a column of electrons extending from the sun to the earth, having in some average sense a ratio of stream to background plasma density $n_s/n_o \sim 10^{-6}$, and a speed $\sim c/3$. Suppose these same electrons have produced wave levels which everywhere reach the marginal value for strong turbulence effects, which we assume in some average sense is $W \sim 10^{-4}$.

Using a background plasma temperature ~ 50 eV, the ratio of wave energy density to stream kinetic energy density is

$$(Wn_o KT_e)/[\tfrac{1}{2} n_s m_e (c/3)^2] \sim .2.$$

Thus, in some average sense only 20% of the stream kinetic energy needs to be converted to Langmuir wave energy, and the stream can continue to propagate. This comparison could be improved by noting that some of the inhomogeneous quasilinear calculations (Magelssen and Smith, 1977) yield wave levels very close to the sun which are far below the threshold for strong turbulence effects; our stream may then need to produce strongly turbulent wave levels only at distances, for example $\gtrsim 3\ R_\odot$. Thus there is no reason, in principle, why strong turbulence effects could not play a role in stream propagation over significant distances.

We conclude this section by briefly considering the possible effects of the weak solar wind magnetic field. The ratio of electron gyrofrequency ω_{ce} to electron plasma frequency ranges between 0.1 and 0.01 from the sun to the earth. Thus, the low frequency wave motions associated with strong Langmuir turbulence occur at frequencies much lower than the electron gyrofrequency, but higher than the ion gyrofrequency; as a result the ions are unmagnetized and the electrons are magnetized relative to these low frequency waves.

When the initial beam excited waves have even a small two-dimensional character (of order one percent or greater), the strongly turbulent situation will consist of three-dimensional collapsing solitons. The basic physical reason is that a collapsing soliton needs to push plasma away in order to dig a density hole; it does not care whether it does this by pushing electrons in all directions, or merely in the direction of the magnetic field. The details of the collapse may be affected by the weak magnetic field. Although this scenario has not yet been calculated in detail, the basic ideas have been set out in Kuznetsov (1974), Zakharov (1975), and Nicholson et al. (1978). Confirmation of these ideas also comes from extending the two-dimensional growth rate calculations of Bardwell and Goldman (1976) to the weak magnetic field case. As pointed out by Tsytovich (1977), the oscillating two-stream instability branch of Bardwell and Goldman (1976) tends to be strongly suppressed by a weak magnetic field. However, Weatherall et. al. (1978) and Nicholson et al. (1978) have shown that the other two branches of Bardwell and Goldman, the decay branch and the stimulated modulational instability branch, are only slightly affected by a weak magnetic field. Thus, the overall three-dimensional parametric instability and collapse picture for an intense monochromatic wave is not substantially affected by a weak magnetic field. Strong magnetic fields ($\omega_{ce}^2/\omega_{pe}^2 \gtrsim 1$) can possibly force all of the physics to become one-dimensional and prevent three-dimensional soliton collapse (Petviashvili, 1975 and Krasnosel'skikh and Sotnikov, 1977).

For a type III burst consisting of a narrow stream of electrons in velocity space, in the presence of the weak solar wind magnetic field, the initial stream excited waves will have substantial spread in wavenumbers in all directions and, if the wave intensities reach the strongly turbulent regime, the above three-dimensional soliton collapse effects will occur. Some bursts may have two components of electrons, a core and a halo (Kurt et al., 1977, see Section 2). In spatial regions which have only core electrons or both core and halo electrons, the excited wavenumbers will have substantial bandwidths in all directions and the resulting turbulence, will be fully three-dimensional whether or not the strongly turbulent regime is reached. In spatial regions which contain only anisotropic halo electrons, the stream plasma instability in the presence of a weak magnetic field may be modified (Papadopoulos and Freund, 1978; R. Smith, Goldstein, and Papadopoulos, 1978; and Goldman, private communication). In the absence of magnetic field effects, one dimensional strongly turbulent situations are quickly unstable to two dimensional perturbations. This fact may remain true in the presence of weak magnetic fields and is currently under investigation.

5. RADIATION

We consider recent developments in the transformation of Langmuir plasma waves into radiation near the fundamental (ω_{pe}) and second harmonic ($2\omega_{pe}$) of the plasma frequency. It is clear from the results of Sections 2-4 that both the weak and strong turbulence regimes should be considered. However, aside from basic dependences on parameters such as W_p, calculations of radiation in the strong turbulence regime are presently not in a sufficiently "clean" state to be reviewed although work is in progress (Breizman and Pekker, 1978; Papadopoulos and Freund, 1978; Reiter et al. 1978). Thus, our discussion is confined primarily to weak turbulence results.

The main new result for fundamental emission is ray tracing including amplification of the radiation in a source with random density inhomogeneities (Smith and Riddle, 1975). The amplification process is considered in Smith (1970, 1974b). Smith and Riddle (1975) showed for the 80 MHz plasma level that density inhomogeneities which are sufficiently large to explain observed source sizes would not permit amplification of fundamental radiation if the plasma wave level is uniform over the source area. The only possibility for fundamental amplification is for the plasma waves to be concentrated in clumps of size similar to the typical scale lengths of the density inhomogeneities, i.e. 10^2-10^3 km, i.e. similar to the minimum scale size of the clumps of plasma waves inferred from observations near 0.45 A.U. (see Section 2). Inhomogeneities of this size in the meter wavelength range were also inferred by Melrose (1975) who argued that they would lead to strong mode-mode coupling which would explain, for example, the lack of polarization reversals in type I bursts. Thus there is strong indirect evidence near 100 MHz and direct evidence near 0.45 A.U. that the plasma wave distribution is spatially concentrated into clumps of size $\sim 10^2$-10^3 km. A possible origin for this clumping is discussed in Section 6.

There are no fundamentally new results for weak turbulence second harmonic radiation, but it has been shown that the radiation observed by Gurnett and Anderson (1976) can be explained by the observed plasma wave levels using weak turbulence theory (Smith, 1977). Because of the importance of this result, both for possible comparison with strong turbulence results and its demonstration of the necessity of rapid weak or strong turbulence wave-wave processes, we summarize the main points of this calculation. The process for second harmonic radiation which is the principle radiation in the hectometric and kilometric wavelength ranges is the combination of two excited plasma waves, p and p', which can be represented schematically by

$$p + p' \rightarrow t(2\omega_p) \ . \tag{5.1}$$

Here t is the transverse electromagnetic wave at twice the plasma wave frequency, $2\omega_p$. It is well known that for $v_{ph}/c \ll 1$, process (5.1) can only occur if the plasma waves meet almost "head-on" and both of the plasma waves must have high effective temperatures to produce radiation with a high effective temperature (Melrose, 1970; Smith and Sturrock, 1971). Since the waves produced by the electron stream have a narrow spread of angles (Section 3), some type of scattering process is required to produce waves in the backward cone. Alternately, for $v_{ph}/c \sim 1$, the scattering process could produce such small wavenumbers that they could combine in the same cone to produce radiation.

The most effective weak turbulence process for this purpose is scattering on the polarization clouds of ions or the parametric decay instability (Smith, 1970, 1974a, Section 4) which can be represented schematically by

$$p + i \rightarrow p' + i' \ , \tag{5.2}$$

where i and i' are the ion states before and after the scattering, respectively. Process (5.2) can be either spontaneous or induced, but it is shown in Smith (1977) that only the induced process is important for type III bursts in the solar wind. The maximum growth rate for this process is (Smith, 1977)

$$\gamma(k_2) = 1.9 \times 10^{-1} \omega_{pe} W, \qquad (5.3)$$

where k_2 is the wave vector of the scattered plasma wave. When the wavenumber k_o of the initial plasma wave satisfies $k_o < 2\omega_{pe} v_i/(3v_e^2)$, where v_i is the ion thermal velocity, waves in the forward cone relative to k_o are amplified and when $k_o > 2\omega_{pe} v_i/(v_e^3)$, waves in the backward cone are amplified.

We now apply these results to the type III burst of 1976 March 31, 18:10 U.T. which is the strongest burst observed to date (Gurnett and Anderson, 1976). This burst had an electric field strength for plasma waves of 14.8 millivolts m^{-1} which leads to $W = 1.4 \times 10^{-5}$ for $n_o = 42$ cm^{-3} and $T_e = 1.2 \times 10^5$ K. The corresponding radiation intensity near $2\omega_p$ was 1.15×10^{-17} Wm^{-2} Hz^{-1} which leads to a volume emissivity assuming isotropic emission of $J(2\omega_p) = 1.55 \times 10^{-23}$ ergs cm^{-3} s^{-1} ster^{-1} (Smith, 1977). For a spectrum of plasma waves we take

$$W_p(k_o) = \text{const. for } k_o \leq 1.9 \times 10^{-2} \lambda_e^{-1} \qquad (5.4)$$

distributed isotropically consistent with the numerical results of Magelssen and Smith (1977), the results of induced scattering and the fact that a burst lasts about $10^8 \omega_{pe}^{-1}$ at 0.45 A.U. Thus from Eq. (5.3) with $W = 1.4 \times 10^{-5}$, $\gamma(k_2)/\omega_{pe} = 2.6 \times 10^{-6}$ so that 260 e-folds would occur during the lifetime of the burst. Long before this time, however, in about 20 e-folds, the energy density in scattered waves would reach the energy density in initial waves, justifying the use of an isotropic spectrum. Numerical evaluation of several spectra like (5.4) in Smith (1970) show that a representative value for the volume emissivity is

$$J(k) \simeq 0.1 \frac{\omega_{pe} W_p^2}{n_o m_e c^2} \left(\frac{k}{k_1}\right)^3, \qquad (5.5)$$

where $k = (3)^{\frac{1}{2}} \omega_{pe}/c$ is the wavenumber of the emitted transverse wave and k_1 is the maximum wavenumber in the spectrum. For spectrum (5.4)

$$J(k) = 8 \times 10^{-3} \frac{\omega_{pe} W_p^2}{n_o m_e c^2} \qquad (5.6)$$

and with the above parameters $W_p = 9.5 \times 10^{-15}$ ergs cm^{-3} so that $J(k) = 7.6 \times 10^{-21}$ ergs cm^{-3} s^{-1} ster^{-1}. Comparing this value with the required $J(2\omega_p) = 1.55 \times 10^{-23}$ ergs cm^{-3} s^{-1} ster^{-1}, we see that the derived emission is nearly three orders of magnitude larger than the required emission.

However, if we interpret the three-orders-of-magnitude difference between the peak and average electric fields in the burst of 1976 March 31, 18:10 U.T. (Gurnett et al., 1978) as due to clumps of intense plasma waves which occupy only 10^{-3} of the volume of the source, we can still obtain the observed radiation. In fact it should be emphasized that any radiation theory must obtain about three orders of magnitude more radiation from the clumps of intense plasma waves to explain the observed radiation in this burst since, due to its high group velocity, the radiation is distributed uniformly over the clump and interclump regions on time scales much shorter than the shortest observing time. Thus we conclude that weak turbulence theory can explain the observed radiation, but only with a quasi-isotropic plasma wave distribution during most of the burst due to process (5.2) which again calls into question the one-dimensional quasilinear results of Section 3. Moreover, the value of W is so high that for sufficiently small Δk_o, the relevant criterion for collapse $W > (\Delta k_o \lambda_e)^2$ will be satisfied (see Section 4). This leads us to the question of what is the expected second harmonic radiation from collapsing solitons. This question was considered by Breizman and Pekker (1978), Papadopoulos and Freund (1978) and Reiter et al. (1978). As noted above, however, this field is not yet in a state to be reviewed.

6. DISCUSSION

We have reviewed one-dimensional quasilinear theory and found that while it leads to adequate agreement with observations, it is inconsistent with both weak and strong turbulence theories for many type III bursts. A two-dimensional analysis including weak and strong turbulence wave-particle and wave-wave processes was called for in Section 3. However, we do not feel that the resources required for such an undertaking can be justified until our knowledge of the steady-state characteristics of the strongly turbulent regime becomes more complete as noted in Section 4.

In the second harmonic radiation problem the same difficulty occurs. Thus we consider determination of the steady-state characteristics of strong Langmuir turbulence as a key problem in need of solution before the area of nonlinear processes involved in the generation of type III bursts can proceed much further. Once this determination is made, a two-dimensional analysis (assuming axial symmetry) including weak and strong turbulence wave-particle and wave-wave processes should by made. Such an analysis should be possible with the new generation of advanced computers which are presently coming into service.

A subsidiary problem is the origin of the plasma wave clumping on scales of 10^2-10^3 km as required for fundamental amplification at meter wavelengths and as observed directly in the solar wind. We offer the following possible explanation which is presently being pursued by one of us (D.S.) to explain the observations near 0.45 A.U. It is known from interplanetary scintillations that the solar wind is inhomogeneous in density on scales from 30-several thousand kilometers. If a stream tries to amplify plasma waves in this medium, the plasma waves will still be subject to the laws of reflection and refraction and thus tend to regions of relatively lower density and be partially confined there. Succeeding parts of the amplifying stream will find higher levels of plasma waves in the regions of relatively lower density leading to still higher levels since the amplification is starting from a much higher level relative to regions of relatively higher density where plasma waves are refracted out. The basic principle of this idea has already been tested by modifying the program of Smith and Riddle (1975) and found to be sound. In that program transverse waves are amplified by a uniform distribution of plasma waves. Whether or not plasma waves amplified by a uniform stream with the level of density inhomogeneities present near 0.45 A.U. will show the same behavior is presently being investigated. If this turns out not to be the case, then other possibilities such as spatially filamented streams will have to be considered.

Thus, while the origin of the plasma wave clumps is not yet clear, some promising possibilities exist. When the formation mechanism has been identified, it will have to be included in complete theories of type III sources since it will affect collapse criteria, etc.

In summary, the field of nonlinear processes involved in the generation of type III bursts has seen a rapid development in the last four years due both to direct observations and theoretical developments. We anticipate an even more rapid development in the near future as better observational techniques, larger computers, and more powerful analytical tools are applied to the problem.

ACKNOWLEDGEMENTS

We thank M. Goldman, R. Lin, D. Melrose, G. Reiter, and J. Weatherall for useful comments on this manuscript. The work of one of us (DRN) was supported by the Atmospheric Research Section, National Science Foundation, ATM 76-14275.

REFERENCES

Akhiezer, A.I., Akhiezer, I.A., Polovin, R.V., Sitenko, A.G., and Stepanov, K.N.: 1967 "Collective Oscillations in a Plasma," MIT Press, Cambridge, Massachusetts.
Bardwell, S., and Goldman, M.V.: 1976, Astrophys. J., 209, 912.
Breizman, B.N., and Pekker, L.S.: 1978, Phys. Lett. 65A, 121.
Fitzenreiter, R.J., Evans, L.G., and Lin, R.P.: 1976, Solar Phys. 46, 437.
Galeev, A.A., Sagdeev, R.Z., Sigov, Yu. S., Shapiro, V.D. and Shevchenko, V.I.: 1975, Sov. J. Plasma Phys. 1, 5.

Goldman, M.V., and Nicholson, D.R.: 1978, Phys. Rev. Lett. 41.
Gurnett, D.A., and Anderson, R.R.: 1976, Science 199, 1159.
Gurnett, D.A., and Anderson, R.R.: 1977, J. Geophys. Res. 82, 632.
Gurnett, D.A., Anderson, R.R., Scarf, F.L., and Kurth, W.S.: 1978, J. Geophys. Res. (submitted).
Hubbard, R.F., and Joyce, G.: 1976, Plasma Phys. 18, 681.
Krasnosel'skikh, V.V. and Sotnikov, V.I.: 1977, Sov. J. Plasma Phys. 3, 491.
Kurt, V.G., Mishin, E.V. and Pissarenko, N.F.: 1977, Pis'ma v Astron. Zh. 3, 170.
Kuznetsov, E.A.: 1974, Sov. Phys. JETP 39, 1003.
Landau, L.D., and Lifshitz, E.M.: 1960, Mechanics, Addison-Wesley, Reading, MA, pp. 93-95.
Lin, R.P.: 1974, Space Sci. Rev. 16, 189.
Lin, R.P.: 1976, Solar Phys. 46, 433.
Lin, R.P., Evans, L.G., and Fainberg, J.: 1973, Astrophys. Letters 14, 191.
Magelssen, G.R., and Smith, D.F.: 1977, Solar Phys. 55, 211.
Melrose, D.B.: 1970, Aust. J. Phys. 23, 885.
Melrose, D.B.: 1974, Solar Phys. 38, 205.
Melrose, D.B.: 1975, Solar Phys. 43, 79.
Nicholson, D.R., Goldman, M.V., Hoyng, P., and Weatherall, J.: 1978, Astrophys. J. 223, in press.
Nicholson, D.R., and Goldman, M.V.: 1978, Phys. Fluids 21, in press.
Papadopoulos, K.: 1975, Phys. Fluids 18, 1769.
Papadopoulos, K., Goldstein, M.L., and Smith, R.A.: 1974, Astrophys. J. 190, 175.
Papadopoulos, K., and Freund, H.P.: 1978, NRL preprint.
Petviashvili, V.E.: 1975, Sov. J. Plasma Phys. 1, 15.
Reiter, G.F., Goldman, M.V., and Nicholson, D.R.: 1978, in preparation.
Rowland, H.L., and Papadopoulos, K.: 1977, Phys. Rev. Lett. 39, 1276.
Smith, D.F.: 1970, Adv. Astron. Astrophys. 7, 147.
Smith, D.F.: 1974a, Space Sci. Rev. 16, 91.
Smith, D.F.: 1974b, Solar Phys. 34, 393.
Smith, D.F.: 1977, Astrophys. J. 216, L53.
Smith, D.F., and Sturrock, P.A.: 1971, Ap. Space Sci. 12, 411.
Smith, D.F., and Riddle, A.C.: 1975, Solar Phys. 44, 471.
Smith, R.A., Goldstein, M.L., and Papadopoulos, K.: 1976, Solar Phys. 46, 515.
Smith, R.A., Goldstein, M.L, and Papadopoulos, K.: 1978, Goddard preprint no. 78079.
Takakura, T.: 1977, Solar Phys. 52, 429.
Takakura, T., and Shibahashi, H.: 1976, Solar Phys. 46, 323.
Tsytovich, V.N.: 1970, Nonlinear Effects in a Plasma, Plenum Press, New York.
Tsytovich, V.N.: 1977, Research in Geomagnetism, Aeronomy and Solar Physics 42, 3 (in Russian).
Weatherall, J.C., Goldman, M.V., and Nicholson, D.R.: 1978, in preparation.
Zaitsev, V.V.: 1977, Radiofizika 20, 1379.

Zaitsev, V.V., Mityakov, N.A., and Rapoport, V.O.: 1972, Solar Phys. 24, 444.
Zaitsev, V.V., Kunilov, M.V., Mityakov, N.A., and Rapoport, V.O.: 1974, Soviet Astron.-AJ 18, 147.
Zakharov, V.E.: 1972, Soviet Phys. JETP 35, 908.
Zakharov, V.E.: 1975, JETP Lett. 21, 221.

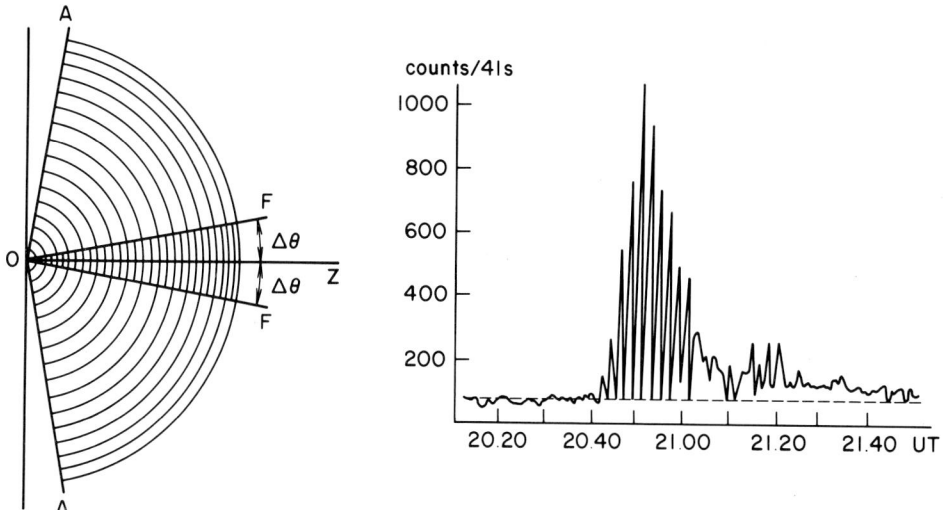

Fig. 1 (from left to right) Schematic division of the electron stream into a core component FOF and a halo component AOA along with the time profile of the intensity of electrons with energies E > 40 keV measured by the Prognoz satellite on 27 April 1972. The 100% modulation around 21.50 U.T. is caused by the spacecraft spin and the small angular spread $\Delta\theta$ of the core. (Adapted from Figs. 1 and 2 of Kurt et al., 1977)

A THEORY OF SOLAR TYPE III RADIO BURSTS

Melvyn L. Goldstein
NASA/Goddard Space Flight Center
Greenbelt, MD 20771

Konstantinos Papadopoulos
Naval Research Laboratory
Washington, DC 20390

Robert A. Smith
JAYCOR
Alexandria, VA 22304

ABSTRACT

A theory of type III bursts is reviewed. Energetic electrons propagating through the interplanetary medium are shown to excite the one dimensional oscillating two stream instability (OTSI). The OTSI is in turn stabilized by anomalous resistivity which completes the transfer of long wavelength Langmuir waves to short wavelengths, out of resonance with the electrons. The theory explains the small energy losses suffered by the electrons in propagating to 1 AU, the predominance of second harmonic radiation, and the observed correlation between radio and electron fluxes.

INTRODUCTION

Solar type III radio bursts have been studied for more than 30 years. The persistent interest in this phenomenon has been due in no small part to the theoretical difficulties encountered in constructing a convincing interpretation of many of the most striking properties of the bursts. Several basic questions were posed by Sturrock fifteen years ago, and are only now beginning to be answered. Among the issues he raised (Sturrock, 1964) are three that will be discussed in some detail in this brief review. First, why is the electron beam that excites the bursts not significantly decelerated; second, why is the radiation predominately emitted at the second harmonic of the local plasma frequency, $f_{pe} = \omega_e/2\pi$; and third, why does the beam have such a well defined velocity, typically between 0.3 and 0.2c.

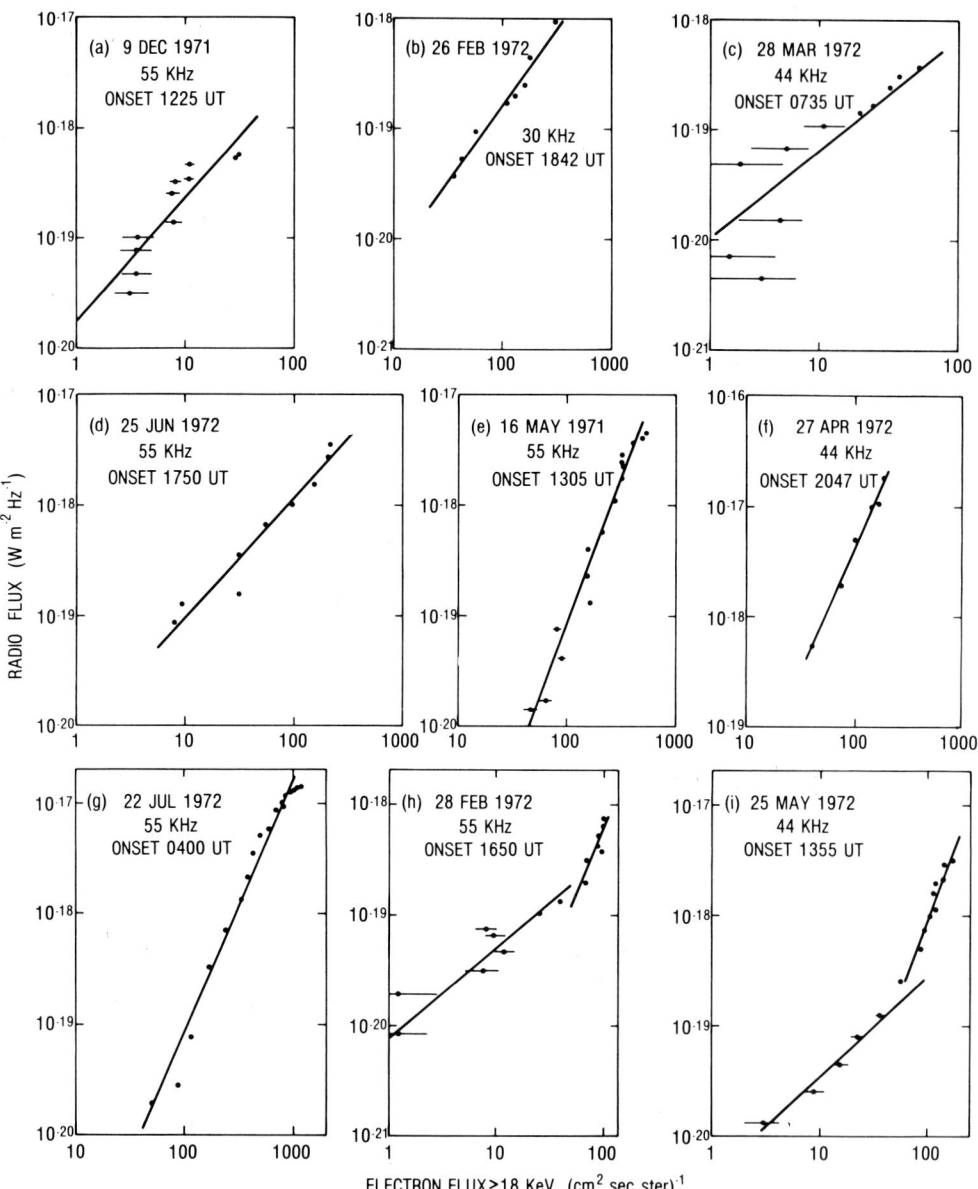

Figure 1. After Fitzenreiter et al. (1976), showing the correlation between radio and electron fluxes for nine type III bursts observed at 1 AU. The dots represent data taken approximately two minutes apart.

In 1976 yet another curious observation was reported by Fitzenreiter, Evans, and Lin (1976). In looking at simultaneous observations of both the electron and radio fluxes of type III bursts that had traveled out to 1 AU, they found that for electron fluxes less than about 100 $(cm^2 \cdot s \cdot ster)^{-1}$, the radio intensity I and the electron flux J_E were approximately linearly proportional. For larger electron fluxes I $\propto J_E^{2.4}$. Figure 1 shows this power law dependence for nine events analyzed by Fitzenreiter, et al. (1976).

In 1974 we proposed a theory (Papadopoulos, Goldstein, and Smith 1974) which demonstrated that effects of strong plasma turbulence can readily account for the observed fact that the electron streams associated with type III bursts are able to travel large distances without significant deceleration. In contrast, conventional weak turbulence plasma theory predicts that all the streaming energy should be dissipated within a few kilometers of the injection site.

The strong turbulence theory also suggested an explanation for the dominance of second harmonic radiation. During the last several years, that theory has been expanded in a series of papers (Smith, Goldstein, and Papadopoulos 1976, 1978; and Goldstein, Smith, and Papadopoulos 1978). In its present version the theory not only accounts for the minimal energy losses suffered by the electrons, but also is able to account for the observed intensities of electromagnetic radiation (at $2\omega_e$), and the correlation between the radio and electron fluxes.

The essential features of the theory will be reviewed in this paper. The reader is referred to the original papers mentioned above for additional details, and to the review by Nicholson and Smith (1979) for a discussion of other aspects of type III burst observations and theory.

ELEMENTS OF THE THEORY

Following the injection into the solar atmosphere of a power law distribution of electrons, the faster particles will begin to out pace the slower ones as the beam propagates to higher altitudes. Sufficiently far from the injection point only the fastest particles will at first be detected. The distribution function f_T will have the form of a "bump" on the tail of the thermal component, and will be unstable to the excitation of Langmuir waves with phase velocities equal to the velocities of the fast electrons. In the conventional weak turbulence theory, as the slower particles arrive, they continue to amplify the Langmuir waves ("oscillation pileup") until the energy flux of the waves equals that of the beam $[v_g(E^2/8\pi) \simeq (1/2)n_b m v_b^3 (\Delta v_b/v_b)]$, where v_g and E are the group velocity and electric field of the Langmuir waves; and n_b, v_b, and Δv_b are the density, velocity, and velocity spread of the energetic electrons, respectively. In the solar corona a beam could propagate only some 60 km before losing all of its energy to Langmuir waves. Clearly this does not happen because energetic electrons are often observed at 1 AU in association with type III bursts (Figure 2).

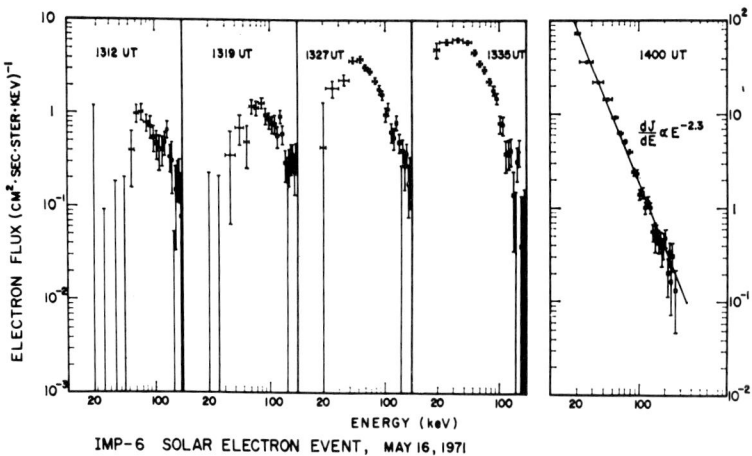

Figure 2. After Lin et al. (1973). Electron spectra are shown at various times during the type III burst of May 16, 1971.

The strong turbulence theory can be formulated in terms of a two-fluid hydrodynamic model of the plasma in which the motions are separated into fast plasma oscillations and quasineutral slow ion oscillations. If one writes the ion and electron densities as $n_i = n_o + \delta n_i$ ($\delta n_i/\delta n \ll 1$), and $n_e = n_i + \delta n_e$ ($\delta n_e/n_i \ll 1$), then the fundamental equations describing the coupling of high frequency plasma waves to low frequency density fluctuations are

$$\frac{\partial \delta n_e}{\partial t} - n_o \nabla \cdot \delta \underline{v}_e = -\nabla \cdot (\delta n_i \delta \underline{v}_e)$$

$$n_o \frac{\partial}{\partial t} \delta \underline{v}_e + \frac{\gamma_e T_e}{m} \nabla \delta n_e = \frac{e}{m} n_o \underline{\varepsilon} - \nu_e n_o \delta \underline{v}_e \qquad (1)$$

where $\underline{\varepsilon}$ satisfies $\nabla \cdot \underline{\varepsilon} = -4\pi e \delta n_e$, γ_e is the polytropic index, and ν_e is a phenomenological damping decrement. In writing equations (1) we have assumed that $(k\lambda_e)^2 \ll 1$ and $\nu_e/\omega_e \ll 1$. C_s is the speed of sound, defined by $C_s^2 = (\gamma_e T_e + \gamma_i T_i)/M$. The brackets, $\langle \; \rangle$, denote averages over the fast time scale. Introducing the slowly varying quantity $\underline{E}(x,t)$

$$\underline{\varepsilon}(\underline{x},t) = \frac{1}{2} \{\underline{E}(\underline{x},t)e^{-i\omega_e t} + \underline{E}^*(\underline{x},t)e^{i\omega_e t}\}$$

results in the following simplification

$$\{i \frac{\partial}{\partial t} + \frac{\gamma_e T_e}{m\omega_e} \nabla\nabla + i\nu_e\}\underline{E} = \frac{\omega_e}{2n_o} \delta n_i \underline{E}$$

$$\{\frac{\partial^2}{\partial t^2} + \nu_i \frac{\partial}{\partial t} - C_s^2 \nabla^2\}\delta n_i = \frac{1}{16\pi M} \nabla^2 |\underline{E}|^2 \tag{2}$$

Details of this derivation can be found in Papadopoulos, et al. (1974), Papadopoulos (1975), Manheimer and Papadopoulos (1975), and Smith et al. (1978). Equations (2) describe modulational instabilities including the OTSI, or soliton formation, depending on ones point of view.

Upon taking the Fourier transform of equations (2) the general dispersion relation for the OTSI and other modulation instabilities can be written as

$$[-\omega(\omega+i\nu_i) + k^2 C_s^2] = - \frac{k^2 \omega_e}{32\pi M n_o} \int d\underline{k}' |\underline{E}(\underline{k}')|^2 \times$$

$$\frac{[\underline{k}' \cdot (\underline{k}-\underline{k}')]^2}{|\underline{k}'|^2 |\underline{k}-\underline{k}'|^2}[- (\omega+i\nu_{ek}) - \frac{3}{2}\omega(k'\lambda_e)^2 + \frac{3}{2}(\underline{k}-\underline{k}')^2\lambda_e^2\omega_e]^{-1} +$$

$$\frac{[\underline{k}' \cdot (\underline{k}-\underline{k}')]^2}{|\underline{k}'|^2 |\underline{k}+\underline{k}'|^2}[- (\omega+i\nu_{ek}) - \frac{3}{2}\omega(k'\lambda_e)^2 + \frac{3}{2}(\underline{k}+\underline{k}')^2\lambda_e^2\omega_e]^{-1} \tag{3}$$

where we have assumed that a zero-order spectrum of "pump" Langmuir waves exists between $\underline{k}_1 < \underline{k} < \underline{k}_2$. Outside that region the waves are taken to be small perturbations, $\delta\underline{E}(\underline{k},t)$ and $\delta n(\underline{k},t)$. The analysis can be further simplified by noting that a full three dimensional treatment is not necessary (contrast Bardwell and Goldman 1976 and Nicholson et al. 1978). In reducing these equations to one dimension, one must separately consider the question of whether the pump waves are predominately one dimensional, whether the parametric instabilities preferentially give rise to daughter waves also aligned predominately along the magnetic field direction, and whether the decay instability has a lower threshold than the OTSI.

The pump waves are excited by the beam plasma instability which in the presence of a magnetic field, for arbitrary propagation direction, has a growth rate given by:

$$\gamma_b = \frac{n_b}{n_o} \sqrt{\pi} \exp(-\lambda) I_o(\lambda) \left(\frac{\omega - \underline{k} \cdot \underline{V}_b}{k \Delta V_b}\right) \times \exp\left(\frac{\omega - \underline{k} \cdot \underline{V}_b}{k \Delta V_b}\right)^2$$

where $\lambda \simeq k_\perp \Delta v_b / \Omega_e$ and we assume that the beam velocity spread, Δv_b, is isotropic. Because $\gamma_b(k_\perp) \neq 0$ only if $\lambda \ll 1$, the angular spread of the spectrum θ will be given by $\sin(\theta) < (\Omega_e/\omega_e) \cdot (V_b/\Delta V_b) \simeq (1/3) \cdot (\Omega_e/\omega_e)$. With $\Omega_e/\omega_e \simeq 10^{-2}$ the angular spread of the instability spectrum is less than $1°$. Thus the pump waves are one dimensional.

One can now approximate the pump spectrum, $W(k')$ in equation (3) as a one dimensional Lorentzian in which the pump wave spectrum is centered at k_o with width Δk_o. The dispersion relation then becomes

$$[\omega(\omega + i\nu_i) - \mu k^2 T] + \frac{\frac{3}{4} \mu k^2 W_o F(k, k_o, \psi)}{\frac{9}{4} k^4 - [(\omega + i\nu_e) + 3i|k|\Delta k_o \cos\psi - 3|k|k_o \cos\psi]^2} = 0 \quad (4)$$

where, for convenience, we have used a dimensionless notation in which $k \to (k\lambda_e)$, $\omega \to (\omega/\omega_e)$, $\nu \to \nu/\omega_e$, $\mu \equiv m/M$, $W_o \equiv \int dk'^2 E(\underline{k}')|^2/8\pi n T_e$ and $T \equiv 1 + (3/2) \cdot (T_i/T_e)$. In equation (4) ψ is the angle between $\underline{k}$ and the magnetic field direction, $F(\psi) \simeq \cos^2\psi$, $k \gg k_o$ and $F(\psi) \simeq 1 - k^2 \sin^2\psi/k_o^2 + 4k^2\cos^2\psi/k_o^2$ for $k \ll k_o$. In the limit $(\Delta k_o/k_o) \to 0$ the dispersion relation reduces to the one found by Papadopoulos (1975) and Bardwell and Goldman (1976). In the interplanetary medium $W_o < \mu$ and $\omega(\omega + i\nu_i) < \mu k^2 T$ with the result

$$\frac{9}{4} k^4 - [\omega+3i|k|\Delta k_o \cos\psi - 3|k|k_o \cos\psi]^2 - \frac{3}{4} k^2 \frac{W_o F(\psi)}{T} = 0$$

The maximum growth rate as a function of angle is given by $\partial(\mathrm{Im}\omega)/\partial\psi = 0$, or

$$-3|k|\Delta k_o \sin\psi + \frac{\sqrt{3}}{4} \left\{ \frac{W_o F(\psi)}{T} - 3k^4 \right\}^{-(1/2)} \partial F(\psi)/\partial\psi = 0$$

which can be satisfied only if $\psi = 0$ for both $k \lessgtr k_o$. Note that we have neglected ion inertia so that one must have $k^2 < \mu$ as well as $W_o < \mu$; but both conditions are easily satisfied in the interplanetary medium. Therefore we conclude that the daughter waves are also one dimensional and the one dimensional treatment of parametric instabilities developed in Papadopoulos et al. (1974) is completely justified for type III bursts.

Finally, the question arises as to whether the decay instability is important when $W_o \simeq 10^{-5}$, the threshold for the OTSI. The decay instability is a three wave interaction whose threshold is found by setting $F(\psi) = 1$ and $\cos\psi = 1$ in equation (4). If the decay instability were excited, the daughter waves would have smaller wave numbers than the pump wave. Stabilization of the beam plasma instability could not be achieved in that case. The threshold condition is $W_{decay}^{thr} \simeq (8\nu_i/\mu kT)$ $\cdot(3|k|\Delta k_o)$. For $T_e/T_i \simeq 1$, $\nu_i/\mu kT \simeq 2$ and $W_{decay}^{thr} \simeq 50 |k|k_o \gg 50 k_o \Delta k_o \simeq 10 k_o^2 \gg 10^{-5}$, and thus the OTSI has the lower threshold.

The foregoing discussion has shown that both the pump and daughter waves are produced with wavenumbers aligned along the direction of the electron beam. In addition, none of the competing modulational instabilities will have lower thresholds than the OTSI, at least for the parameter range appropriate for type III bursts. The derivation of the OTSI growth rate, γ_{OTS}, has been given in Papadopoulos et al. (1974) and Smith et al. (1976). If $\delta \equiv \omega_{ek} - \omega_{ek'}$ is the frequency shift between the pump wave with wavenumber k_o and the daughter wave with wavenumber k', then for the OTSI $\delta < 0$ and γ_{OTS} is given by

$$\gamma_{OTS}^2 (k_o, k') = -(1/2)\cdot(\omega_A^2 + \delta^2) +$$

$$\{(\omega_A^2 + \delta^2)^2 - 4\delta^2 \omega_A^2[1 + W(k_o)\omega_e^2/(2\delta\omega_{ek'})]\}^{(1/2)} \qquad (5)$$

where $(\omega_A/\omega_e)^2 = (m/M)\cdot(k'\lambda_e)^2/[1+(k'\lambda_e)^2]$. In the absence of collisions the threshold for the OTSI is $W_T = -2\delta_{ek}'/\omega_e^2$.

Once the OTSI is excited, the pump waves in resonance with the electron beam will couple to daughter waves having larger wavenumbers. These daughter waves will no longer resonate or exchange energy with the electrons. It is important to note that other modulational instabilities, e.g. direct collapse, do not have this property of scattering the pump waves out of resonance with the electrons. Thus the beam plasma instability would continue to amplify the pump waves until the threshold for the OTSI is reached. Once the OTSI begins to dominate the wave-wave interactions, stabilization of the electrons against catastrophic energy loss will proceed as outlined.

The OTSI is not by itself adequate to explain the lack of energy loss of the electron streams. Excitation of the OTSI itself would result in a marginally stable situation in which wave energy densities of order $W_T \simeq 10^{-5}$ remain resonant with the beam, and over large distances could decelerate it. However, in the presence of such energy densities, an approximately isothermal plasma will produce density depressions $\delta n/n \simeq \widetilde{W}/nT$ due to pressure balance ($nT + \widetilde{W}$ = const.). (The $\sim$ denotes dimensioned variables.) These will modify the local plasma frequency, and hence the local Bohm-Gross dispersion relation of both the pump and daughter Languir waves. The nonlinear dispersion relation will be approximately given by $\omega_k = \omega_e[1 +(3/2)k^2\lambda_e^2 - (1/2)\delta n/n] \simeq \omega_e[1 + (3/2)k^2\lambda_e^2 - (1/2)\widetilde{W}/nT]$. When $W \simeq (k\lambda_e)^2$ this correction term will become important and will cause δ and W_T to decrease. Then virtually all the energy initially resonant with the energetic electrons will be rapidly transferred to larger wavenumbers, out of resonance. A more rigorous derivation of these effects can be found in Kaw, Lin, and Dawson (1967) and Smith et al. (1978).

An additional effect which must be included is anomalous resistivity (Dawson and Oberman 1963, and Dawson 1968). Once nonthermal levels of ion fluctuations are excited, a high frequency anomalous resistivity is produced which causes long wavelength Langmuir waves to cascade to shorter wavelengths. In the presence of a three-dimensional spectrum of correlated, nonthermal ion fluctuations, the scalar impedance at frequencies near ω_e is

$$Z(\omega \simeq \omega_e) \simeq \frac{(2\pi)^3(\delta n_i/n_o)^2}{36\pi k_i^2 \Delta k} \frac{\omega(k_i\lambda_e)}{\omega_e^2 \lambda_e^3}$$

The effective electron-ion collision time, τ_c, is related to Z by

$$Z(\omega) = \frac{4\pi i \omega}{\omega_e^2}(1 - i/\omega\tau_c)$$

so that

$$\gamma_{NL}/\omega_e \equiv 1/(\omega_e \tau_c) \simeq \frac{\pi}{2} \frac{(\delta n_i/n_o)^2}{(k_i \lambda_e)^2}$$

$$= \Sigma \frac{[S(k_i) - S_o(k_i)]}{(k_i \lambda_e)^4}$$

where $S_o(k)$ is the thermal noise level of the ion spectrum and $S(k)$ is the total electric field energy density in ion waves at wavenumber k.

Landau damping of Langmuir waves by the thermal solar wind electrons has also been included. The total electron distribution function, f_T, is the sum of the solar wind and beam components. Thus Landau damping, the linear beam plasma instability, and reabsorption caused by evolution of the electron beam to lower velocities can all be described in terms of a single damping decrement, γ_L, where

$$\frac{\gamma_L}{\omega_e} = \pi \frac{\beta^3}{|\beta|} \frac{\partial f_T}{\partial \beta} \bigg|_{\beta = \omega/kc}$$

The evolution of the growth, spectral transfer, and damping of the Langmuir and ion waves in the presence of the time evolving electron beam can now be described in terms of the various transfer rates γ_L, γ_{OTS}, and γ_{NL}. The complete set of rate equations has been given before (Smith et al. 1976, 1978) and we will not repeat them here.

Before discussing the results of our calculations, it is necessary to briefly review some properties of the model we constructed for the electron beam (v. Smith et al. 1976, 1978 for more details). The beam distribution model is a semi-empirical one based on in situ particle observations at 1 AU (Lin 1974, Lin, Evans, and Fainberg 1973). In general it is very misleading to attempt to construct a beam model based solely on its interaction with the self-consistently produced Langmuir turbulence because of the importance of scattering by magnetic irregularities even for events called "scatter-free". One such event was observed on May 16, 1971 (Figure 2). Of particular importance is the fact that it took more than 15 minutes for the peak of the spectrum to evolve from 80 eV to 33 keV, implying that reabsorption of Langmuir waves by the electron stream is unimportant. Theories which do not accurately model this behavior will, of necessity, conclude that reabsorption is important. As an example of the latter, Magelssen and Smith (1977) use a model for the electron beam evolution for the May 16 event which takes less than 5 minutes to evolve from 80 to below 20 keV. We will return to this subject below when discussing the results of our numerical calculations.

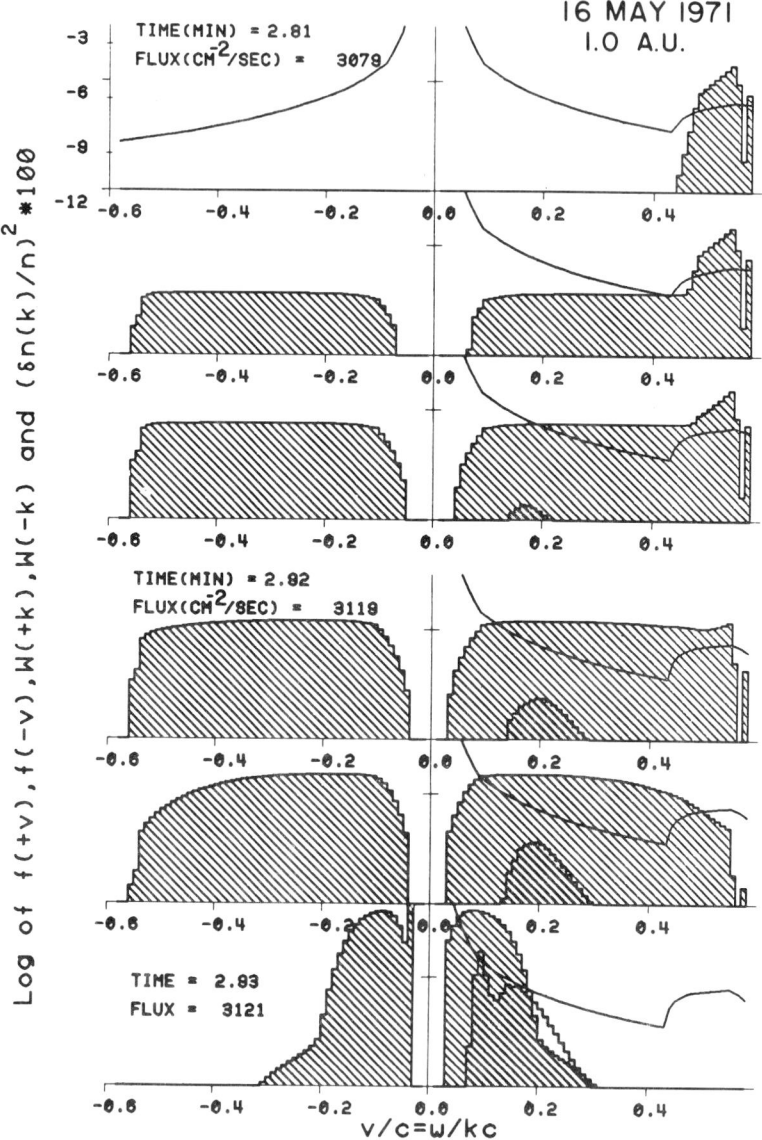

Figure 3. Results of a numerical solution of the rate equations that describe the OTSI. Parameters were chosen to model the May 16, 1971 event at 1 AU. The top panel (3a) shows the distribution function f_T of the solar wind plus the linerarly unstable beam. Langmuir waves (diagonally striped histograms) are shown near W_T (3a), and during subsequent stages of excitation and stabilization of the OTSI (3b-3f). Ion oscillations are depicted by the gray shading. Times computed from the start of the numerical calculations and the calculated values of the electron flux are given in 3a, 3d, and 3f.

NUMERICAL CALCULATIONS

The numerical evaluation of the rate equations can be carried out at any point in the interplanetary medium at which the density and temperature of the ambient solar wind can be estimated. Typically, we chose distances between 0.1 and 1.0 AU, and assumed that the ambient density varied as r^{-2}. At a given location the calculation began (t = 0) with the arrival of energetic electrons with velocities of about 0.7c. The exact velocity distribution being given by the beam evolution model. As an example, consider the burst on May 16, 1971. The local plasma frequency at 1 AU on that date was about 30 kHz and electrons with energies above 100 keV were first observed at 1305 UT when the radiometer on IMP-6 first detected radio noise at 55 kHz ($\simeq 2\omega_e/2\pi$). The radio noise increased in intensity until 1335 UT, and little further evolution was observed in the electron spectrum after that time. From Figure 2 we can see that the distribution function had a positive slope below the peak energy. The other parameters needed for the numerical model were the path length traversed by the electron beam, taken to be 1.5 AU; the ratio of the beam to solar wind density η estimated to be 5×10^{-6}; and the spectral index $\zeta \simeq 4.6$ of the power-law portion of f_T.

The solution is shown in Figure 3, where the logarithms of $f_T(\beta)$, $W(\pm k)$, and $(\delta n/n)^2 = S(k)/(k\lambda_e)^2$ are plotted against $v_{ph} = \omega_{ek}/kc$ at various times.

Initially, the linearly unstable beam produces resonant plasma waves (indicated by cross hatching in Figure 3a) that grow until the OTSI threshold is reached (Figure 3a). Aperiodic ion waves are then excited (gray shading) at the rate γ_{OTS}, as are shorter wavelength "daughter" Langmuir waves (Figure 3b-d). The combined effects of nonlinear changes in the Bohm-Gross dispersion relation and anomalous resistivity then complete the decoupling of the electron beam from the Langmuir turbulence (Figure 3d-f). In our calculations the collapse to short wavelengths ceases when Landau damping by the thermal solar wind electrons balances the spectral transfer. No further energy exchange will then take place. Gradually the ion fluctuations and Langmuir waves will simultaneously decay back to thermal levels whereupon the linear instability will again be excited, and the process will cyclically repeat until the electron beam has merged with the ambient solar wind distribution and no positive slope exists to $f_T(\beta)$.

It is important to note that the total elapsed time between the onset of the OTSI and its final stabilization was little more than 0.1s, during which the electron distribution was essentially constant. Therefore, neither reabsorption nor quasilinear relaxation can be important.

Similar calculations were performed at 0.5 and 0.1 AU and for the type III bursts observed on May 25, 1972 and February 28, 1972; the results are similar to those described here and are reported in

Goldstein et al. (1978). In all cases stabilization and decoupling of
the electron beam from the Langmuir turbulence is due to excitation of
the OTSI.

ADDITIONAL THEORETICAL RESULTS

We now turn to the question of why type III bursts are preferentially observed at the second harmonic of the local plasma frequency. Much of this discussion is based on a recent paper by Papadopoulos and Freund (1978).

From a comparison of Figure 5a and f, one sees that the long wavelength pump waves have collapsed into shorter wavelength daughter waves. In configuration space these short wavelength structures are solitons (Manheimer and Papadopoulos 1975), whose spatial extent in the direction parallel to the magnetic field can be estimated to be about $50\lambda_e$, with an energy density, W, of nearly 10^{-2}. Such structures are very difficult to observe with present spacecraft instrumentation. In a 400 km/s solar wind, a 350 m ($50\lambda_e$) soliton is convected past a 30m dipole antenna in little more than a millisecond. This must be compared to the electronic response times of plasma wave experiments typically no faster than 20 ms (Gurnett, private communication).

Papadopoulos and Freund (1978) found that the total volume emissivity of a soliton, integrated over solid angle is

$$J(2\omega_e) = \frac{3\sqrt{3}}{8} \left(\frac{v_e}{c}\right)^4 \frac{cE_o^2}{8\pi\Delta Z} \left(\frac{1}{k_o L}\right)^2 \tag{6}$$

where Δz is the parallel dimension of the linearly unstable wave-packet, $k_o = \sqrt{3}\omega_e/c$ is the wavelength of the electromagnetic wave at $2\omega_e$, v_e is the thermal electron velocity, E_o is the electric field in the soliton, and L is the dimension of the soliton transverse to the magnetic field. (Papadopoulos and Freund argue that collapse is likely only in the parallel dimension, and that L should be greater than the electron Larmor radius.) Equation (6) is valid for $k_o^2 L^2 \gg 4$, a good approximation throughout the interplanetary medium. The intensity of emission outside a spherical shell of radius R and thickness ΔR centered on the sun is (Gurnett and Frank 1975) $I = JR(2\omega_e/2\pi)$. For the May 16 burst at the time of soliton formation (Figure 3f), $I(2\omega_e) \simeq 1 \times 10^{-17}$ W m^{-2} s^{-1}, close to the peak intensity observed at 55 kHz.

The correlation between the radio and electron fluxes also has a straightforward explanation. In Figure 4 we plot γ_{OTS} against $W(k_o)$--

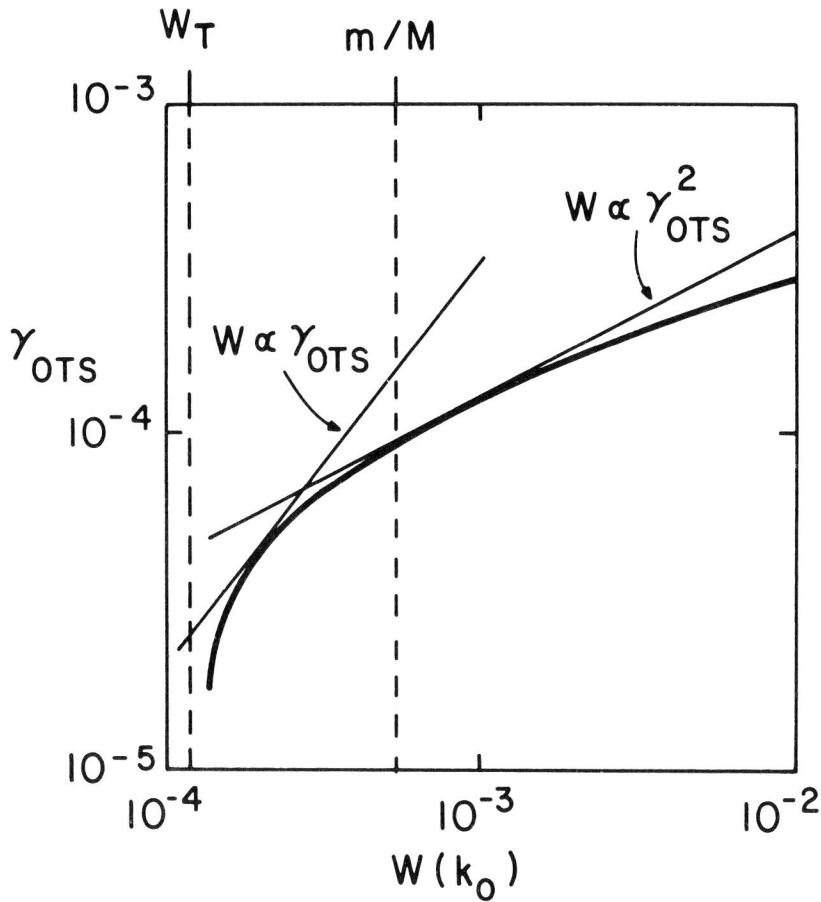

Figure 4. γ_{OTS} versus $W(k_o)$ from equation (5). Note the approximate scaling relations for $W_T < W(k_o) < m/M$ and $W(k_o) > m/M$.

equation (5). Papadopoulos (1975) and Rowland (1977) have shown that one can find an approximate relationship between $W(k_o)$ and γ_{OTS} near and above the threshold:

$W(k_o) \sim \gamma_{OTS}$ $\qquad\qquad\qquad\qquad\qquad m/M > W(k_o) > W_T$

$W(k_o) \sim \gamma_{OTS}^2$ $\qquad\qquad\qquad\qquad\qquad W(k_o) > m/M$

These regimes are noted in Figure 4.

When the OTSI stabilizes the linear instability, $\gamma_{OTS} \simeq \gamma_L$, and $\gamma_L \propto n_b$. In addition, from equation (6), $I(2\omega_e) \propto W(k_o)$ the electron flux, $J_E \propto n_b \langle v \rangle$, where $\langle v \rangle$ is the <u>peak</u> of the electron distribution, so that with $n \propto \langle v \rangle^{-\zeta+1}$ one has

$$I(2\omega_e) \propto J_E^{(1-\zeta)/(2-\zeta)} \qquad m/M > W(k_o) > W_T$$

$$I(2\omega_e) \propto J_E^{2(1-\zeta)/(2-\zeta)} \qquad W(k_o) > m/M \qquad (7)$$

In Figure 5 we compare the observations of Fitzenreiter et al. (1976) with the predictions of the theory from equations (7). The three bursts

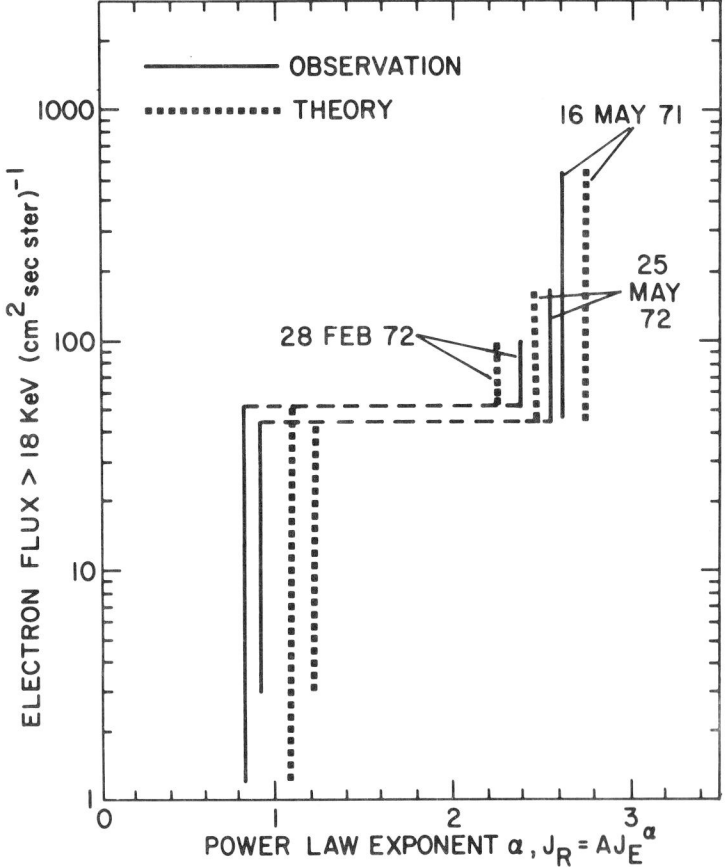

Figure 5. After Fitzenreiter <u>et al</u>. (1976). The electron flux and power law exponent α from the relationship $I \propto J_E^{\alpha}$ are shown for the three events for which numerical calculations could be performed. Observed and computed values of α are plotted.

shown in Figure 5 are the only ones for which the electron observations were sufficiently detailed to permit estimation of ζ, L, and η--the parameters used in our beam evolution model. It should be emphasized that the correlation between J_E and I cannot be explained using weak turbulence theories. The excellent agreement shown in Figure 5 is confirmation that type III bursts are stabilized by the OTSI and that solitons radiate electromagnetic radiation proportional to W and not W^2.

Although the scaling $W(k_0) \sim \gamma_{OTS}$ would seem to imply that α should not be less than 1, Figure 4 indicates that very close to threshold, $W(k_0) \sim \gamma_{OTS}^\nu$ with $\nu < 1$. Thus, for bursts such as the ones on February 28 and May 25, 1972 which initially only weakly excite the OTSI, values of $\alpha < 1$ are quite reasonable.

Thus far we have tacitly assumed that because the electron beam becomes decoupled from the radiation field, no significant energy loss will occur. Smith et al. (1978) have investigated this in some detail; we only summarize that discussion here.

If the beam is injected near the solar surface, the total energy lost by the beam in propagating to the point R is given by

$$\Delta E = \int_{R_0}^{R} dr A(r) \int_{t_1(r)}^{t_2(r)} dt \, \frac{d\widetilde{W}(r,t)}{dt} \tag{8}$$

where $A(r)$ is the source area at r, and $t_1(r)$, and $t_2(r)$ are the times at which the instabilities at r begin and end. Because all the beam energy loss occurs in the resonant region until the onset of the OTSI, one can assume that it takes place at the steady rate $dW/dt = W_T \tau_0$, where W_T is taken to be $W_0 \exp(\gamma_L \tau_0)$.

When equation (8) was evaluated, Smith et al. (1978) found that $\sim 90\%$ of the energy loss occurred in the inner corona, and that $\Delta E = 10^{30}$ W (ergs). With $W \simeq 10^{-4}$, the exciter loses some 10^{26} ergs in leaving the corona. The total energy in the type III exciter has been estimated to be $\simeq 10^{28}$ ergs (Lin 1971). Thus, the exciter will typically lose only a few percent of its energy.

One additional consequence of this energy-loss calculation was that it provides an explanation for why the electron streams appear to have such well-defined velocities, of order c/3 at high frequencies (Wild and Smerd 1972), decreasing to c/2 or less at low frequencies (Evans, Fainberg and Stone 1973).

The peak intensity at any frequency is reached just before the linear beam-plasma instability stops at that frequency, for at that time the density in the energetic electron beam is maximum. It is this <u>peak</u> velocity which is directly deduced from the observed frequency drift rates as being the nominal velocity of the beam.

Smith et al. (1978) found that in the inner corona the peak velocity when the linear instability stopped was v_p = 0.3c, while near 1 AU, because the ambient solar wind is cooler, v_p was about 0.2c. This suggests that the nominal velocity (c/3) is not characteristic of electron acceleration, but rather reflects the evolution of the particle spectrum. In addition, the observations of Evans et al. (1973) do not necessarily imply that the exciter is decelerated between 0.05 AU - 1 AU, but rather reflects the decrease in the temperature of the solar wind with increasing heliocentric distance.

We have reviewed a theory of type III bursts which is able to account for many seemingly diverse aspects of the phenomenon, in particular we have offered an explanation of the small energy losses of the exciter, the predominance of radiation at $2f_{pe}$, the characteristic exciter velocities of 0.2-0.3c, and the correlation between electron and radio fluxes.

REFERENCES

Bardwell, S., and Goldman, M. V.: 1976, Astrophys. J., 209, 912.
Dawson, J.: 1968, "Advances in Plasma Physics", Vol. 1, ed. A. Simon and W. B. Thompson (Interscience: New York), pp. 1-66.
Dawson, J., and Oberman, C.: 1963, Phys. Fluids, 6, 394.
Fitzenreiter, R. J., Evans, L. G., and Lin, R. P.: 1976, Solar Phys., 46, 437.
Goldstein, M. L., R. A. Smith, and Papadopoulos, K.: 1978, NASA TM 79684.
Gurnett, D.A., and Frank, L. A.: 1975, Solar Phys., 45, 477.
Kaw, P. K., Lin, A. T., and Dawson, J. M.: 1973, Phys. Fluids, 16, 1967.
Lin, R. P.: 1973, "High Energy Phenomena on the Sun--Symposium Proceedings", ed. R. Ramaty and R. G. Stone (NASA SP 342), p. 439.
Lin, R. P.: 1974, Sp. Sci. Rev., 16, 189.
Lin, R.P. Evans, L. G., and Fainberg, J.: 1973, Astrophys. Letters, 14, 191.
Lin, R. P., and Hudson, H. S.: 1971, Solar Phys., 17, 412.
Magelssen, G. R. and Smith, D. F.: 1977, Solar Phys., 55, 211.
Manheimer, W. M., and Papadopoulos, K.: 1975, Phys. Fluids, 18, 1397.
Nicholson, D. R., Goldman, M. V., Hoyng, P. and Weatherall, J. C.: 1978, Astrophys. J., 223, 605.
Nicholson, D. R., and Smith, D. F.: 1979, these proceedings.
Papadopoulos, K.: 1975, Phys. Fluids, 18, 1769.
Papadopoulos, K. and Freund, H.: 1978, Geophys. Res. Lett., 5, 881.
Papadopoulos, K., Goldstein, M. L., and Smith, R. A.: 1974, Astrophys. J. 190, 175.
Rowland, H. L.: 1977, Ph.D. Thesis, Univ. of Md., College Park, MD.
Smith, R. A., Goldstein, M. L., and Papadopoulos, K.: 1976, Solar Phys., 46, 415.
Smith, R. A., Goldstein, M. L., and Papadopoulos, K.: 1978, NASA TM 78079.
Sturrock, P. A.: 1964, "AAS-NAS Symposium of the Physics of Solar Flares", ed. W. N. Hess, NASA SP 50, p. 357.
Wild, J. P., and Smerd, S. F.: 1972, Ann. Rev. Astr. and Astrophys 10, 159.

PART IV

IONOSPHERIC F-REGION IRREGULARITIES

A REVIEW OF RECENT RESULTS ON SPREAD F THEORY

S. L. Ossakow
Geophysical & Plasma Dynamics Branch
Plasma Physics Division
Naval Research Laboratory
Washington, D.C. 20375

Ionospheric Spread F was discovered some four decades ago. Yet only in the past few years has significant progress been made in the theoretical explanation of such phenomena. In particular, considerable effort has been expended to explain equatorial Spread F and the attendant satellite signal propagation scintillation phenomena. The present review dwells mainly in this low latitude area. The various linear plasma instabilities thought to initiate equatorial Spread F will be discussed. Recent theoretical and numerical simulation studies of the nonlinear evolution of the collisional Rayleigh-Taylor instability in equatorial Spread F will be reviewed. Also, analytical studies of rising equatorial Spread F bubbles in the collisional and collisionless Rayleigh-Taylor regime will be discussed, as well as the nonlinear saturation of instabilities in these two regimes. Current theories on very small scale ($\leqslant$ 10 meters) size irregularities observed by radar backscatter during equatorial Spread F and their relation to the larger wavelength scintillation causing irregularities will be discussed. Application of turbulence theory to equatorial Spread F phenomena will be reviewed. Remaining problems to be dealt with at equatorial latitudes will be summarized.

I. INTRODUCTION

Spread F, as exhibited by diffuse echoes on ionograms, was discovered some forty years ago by Booker and Wells (1938). Until recently, there has been only much statistical data concerning Spread F. However, in the last few years with advances in radar backscatter measurements, in situ measurements, and theoretical and numerical simulation techniques a clearer picture of the fundamental plasma instability mechanisms causing equatorial Spread F phenomena has been evolving. It should be emphasized that this paper will deal only with equatorial Spread F theory and, in addition, dwell only on those theories using plasma mechanisms as a basis. For experimental results, the works of Farley et al. (1970), Dyson et al. (1974), Kelley et al. (1976), Woodman and La Hoz (1976), Morse et al. (1977), and McClure et al. (1977) should serve as good references for the interested reader. Also, see the review by Kelley in this volume.

In speaking about Spread F one should at least show its basic manifestation on an ionogram (see Fig. 1). After all, the terminology Spread F emerged from the results of ionosonde traces such as those exhibited in Fig. 1. The multiple traces are caused by magnetic field aligned irregularities. If there were no irregularities, a single crisp trace rising to the right would be exhibited on the ionogram. At this point we want to remember the basic equatorial geometry. Figure 2 exhibits the basic equatorial nighttime ionospheric F region geometry, i.e., the geometry under which equatorial Spread F occurs. $N(y)$ represents the background electron density as a function of altitude (y). Gravity, $\underline{g}$, points down, the ambient magnetic field, $\underline{B}$, is horizontal (pointing north) and $\underline{k}$ represents a horizontal perturbation (in the westward direction). The maximum in the electron density profile is the F peak. The underside of the profile steepens at night due to chemical recombination effects and electrodynamic forces. The E region has been severely reduced by chemical recombination and plays a negligible role. To a plasma physicist this geometry is a classical flute mode geometry and one might expect this equatorial ionospheric geometry to be unstable to a variety of plasma instabilities. However, until recently, one of the basic difficulties was getting the unstable irregularities to the topside (i.e., above the F peak) when they were initiated on the bottomside. One must remember that the experimental evidence has exhibited both top and bottomside irregularities.

At this point a general brief review will be given of equatorial Spread F (ESF) theories. First, we will discuss the linear theories. Dungey (1956) was the first to suggest that ESF was initiated on the bottomside by a Rayleigh-Taylor instability. Dagg (1957) suggested that the ESF phenomena was due to E to F region coupling, i.e., irregularities in the E region coupled up to the F region. In 1959, Martyn (1959) was the first to suggest that ESF was a manifestation of the $\underline{E} \times \underline{B}$ gradient drift instability. Calvert (1963) proposed that the downward motion of the neutral atmosphere at night was responsible for ESF. This mechanism is essentially equivalent to the $\underline{E} \times \underline{B}$ instability because of the relative motion between ions and neutrals in determining the instability. All of these previously invoked linear instability mechanisms could only explain the formation of bottomside irregularities. The collisional Rayleigh-Taylor instability with field line averaging was also proposed (Balsley et al., 1972; Haerendel, 1974) as a linear instability mechanism. By averaging (integrating) the density along the magnetic field, the total electron content profile becomes steeper on the bottomside and its peak is raised in altitude with respect to the local electron density peak. This would allow the linear mechanism to operate to slightly higher altitudes ($\sim$ 100 km greater), but still would not explain the

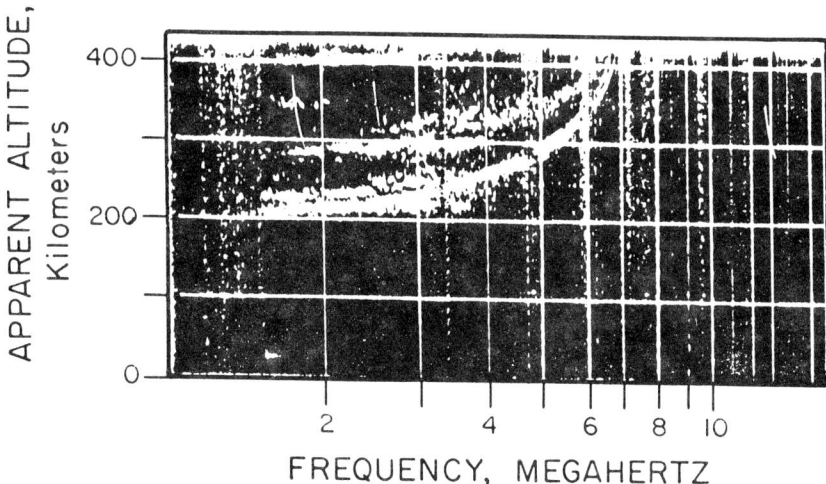

Figure 1. Ionogram recorded at Natal, Brazil (equatorial station), on November 18, 1973 at 2122-2123 UT. This figure was taken from Hudson and Kennel (1975).

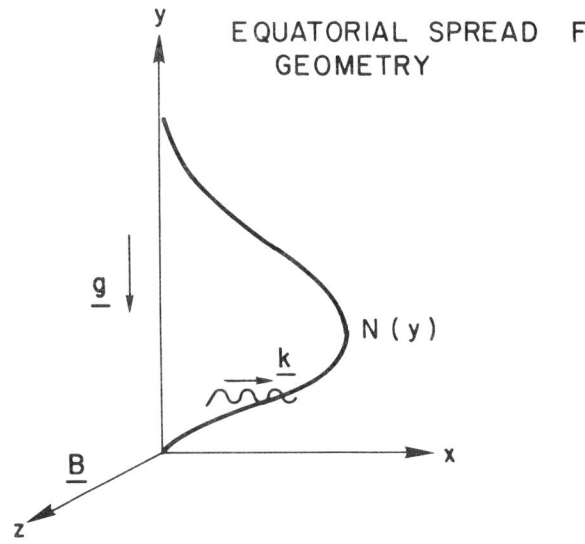

Figure 2. Basic equatorial Spread F geometry.

existence of irregularities above this "new peak". Hudson and Kennel (1975) pointed out the importance of the collisional drift mode in ESF in the wavelength regime 30m - 100m. This mode could be excited on both the top and bottomside but still would not explain the longer wavelengths. In their paper, finite larmor radius (FLR) corrections were also applied to the collisionless and collisional Rayleigh-Taylor instability.

Several nonlinear theories have been invoked to explain the different ESF observations. For example, Hudson et al. (1973) suggested that the very smallest scale ($\lesssim$ 10m) irregularities (e.g., those seen by radar coherent backscatter) were due to a two step process. In this prescription a longer wavelength instability sets up the driving conditions for the shorter wavelengths to become unstable. This is similar in spirit to the successful two step theory (Sudan et al., 1973) proposed for Type II equatorial E region electrojet irregularities. Haerendel (1974) suggested that the range of wavelengths (many kilometers down to centimeters) exhibited by ESF phenomena was due to a multi-step process. This scenario is as follows: (i) the collisional Rayleigh-Taylor (R-T) instability with horizontal wavevectors is driven by gravity and the background, zero order electron density gradient scale length on the bottomside; then (ii) the $\underline{E} \times \underline{B}$ gradient drift instability with vertical wavevectors arises due to the horizontal density, large amplitude variations set up by the collisional R-T instability; then (iii) the inertia (collisionless) dominated R-T instability arises; and finally (iv) kinetic drift waves grow upon these irregularities after they reach large amplitude. Chaturvedi and Kaw (1976) interpreted the k^{-2} measured power spectrum of the ESF plasma density irregularities in terms of a two step theory. In this theory longer wavelength R-T modes couple to kinetic collisional drift waves in such a manner that the mode coupling results in the observed k^{-2} spectrum.

A major breakthrough was made by Scannapieco and Ossakow (1976) who performed a nonlinear numerical simulation of the collisional R-T instability for ESF geometry. The simulation results showed that the collisional R-T instability generated irregularities and bubbles (plasma density depletions) on the bottomside of the F region which subsequently rose beyond the F peak by nonlinear polarization induced $\underline{E} \times \underline{B}$ forces. This was the first theoretical result to explain how long wavelength irregularities could appear on both the bottomside and topside of the F region. The bubble phenomena was in accord with the recent observations (Kelley et al., 1976; McClure et al., 1977, Woodman and La Hoz, 1976) of plasma density depletions. An analytical nonlinear mode-mode coupling theory for the coherent development of the collisional R-T instability was performed by Chaturvedi and Ossakow (1977). This theory suggested that vertical modes would be dominant and result in a k^{-2} power spectrum. Hudson (1978) extended the previous results to the collisionless R-T regime and reached similar conclusions. Analytical models for the rise of collisional and collisionless R-T ESF bubbles, in analogy with fluid bubbles, was presented by Ott (1978). At the same time, Ossakow and Chaturvedi (1978) presented analytical models for the rise of collisional R-T ESF bubbles within the context of the electrical analogy with barium clouds.

Costa and Kelley (1978a,b) suggested that coherent steepened structures and not turbulences would give a k^{-2} power spectrum. Moreover, these sharp gradients could cause small scale sizes ($\sim$ 20m) by collisionless low frequency (much less than the ion gyrofrequency, Ω_i) kinetic drift waves via a two step process. Their analysis was a linear one carried out on a nonlinear state, i.e., one achieves the steepened gradients by nonlinear processes and then one performs linear theory on this state. Kelley and Ott (1978) suggested that the ESF bubbles, in the collisionless R-T regime, generate a wake with vortices. They then applied two dimensional fluid turbulence theory to the model. This resulted in the development of turbulence at shorter and longer wavelengths than the bubble size. This in turn led to a prediction of k^{-1} for the power spectrum (which does not appear to be in agreement with existing experimental observations) in the range $L_S^{-1} < k < L_D^{-1}$, where L_S is the stirring (bubble) size and L_D is a dissipation length cutoff. In a continuation of the numerical simulation work, Ossakow et al. (1978) showed a more rapid ESF development and higher bubble rise velocities resulting from sharper bottomside background electron density gradients and higher altitudes of the F peak. In Huba et al. (1978) very small scale (wavelengths $\sim$ 1m and 36cm) irregularities are reported. A two step process, utilizing high frequency ($\gtrsim \Omega_i$) kinetic drift cyclotron or lower hybrid drift instabilities, is invoked to explain them. Linear theory for these instabilities, in the ESF environment, was performed on the nonlinear state with encouraging results.

The above introductory remarks glaringly point out that much work and significant progress in the theoretical area of equatorial Spread F has been accomplished in the past few years (indeed just look at the number of publications during 1978 along). Notwithstanding the recent successes, much work still needs to be done. Indeed, the theoretical and numerical simulation efforts in ESF are continuing along a hot and heavy path. Section II of this paper presents outlines of some of the theoretical efforts briefly mentioned in the preceding paragraphs. Given the length limitations, it would be exceedingly difficult to outline all of the theoretical works mentioned or even to present all of the details of a few works. Section II, hopefully, will wetten the reader's appetite to read the referenced works. Section III presents a summary concerning ESF theory.

II. THEORY

In this section we present some representative theoretical and numerical simulation works with the appropriate references.

a. General. The basic plasma fluid equations applicable to the equatorial Spread F ionosphere are as follows:

$$\frac{\partial n_\alpha}{\partial t} + \nabla \cdot (n_\alpha \underline{V}_\alpha) = P - \nu_R n_\alpha \tag{1}$$

$$-T_e \nabla n_e - en_e(-\nabla \phi + \frac{\underline{V}_e \times \underline{B}_o}{c}) = 0 \tag{2}$$

$$m_i n_i (\frac{\partial}{\partial t} + \underline{V}_i \cdot \nabla) \underline{V}_i = -T_i \nabla n_i$$

$$+ en_i(-\nabla \phi + \frac{\underline{V}_i \times \underline{B}_o}{c}) + m_i n_i \underline{g} - m_i n_i \nu_{in} \underline{V}_i \tag{3}$$

$$\nabla \cdot \underline{J} = 0 \tag{4}$$

$$\underline{J} = ne (\underline{V}_i - \underline{V}_e) \tag{5}$$

In the above equations the subscript α denotes species (e is electron, i is ion), n is density, $\underline{V}$ is velocity, P is the production, ν_R is the chemical recombination rate, T is temperature, ∇ is the gradient operator, $\underline{B}_o$ is the ambient magnetic field (taken to be uniform), e is the electronic charge, c is the speed of light, m is mass, $\underline{g}$ is gravity, ν_{in} is the ion-neutral collision frequency, $\underline{J}$ is current, and the electrostatic approximation has been made where $\underline{E} = -\nabla \phi$. Equation (1) is the continuity equation, (2) and (3) are the electron and ion momentum equations, respectively, (4) is the divergence of the current and (5) is the current equation. What we have in mind is to apply the set of equations (1) - (5) to the two dimensions perpendicular to $\underline{B}_o$ at the geomagnetic equator, making various approximations.

Assuming a harmonic perturbation dependence of the form exp-i ($\underline{k} \cdot \underline{x}_\perp - \omega t$), where $\perp$ denotes perpendicular to $\underline{B}_o$ ($\underline{k}$ is horizontal) and linearizing equations (1) - (5) we obtain for the linear growth rate

$$\gamma = \left[-\nu_{in} + (\nu_{in}^2 - 4g \cdot \frac{\nabla n_o}{n_o})^{\frac{1}{2}}\right]/2 - \nu_R \qquad (6)$$

$$\omega \equiv \omega_r + i\gamma$$

which reduces to

$$\gamma = \begin{cases} \dfrac{g}{\nu_{in} L} - \nu_R, & \nu_{in}^2 \gg 4g/L \qquad (7a) \\[6pt] \left(\dfrac{g}{L}\right)^{\frac{1}{2}} - \nu_R, & \nu_{in}^2 \ll 4g/L \qquad (7b) \end{cases}$$

$$|L| = \left|\frac{\nabla n_o}{n_o}\right|$$

in the collision dominated and inertia dominated regimes, for $kL \gg 1$. Equations (7a) and (7b) represent the R-T instability, including recombination damping, in the respective regimes. (It should be noted that this only represents instability on the bottomside of the F region where the first term in (7a) and (7b) is positive.)

b. *2D Computer Simulation Results.* In this section we will outline the two dimensional ($\perp B_o$) computer simulation results (Scannapieco and Ossakow, 1976; Ossakow et al., 1978). The simulations follow the nonlinear evolution of the collisional R-T instability; consequently, in eqn. (3) inertial terms (i.e., the left hand side of the eqn.) are neglected. Furthermore, one takes for the F region $\nu_{in}/\Omega_i \ll 1$ ($\Omega_i = eB_o/m_i c$) and the quasi-neutrality assumption is made, i.e., $n_e \approx n_i \approx n$. Equations (1) - (5) then become

$$\frac{\partial n_\alpha}{\partial t} + \nabla \cdot (n_\alpha \underline{V}_\alpha) = -\nu_R (n_\alpha - n_{\alpha o}) \qquad (8)$$

$$\underline{V}_e = \frac{c}{B_o} \underline{E} \times \hat{z} \qquad (9)$$

$$\underline{V}_i = \left(\frac{g}{\Omega_i} + \frac{c}{B_o}\underline{E}\right) \times \hat{z} + \frac{\nu_{in}}{\Omega_i}\left|\frac{g}{\Omega_i} + \frac{c}{B_o}\underline{E}\right| \qquad (10)$$

$$\nabla \cdot \underline{J} = 0, \quad \underline{J} = ne(\underline{V}_i - \underline{V}_e) \qquad (11)$$

where one has taken $T_e = T_i = 0$ for simplicity (see Ossakow et al., 1978), $\underline{B}_o = B_o \hat{z}$ and the subscript o refers to equilibrium quantities.

Making the electrostatic assumption

$$\underline{E} = -\nabla\phi \tag{12}$$

and breaking the potential into an equilibrium and a perturbed quantity,

$$\phi = \phi_o + \tilde{\phi} \tag{13}$$

eqns. (8) - (11) become

$$\frac{\partial n}{\partial t} - \frac{c}{B_o}(\nabla\tilde{\phi} \times \hat{z}) \cdot \nabla n = -\nu_R(n-n_o) \tag{14}$$

$$\nabla \cdot (\nu_{in} n\nabla\tilde{\phi}) = \frac{B_o}{c}(\underline{g} \times \hat{z}) \cdot \nabla n \tag{15}$$

where eqns. (14) and (15) are taken to be two dimensional ($\perp \underline{B}_o$). Linearizing eqns. (14) and (15), taking a horizontal perturbation results, as in eqn. (7a) but in a more illustrative form, in the linear growth rate

$$\gamma = -\frac{g}{\nu_{in}} \cdot \frac{\nabla n_o}{n_o} - \nu_R$$

This clearly shows that only the bottomside of the F region where ∇n_o is positive can be linearly unstable (and only if the first term $>\nu_R$).

Equations (14) and (15) were solved numerically using a vertical mesh spacing of $\Delta y = 2$km and total y extent of 200km, and an east-west, horizontal mesh spacing $\Delta x = 200$m and a total horizontal extent of 8km. Realistic profiles of ν_{in} and ν_R as a function of altitude were utilized. The system was initialized with a perturbation of a few percent in the horizontal (x), east-west direction with a wavelength ~ 3km and the evolution in time of (14) and (15) was followed for different background electron density profiles. Figure 3 shows the results for a background electron density, n_o, profile with an F peak at 354km and a minimum bottomside background electron density gradient scale length, $L \sim 10$km. In this case at $t = 4000$ sec a bubble (plasma density depletion) is clearly forming and beginning to rise in the central portion of the mesh (note $n \equiv n_o + \tilde{n}$). The isodensity contours are such that the maximum absolute value of the enhancement or depletion is in the center and the contours decrease (in absolute percentage) as one goes toward the outer contours. At $t = 4000$ sec the maximum depletion within the bubble is 54% and the maximum enhancement over the mesh is 84%. At $t = 8000$ sec we notice that the bubble has reached the altitude of the F peak, with the innermost contour of the rising bubble representing a 41% depletion. At $t = 10^4$ sec the main bubble is

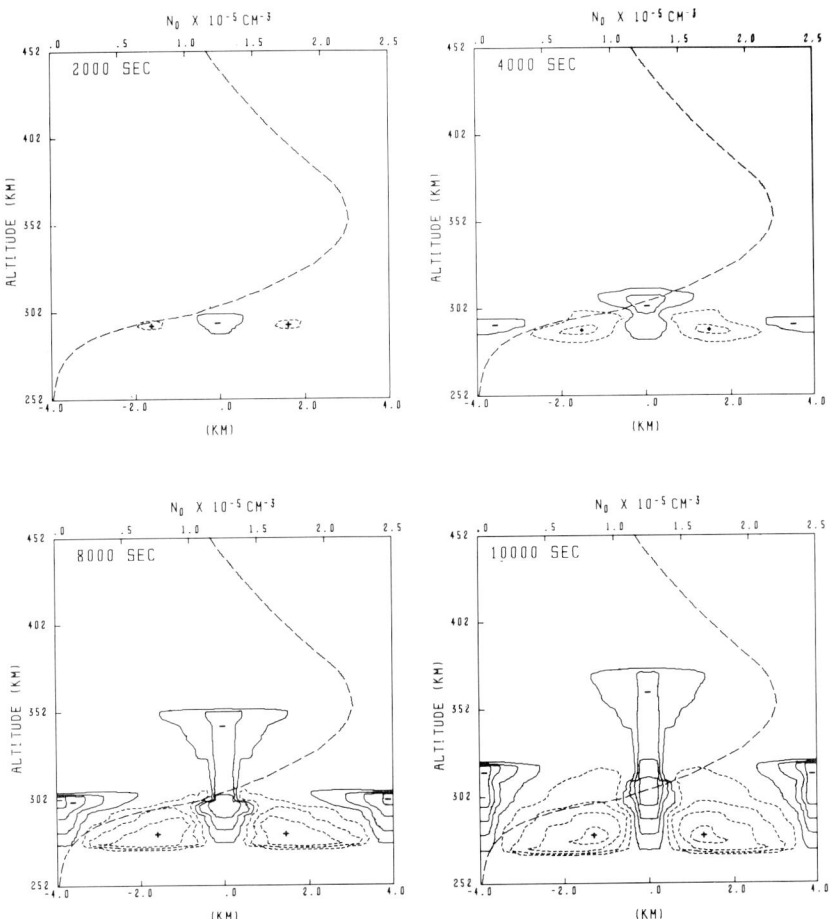

Figure 3. Contour plots of constant $\tilde{n}/n_o$ for the simulation with an F peak at 354km at t = 2000, 4000, 8000, and 10,000 sec. The small dashed contours with a plus sign inside and the solid contours with a minus sign inside indicate enhancement and depletions over the ambient electron number density. The large dashed curve depicts the ambient electron number density (values on upper horizontal axis), n_o, as a function of altitude. The vertical y axis represents altitude, the lower horizontal x axis is east-west range, and the ambient magnetic field is along the z axis, out of the figure. Taken from Ossakow et al. (1978).

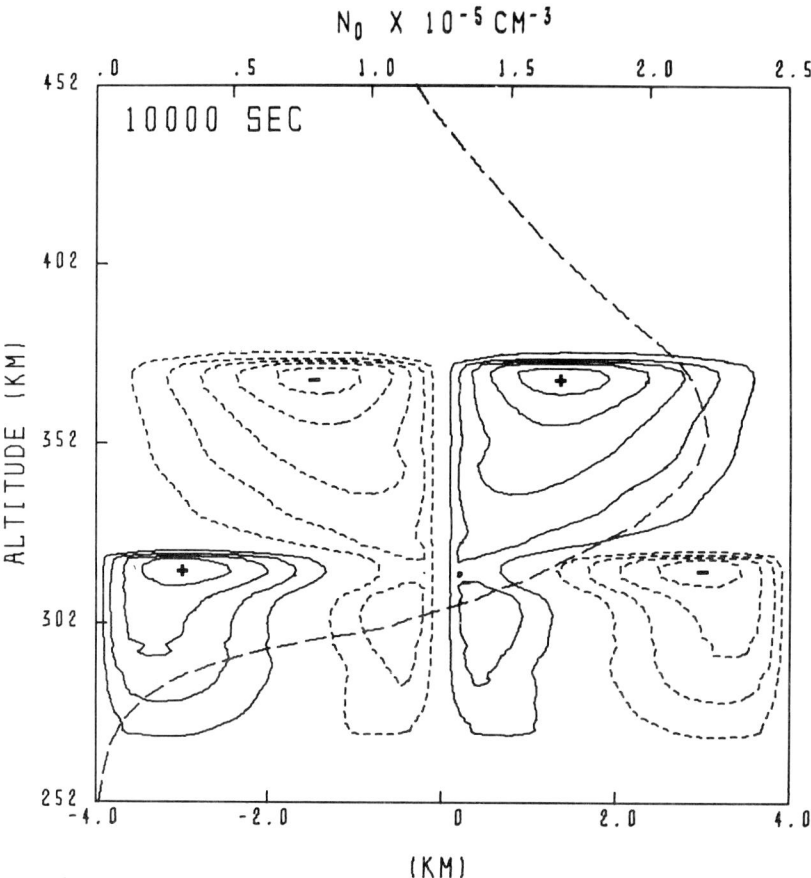

Figure 4. Contours of constant induced (polarization) potential, $\tilde{\phi}$, over the mesh (see Fig. 3) at t = 10,000 sec. Plus and minus denote positive and negative values, with values decreasing in magnitude as one goes from the innermost to the outermost contours. The large dashed curve is n_o. Taken from Ossakow et al. (1978).

clearly through the F peak with an innermost depletion contour of 41%. However, in the ionosphere below the bubble near x = 0 there is a 71% depletion contour, similarly in the wings near $|x|$ = 4km. The innermost enhancement contour, at this time, represents a 236% enhancement with a maximum inside this contour of 294%. Note that the top of the main bubble is at an altitude of 375km while the bottom trail of the bubble is at an altitude of 270km. Between t = 8000 and 10^4 sec the bubble has risen $\sim$ 24km which represents a rise velocity $\sim$ 12m/sec. Also note that the bubble is $\sim$ 1km wide. Figure 4 depicts contours of constant induced potential $\tilde{\phi}$ at t = 10^4 sec. This shows that the more isolated part of the high altitude bubble depicted in Fig. 3 is acted on by an induced electric field which points from west to east and is dipolar in nature. This causes the bubble to rise with a $(-c/B_o)\nabla\tilde{\phi} \times \hat{z}$ velocity. However, the lower portion of the mesh is acted on by an induced electric field which points from east to west. This field is much weaker than the induced field acting on the isolated portion of the central bubble. The lower altitude electric field causes the enhancements and depletions to move downward. Thus, the lower altitude portion of the central bubble becomes captured by the enhancements.

Figure 5 presents for comparison a case in which the background electron density profile (shape) was kept the same, but the entire profile moved up in altitude so that the F peak was at 434km. All other parameters are the same as in Figure 3, except ν_{in} and ν_R are taken for the altitude range 332km to 532km (those used in this simulation). One can immediately note the more rapid time evolution of the Spread F process with respect to that presented in Fig. 3. At t = 700 sec, a rising bubble with an innermost contour of 79% and a maximum depletion inside this contour of 84% were noted. At t = 1000 sec the bubble has reached the peak and at this time the innermost depletion contour is 85%. The long trail associated with the high altitude bubble has a 100 km extension to lower altitudes. At t = 1400 sec the top of the main bubble is at an altitude $\sim$ 500km and has a long trail connecting to an altitude of 357km. There is a maximum 70% depletion within the innermost contour of the high altitude bubble. Between t = 10^3 and 1400 sec the top part of the bubble rose $\sim$ 65km and this represents a rise velocity $\sim$ 160m/sec. Potential contour results for this simulation show similar patterns to those exhibited in Fig. 4. Naturally, the induced electric fields causing the bubble to rise in the present case are stronger. This spread in bubble rise velocities has been observed by AE satellite data (McClure et al., 1977).

Other numerical simulations in this series have been performed and the paper by Ossakow et al. (1978) should be consulted for more details. The basic conclusions reached from these simulations are as follows: (i) the collisional R-T instability causes linear growth on the bottomside of the equatorial Spread F region; (ii) plasma density depletions (bubbles) steepen on their top and nonlinearly rise to the topside ionosphere, beyond the F peak, by polarization (induced) $\underline{E} \times \underline{B}$

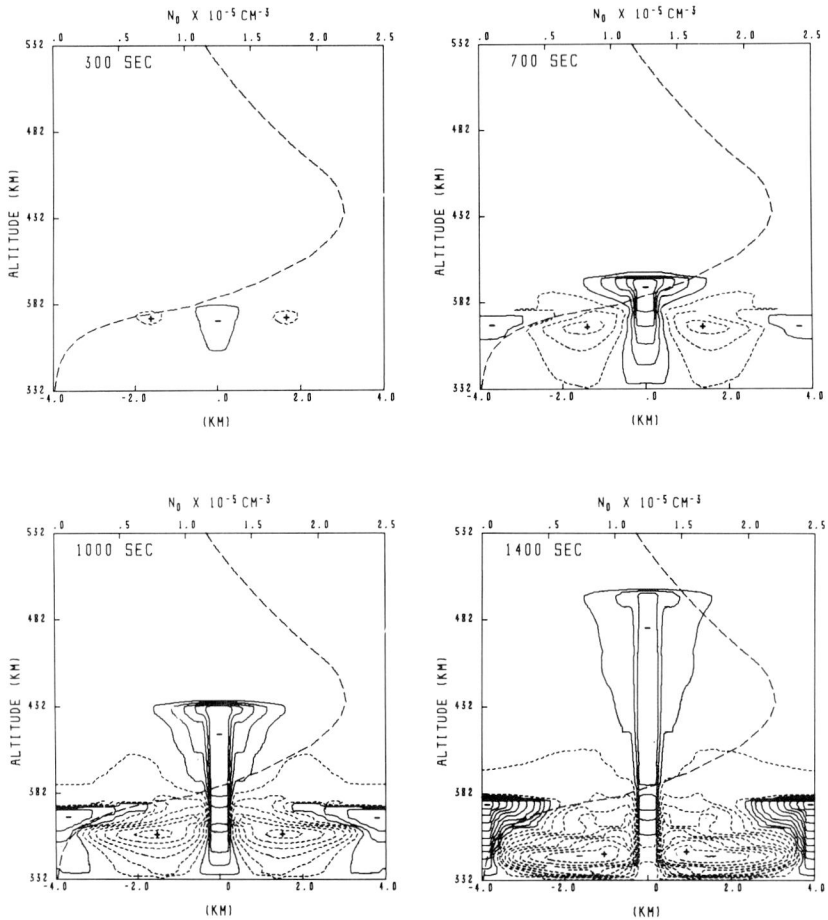

Figure 5. Contour plots of constant $\tilde{n}/n_o$ for the F peak at 434km simulation at t = 300, 700, 1000, and 1400 sec. All other nomenclature is same as Fig. 3. Taken from Ossakow et al. (1978).

forces; and (iii) high altitude of the F peak, small bottomside background electron density gradient scale lengths, and large percentage depletions yield large vertical bubble rise velocities, with the first two conditions favoring collisional R-T linear growth (instability). In addition, large spatial bubbles with similar rise velocities to those presented here, but with almost 100% depletions, have been produced by numerical simulations (Zalesak et al., 1978). In these cases the horizontal mesh covered 200km in extent (intermesh spacing Δx = 5km) and a long wavelength ($\sim$ 75km) initial perturbation was used. Bubbles with horizontal dimensions $\sim$ 50km resulted. The NRL group has also added neutral wind effects to these simulations and found that an eastward neutral wind results in a westward motion of the bubbles in addition to its rise. This is also in agreement with observations of bubble motion (see McClure et al., 1977).

c. Analytical 2D Coherent Mode Coupling Results: A two dimensional nonlinear quasi-final state of the collisional R-T instability was investigated by Chaturvedi and Ossakow (1977) using analytical means and considering coherent mode coupling as the saturation mechanism. Equations (14) and (15) were utilized with $n = n_o + \tilde{n}$, etc. This yields the following coupled nonlinear equations

$$\frac{\partial \tilde{n}}{\partial t} - \frac{c}{B_o} \nabla \tilde{\phi} \times \hat{z} \cdot \nabla n_o = - \nu_R \tilde{n} + \frac{c}{B_o} \nabla \tilde{\phi} \times \hat{z} \cdot \nabla \tilde{n} \qquad (16)$$

$$\frac{g \times \hat{z}}{\Omega_i} \cdot \nabla \tilde{n} - \frac{c}{B_o \Omega_i} \left[n \nu_{in} \nabla^2 \tilde{\phi} + \nabla \tilde{\phi} \cdot \nabla (\tilde{n} \nu_{in}) \right] = 0 \qquad (17)$$

where the second term on the LHS of (16) represents growth, the first term on the RHS represents damping, and the second term on the RHS is the nonlinear term. Comparing the nonlinear term in (16) with the last term in (17) results in

$$\frac{c}{B_o} \nabla \tilde{\phi} \times \hat{z} \cdot \nabla \tilde{n} : \frac{c}{B_o \Omega_i} \nabla \tilde{\phi} \cdot \nabla (\tilde{n} \nu_{in}) \approx \frac{\Omega_i}{\nu_{in}} >> 1$$

Therefore, eqn. (17) is treated linearly and the nonlinearity is retained in (16). A perturbation of the form

$$\frac{\tilde{n}}{n_o} = A_{1,1} \sin (k_y y - \omega t) \cos k_x x + A_{2,0} \sin 2k_x x \qquad (18)$$

is chosen. In this analysis x is vertical (altitude) and y horizontal (east-west). This is the way it appears in the reference. For the convenience of the reader, in referring to the original work, we have kept the coordinate system of the reference.

The coupled mode equations for the amplitudes $A_{1,1}$ and $A_{2,0}$ become

$$\frac{\partial A_{1,1}}{\partial t} = \gamma_{1,1} A_{1,1} - 2\alpha A_{1,1} A_{2,0} \qquad (19)$$

$$\frac{\partial A_{2,0}}{\partial t} = \gamma_{2,0} A_{2,0} + \frac{\alpha}{2} A_{1,1}^2 \qquad (20)$$

where the coupling coefficient $\alpha = k_x k_y^2 g/k^2 \nu_{in}$, and $\omega \equiv \omega_r + i\gamma$. It should be noted that the linear growth rate $\gamma_{2,0}$ is negative, i.e., the purely vertical mode $A_{2,0}$ is linearly damped. Also $\gamma_{1,1}$ is positive and represents linear growth of the mode $A_{1,1}$. In the saturated steady state one has $\partial A_{1,1}/\partial t = \partial A_{2,0}/\partial t = 0$ and from (19) and (20) this results in

$$A_{2,0} = \frac{\gamma_{1,1}}{2} \approx \frac{1}{2k_x L} \qquad (21)$$

$$A_{1,1} \approx \left(\frac{k^2}{k_y^2} \frac{\nu_R \nu_{in}}{gLk_x^2}\right)^{\frac{1}{2}} \approx \left(\frac{-2\gamma_{2,0}}{\alpha} A_{2,0}\right)^{\frac{1}{2}} \qquad (22)$$

For typical values of the parameters, $A_{2,0} \gg A_{1,1}$ and shows $A_{2,0} \propto k_x^{-1}$ (i.e., the power spectrum would $\propto k_x^{-2}$). This represents a coherent nonlinear evolution where the linearly damped mode $A_{2,0}$ is generated nonlinearly by the linearly unstable mode $A_{1,1}$, by a harmonic generation ($A_{1,1}$ contains k_x and $A_{2,0}$ contains $2k_x$). More detailed information can be found in Chaturvedi and Ossakow (1977).

d. _Analytical Models for ESF Bubbles._ First we will discuss collisional R-T bubbles (Ossakow and Chaturvedi, 1978) in which the electrical analogy with plasma density enhancement (e.g., barium clouds) was utilized and results obtained for general 2D ($\perp B_0$) bubble shapes. Here eqns. (14) and (15) are utilized with the further simplifying assumptions of neglecting recombination chemistry ($\nu_R = 0$) and the explicit altitude dependence of the ion-neutral collisional frequency, ν_{in}. The following set of equations result.

$$\frac{\partial n}{\partial t} - \frac{c}{B_0}(\nabla \tilde{\phi} \times \hat{z}) \cdot \nabla n = 0 \qquad (23)$$

$$\nabla \cdot (n\nabla \tilde{\phi}) = \underline{E}^* \cdot \nabla n \qquad (24)$$

$$E^* = E_o + \frac{\Omega_i}{\nu_{in}} \frac{m_i}{e} \underline{g} \times \hat{z} \tag{25}$$

where an ambient horizontal electric field E_o has been included to show more generality. Equation (24) can be thought of, in general, as a potential equation for a dielectric immersed in the applied electric field, E^*. The plasma density depletion (bubble) is analogous to the case of a cavity immersed in a dielectric with a uniform electric field (E^*). In general (23) and (24) do not admit of a two dimensional ($\perp B_o$), steady state solution. However, for the case of a constant density inside the depletion and a constant density outside the depletion, two dimensional steady state solutions can be obtained.

For a constant density depletion with an elliptical shape one has

$$n(x,y) = n_o - n_D(x,y)$$

$$= n_o \left[1 - \frac{\delta n}{n_o} F(x,y)\right] \tag{26}$$

$$n_D = \delta n \, H \left[1 - \left(\frac{x}{a}\right)^2 - \left(\frac{y}{b}\right)^2\right] \tag{27}$$

$$H(x) = \begin{cases} 1, & x > 0 \\ 0, & x < 0 \end{cases}$$

where H is the Heaviside function (note the geometry is taken such that x is vertical and y horizontal). For this elliptical shape, the solution to (24), neglecting E_o, is

$$\frac{\partial \tilde{\phi}}{\partial y} = \frac{\Omega_i}{\nu_{in}} \frac{m_i g}{e} \left(\frac{a}{b + a(1 - \delta n/n_o)} \delta n/n_o\right) \tag{28}$$

Using $-(c/B_o) \nabla \tilde{\phi} \times \hat{z}$ this further results in the nonlinear vertical bubble rise velocity, V_B, given by

$$V_B = \frac{g}{\nu_{in}} \left(\frac{a}{b + a(1 - \delta n/n_o)} \delta n/n_o\right) \tag{29}$$

Limiting cases of (29) for sheet, cylindrical and slab bubble geometries are given by

$$\frac{V_B}{g/\nu_{in}} = \begin{cases} (\delta n/n_o)(1 - \delta n/n_o)^{-1}, & b \ll a \quad (30a) \\ (\delta n/n_o)(2 - \delta n/n_o)^{-1}, & b = a \quad (30b) \\ 0, & b \gg a \quad (30c) \end{cases}$$

For typical values of ν_{in} as a function of altitude, Fig. 6 exhibits rise velocities given by (30b) for various percentage depletions. Table 1 shows the rise velocities, in units of g/ν_{in}, for various shapes. (Note that the linear case comes from linearizing (23) and (24)). All of the geometry results can be expressed in a concise formula

$$V_B = \frac{g}{\nu_{in}} f(\frac{\delta n}{n_o}) \qquad (31)$$

where $f(\delta n/n_o)$ is an increasing function of the percentage depletion $\delta n/n_o$. Basically, the results predict that high altitudes and/or large percentage depletions yield high vertical rise velocities for the bubbles (in agreement with experimental observations).

Collisional and collisionless (inertia dominated) two dimensional cylindrical R-T bubbles have been studied analytically by Ott (1978). This study is based on the analogy with fluid dynamic flows and brings forth some of the work done on two-dimensional fluids. This study begins with the basic equations (1) - (3), considers two dimensions ($\perp \underline{B}_o$), sets $n_e \approx n_i \approx n$, and makes the assumption that

$$\Omega_i \gg \frac{\partial}{\partial t}, \underline{V}_i \cdot \nabla, \nu_{in} \qquad (31)$$

To lowest order, using (31), one obtains from (3) a lowest order ion velocity (with $\hat{z} = \underline{B}_o/|\underline{B}_o|$),

$$\underline{V}_i^{(o)} = \frac{\hat{z} \times \nabla\phi}{B_o} + \frac{\hat{z} \times \nabla p_i}{neB_o} + \frac{m_i \underline{g} \times \underline{B}_o}{eB_o^2} \qquad (32)$$

(Note that mks units are being used here to coincide with the units used by Ott (1978)). Quasi-neutrality, i.e., $\nabla \cdot \underline{J} = 0$, with the assumption of two dimensionality implies that $\underline{J}$ can be specified in terms of a single scalar potential function, ψ, such that

$$\underline{J} = - \hat{z} \times \nabla\psi \qquad (33)$$

Using eqn. (2) for the electron velocity, and a next order ion velocity equation obtained by putting (32) into (3), the following ion velocity equation is obtained

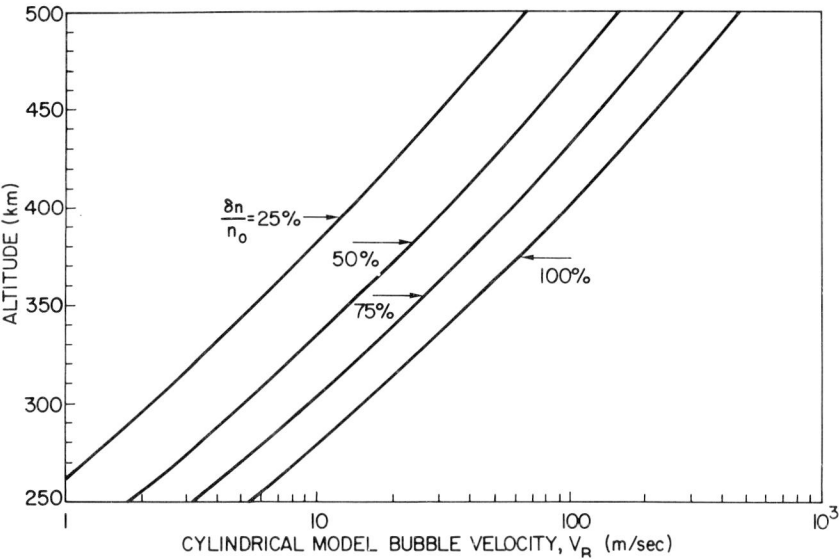

Figure 6. Depletion (bubble) vertical rise velocity V_B as a function of altitude for various values of the percentage depletion $\delta n/n_o$ for the cylindrical bubble model. Taken from Ossakow and Chaturvedi (1978).

V_B' \ $\dfrac{\delta n}{n_o}$	.25	.5	.75	.9	1
Linear	.25	.5	.75	.9	1
Sheet	.33	1	3	9	∞
Cylindrical	.14	.33	.6	.82	1
Elliptical (5:1)	.26	.71	1.67	3	5
Elliptical (10:1)	.29	.83	2.14	4.5	10

Table 1. Bubble rise velocity (in units of g/ν_{in}), V_B', as a function of fractional depleted density, $\delta n/n_o$, for various bubble shapes.

$$nm_i \left[\frac{\partial \underline{V}_i^{(o)}}{\partial t} + \underline{V}_i^{(o)} \cdot \nabla \right] \underline{V}_i^{(o)} = - \nabla \tilde{p} - nm_i \nu_{in} \underline{V}_i^{(o)}$$
$$+ nm_i \underline{g} \qquad (34)$$

with $\tilde{p} = p_e + p_i + B\psi$. Using (32) and the assumption of either isothermal or adiabatic ions one has

$$\nabla \cdot \underline{V}_i^{(o)} = 0 \qquad (35)$$

Using $\underline{V}_i^{(o)}$ in the ion continuity equation (1) with $\nu_R = P = 0$, eqn. (1) becomes

$$\frac{\partial n}{\partial t} + \underline{V}_i^{(o)} \cdot \nabla n = 0. \qquad (36)$$

Equations (34) - (36) form a complete set of equations sufficient to determine the unknown quantities $\underline{V}_i^{(o)}$, n, and $\tilde{p}$. In the limit of $\nu_{in} \to 0$, eqns. (34) - (36) are identical to those of an ideal incompressible fluid. At this point the philosophy taken is that there is much to be learned concerning bubbles in the ESF ionosphere from the extensive studies of bubbles in fluids. One then uses a stream function ξ such that

$$\underline{V}_i^{(o)} \equiv \hat{z} \times \nabla \xi \qquad (37)$$

After a series of manipulations one can obtain an equation for the stream function. In the collisional and collisionless case one finds that for certain values of the bubble rise velocity ξ will satisfy the equation, for a cylindrical shape (see Ott, 1978 for more details).

The results of this study by Ott (1978) predicts the bubbles to be cylindrical (circular cap at top) in two dimensions. The bubble rise velocity in the collision-dominated regime, for a 100% depletion, is given by

$$V_B = \frac{g}{\nu_{in}} \qquad (38)$$

which is altitude dependent. This result is the same as that predicted by (30b) for $\delta n/n_o = 1$, i.e., a 100% depletion. In the inertia (collisionless) dominated regime, for a 100% depletion, the bubble rise velocity is given by

$$V_B \simeq \tfrac{1}{2} (Rg)^{\tfrac{1}{2}} \qquad (39)$$

where R is the radius of curvature at the top of the bubble so that (39) is size dependent.

e. <u>Analytical Work on the Very Small Scale ($\lesssim$ 10m) Irregularities</u>. The work in this section represents essentially multilinear calculations using kinetic theory resulting in plasma kinetic drift wave modes. It is multilinear because it depends on say a two-step process whereby linear theory is performed on the non-linear state. The driving density gradients, in these calculations, are thought to arise through a primary instability, driven by the zero order background ionospheric equatorial F region electron density gradient, achieving a large amplitude state. The zero order ionospheric electron density gradient is of larger scale size than the primary instability electron density gradient scale size, which would be of the order of the instability wavelength. Because the calculations are kinetic, they employ particle distribution functions. Kinetic drift waves have been investigated for laboratory plasma fusion conditions for over twenty years, so a well developed formalism exists.

Before proceeding with specific calculations let us present some general concepts regarding kinetic drift waves which will be useful for both types of calculations presented in this section. The basic geometry is such that

$$\underline{B} = B_0 \hat{z} \tag{40a}$$

$$n_o = n_o(x) \tag{40b}$$

(similar to the zero order background equatorial ionospheric geometry) where here $n_o(x)$ arises due to the primary instability. The orthogonal coordinate system is completed by the ion and electron diamagnetic velocities being along the y axis,

$$\underline{V}_d = (V_{di} - V_{de}) \hat{y} \tag{41a}$$

$$V_{di} = (V_i^2/2\Omega_i) (\partial \ln n_o/\partial x) \tag{41b}$$

$$V_{de} = - (V_e^2/2\Omega_e) (\partial \ln n_o/\partial x) \tag{41c}$$

where $V_{i,e} = (2T_{i,e}/m_{i,e})^{\frac{1}{2}}$ and the larmor radius is defined by $r_\alpha = V_\alpha/\Omega_\alpha$. The linear analysis is then performed with this in mind.

First the low frequency ($\omega \ll \Omega_i$) collisionless drift wave calculations of Costa and Kelley (1978a,b) for $kr_i \sim 1$ (r_i is the ion gyroradius), will be presented. These calculations were meant to provide a basis within which to try and account for the 3 meter radar backscatter observations of Jicamarca (see Woodman and La Hoz, 1976). A linear kinetic dispersion relation is derived using perturbations of the form exp i $[\underline{k} \cdot \underline{x} - \omega t]$, where $\underline{k} = k_\parallel \hat{z} + k_\perp \hat{y}$ and $\omega = \omega_r + i\gamma$. Furthermore, assumptions are made such that $\omega \ll \Omega_i$, $V_i \ll |\omega/k_\parallel| \ll V_e$ and $\omega > \nu_{in}$, ν_{ie}, ν_{ii}, which are the ion-neutral, ion-electron, and ion-ion collision frequencies, respectively. This analysis depends on having a $k_\parallel$, a component of the wavevector along the ambient geomagnetic field. Figure 7 shows some growth rate results from these calculations. The growth rate is in units of the ion thermal velocity divided by the electron density gradient scale length ($L = n_o (dn_o/dx)^{-1}$). Figure 8 depicts some measured inverse electron density gradient scale lengths during ESF and maximum growth rates as a function of these inverse scale lengths. The basic results of these calculations show maximum growth for $k_\perp r_i \approx 1.5$, i.e., $\lambda_\perp \approx 20$m (for typical ESF parameters) with growth rates $\gtrsim 1$ sec^{-1}. For more detailed analysis concerning these low frequency drift waves applied to ESF see Costa and Kelley (1978a,b).

Now we present the high frequency drift wave analysis of Huba et al. (1978). In this reference radar backscatter observed irregularities with wavelengths of 1 meter and 36 cm at Kwajalein during ESF conditions are shown. In an effort to explain these very short wavelength irregularities high frequency ($\omega \gtrsim \Omega_i$) drift waves were analyzed for ESF conditions. These waves are the so-called drift cyclotron (DC) or lower hybrid drift (LHD) instabilities with maximum growth rates for $k_\perp r_e \sim 1$ (and $k_\parallel = 0$). No $k_\parallel$ is required for these instabilities. The parameter determining which instability operates in a collisional plasma is

$$C_f = (\nu_{ii}/\Omega_i)(k_\perp r_i)^2 \quad (42)$$

Utilizing the linear dispersion relation for high frequency drift waves for $C_f \ll 1$, instability results for

$$L/r_i < (1/2\ell)(m_i/m_e)^{\frac{1}{2}} \quad (43)$$

where ℓ is the harmonic number ($\omega_r \approx \ell\Omega_i$). For the O$^+$ ESF ionospheric plasma this requires the electron density gradient scale length $L < 340$m, which is satisfied (see Fig. 8). Growth rates for these instabilities are given by

$$\gamma \approx (m_e/m_i)^{\frac{1}{4}} \ell \Omega_i \quad (44)$$

However, the condition $C_f \ll 1$ implies that $(k_\perp r_i)^2 n \ll 2 \times 10^7$ and for $k_\perp r_e \sim 1$ this means $n \ll 10^3$ cm^{-3}, which is quite restrictive. Longer wavelengths, i.e., smaller values of $k_\perp r_e$ would raise the density restriction somewhat, but still be restrictive for ESF conditions.

For $C_f \gtrsim 1$, the lower hybrid drift instability is operative and there is no threshhold condition on L. Basically, the collisions which

SPREAD F THEORY 285

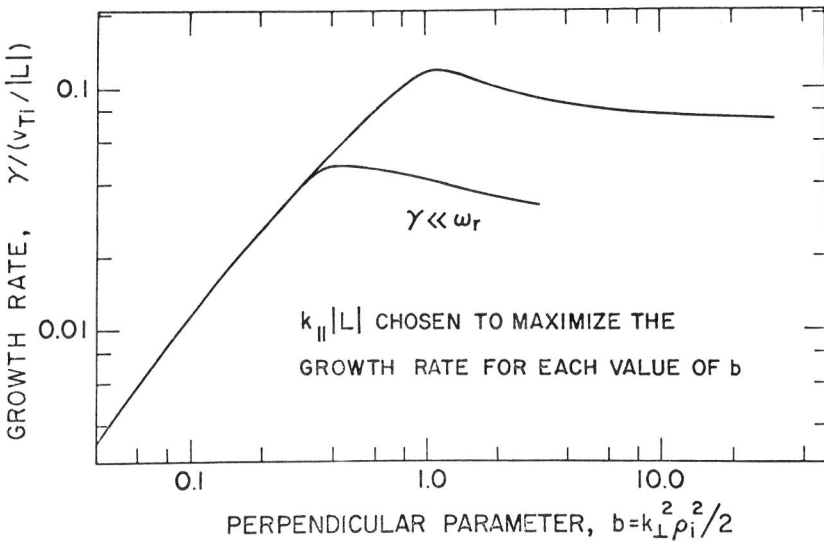

Figure 7. Growth rates as a function of b for the low frequency collisionless drift waves (ρ_i is the ion gyroradius). The curve $\gamma \ll \omega_r$ is obtained from a small growth rate approximation expression. Taken from Costa and Kelley (1978b).

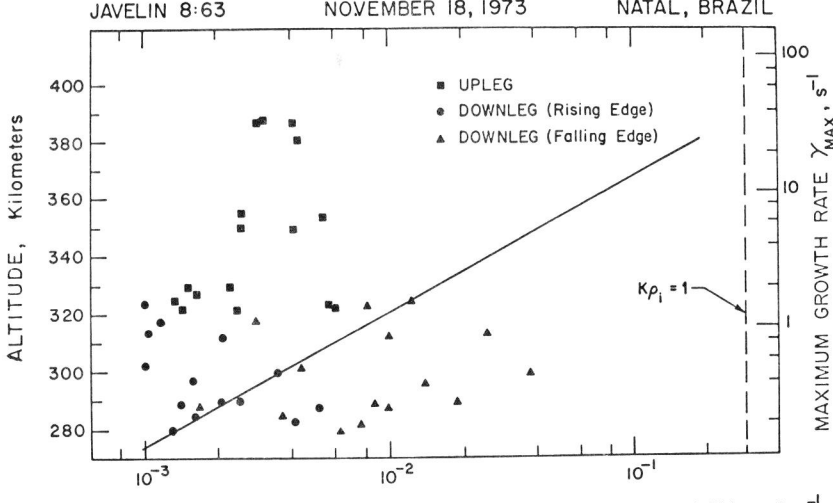

Figure 8. Observed inverse gradient scale lengths $(1/n)(dn/dx)$ as a function of altitude during the Natal rocket flight. Also indicated by the straight line is the maximum growth rate of low frequency drift waves as a function of $(1/n)(dn/dx)$. Note that ρ_i is the ion gyroradius. Taken from Costa and Kelley (1978a).

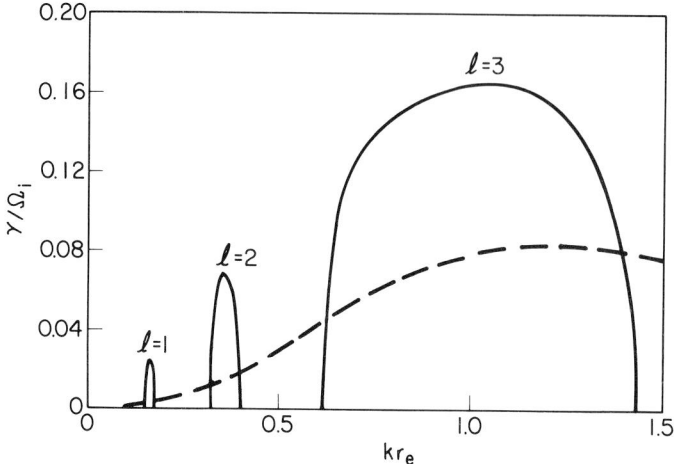

Figure 9. Growth rate as a function of perpendicular wavenumber for the DC instability (solid line) and LHD instability (dashed line) for an O^+ plasma with $T_e = T_i$, $\omega_{pe}/\Omega_e = 10$ (where ω_{pe} is the electron plasma frequency), and $V_{di}/V_i = 0.037$. Growth also occurs for $kr_e > 1.5$, but has not been plotted. Taken from Huba et al. (1978).

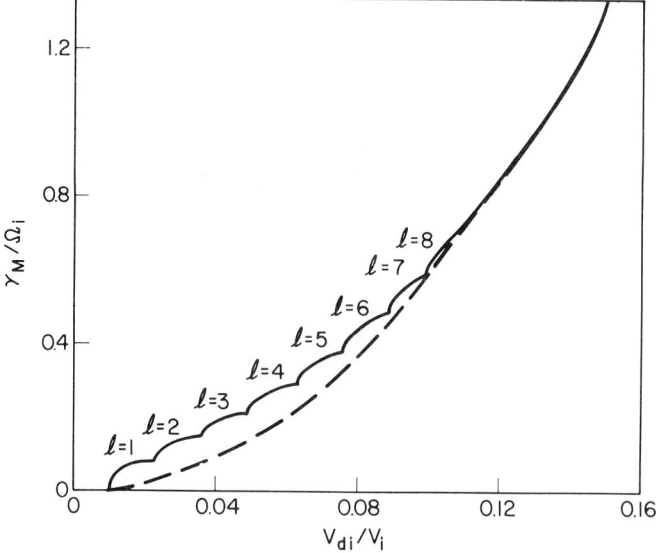

Figure 10. Maximum growth rate as a function of ion diamagnetic drift velocity for the DC instability (solid line) and the LHD instability (dashed line). The other parameters are the same as in Fig. 9. Note that the DC instability goes into the LHD instability for $V_{di}/V_i \gtrsim 0.11$. Taken from Huba et al. (1978).

increase C_f destroy the ion gyroresonances needed for the DC instability to operate. The real and imaginary part of the frequency for the LHD instability are given by

$$\frac{\omega_r}{\Omega_i} \sim \frac{r_i}{L}\left(\frac{m_i}{m_e}\right)^{\frac{1}{2}}, \quad \frac{\gamma}{\Omega_i} \sim \left(\frac{r_i}{L}\right)^2 \left(\frac{m_i}{m_e}\right)^{\frac{1}{2}} \tag{45}$$

In the collisionless limit it should be noted that the DC instability transforms into the LHD instability for high enough ion diamagnetic velocities such that

$$L/r_i \leqslant (m_i/m_e)^{\frac{1}{4}}$$

which for ESF conditions implies that $L \leqslant 30m$. Figures 9 and 10 show some typical results from the analysis of the high frequency drift wave linear dispersion relation.

The results of this analysis predict that the lower hybrid drift instability is dominant for most typical ESF ionospheric parameters. Also, maximum growth of the instability occurs for $k_\perp r_e \sim 1$ ($\lambda_\perp \sim 21$ cm), although good growth rates can occur for $\lambda_\perp \sim 1$ m. Finally, from this instability, large growth rates ($\gamma \leqslant \Omega_i$) resulting in growth times, $\tau = \gamma^{-1}$, less than a second can occur. For more details of this work see Huba et al. (1978).

III. SUMMARY

Although much progress has been made in the theoretical efforts directed toward the equatorial Spread F ionosphere, especially in the past three years, more has to be done. Also the burden cannot be placed on the theoretician alone. Correlative measurements have to be made prior to and during ESF conditions. It is not sufficient to make a single measurement with one instrument and then expect a complete theoretical description of ESF. One needs to know the state of the ionosphere with respect to driving parameters such as background electron density profiles (bottomside electron density gradient scale lengths and height of the F peak), d.c. electric fields, neutral winds, and ionic mass composition in order to build a predictive model. In addition, in order to compare results from the predictive model, measurements of the in situ fluctuating component of the electric field and plasma density have to be made, as well as radar backscatter measurements of the very small scale irregularities ($\leqslant 10m$) and ground measurements of satellite signal propagation amplitude and phase scintillations.

Some of the remaining theoretical problems could be listed as follows. (1) What are the effects of changing initial conditions, including ion inertia, and including neutral winds in the numerical simulations? (2) An analytical description of many bubbles rising. (3) How do bubbles decay and what role does diffusion, etc. play? (4) How does a turbulent development occur? (5) What are the effects of other regions of the ionosphere (e.g., E region) on ESF? (6) A more complete study of collisional effects on drift waves is needed. What determines when the small scale irregularities should occur (a more quantitative description)? (7) A determination of the nonlinear saturation of small scale irregularities (instabilities) is needed. (8) What are the effects of $k_\parallel$? Indeed points (6) and (7) are tied to the more general question of what is the relation between the very small scale ($\lesssim 10$m) irregularities (which radar backscatter observes) and the longer wavelength fluid type (e.g., R-T) irregularities (which are primarily responsible for Spread F seen on ionograms)? I am sure that some more questions and points could be raised. However, the above list should keep the theoreticians busy for a reasonable time.

ACKNOWLEDGEMENTS

The author is grateful to the dedication of his colleagues in their efforts to provide an understanding of equatorial Spread F. In particular, I am deeply indebted to Pradeep Chaturvedi, Joe Huba, Mike Keskinen, Ed McDonald, Tony Scannapieco, and Steve Zalesak whose interest and work in ionospheric irregularities is the sine qua non without which this article could not have been written. In addition, this work was supported by the Defense Nuclear Agency and the Office of Naval Research.

REFERENCES

Balsley, B. B., Haerendel, G., and Greenwald, R. A.: 1972, J. Geophys. Res. 77, p. 5625.
Booker, B. B., and Wells, H. W.: 1938, Terres. Magn. 43, p. 249.
Calvert, W.: 1963, J. Geophys. Res. 68, p. 2591.
Chaturvedi, P., and Kaw, P.: 1976, J. Geophys. Res. 81, p. 3257.
Chaturvedi, P. K., and Ossakow, S. L.: 1977, Geophys. Res. Letts. 4, p. 558.
Costa, E., and Kelley, M. C.: 1978a, J. Geophys. Res. 83, p. 4359.
Costa, E., and Kelley, M. C.: 1978b, J. Geophys. Res. 83, p. 4365.
Dagg, M.: 1957, J. Atmos. Terres. Phys. 11, p. 139.
Dungey, J. W.: 1956, J. Atmos. Terres. Phys. 9, p. 304.
Dyson, P. L., McClure, J. P., and Hanson, W. B.: 1974, J. Geophys. Res. 79, p. 1497.
Farley, D. T., Balsley, B. B., Woodman, R. F., and McClure, J. P.: 1970, J. Geophys. Res. 75, p. 7199.
Haerendel, G.: 1974, preprint, Max-Planck Institute fur Physik und Astrophysik, Garching, West Germany.
Huba, J. D., Chaturvedi, P. K., Ossakow, S. L., and Towle, D. M.: 1978, Geophys. Res. Letts. 5, p. 695.

Hudson, M. K.: 1978, J. Geophys. Res. 83, p. 3189.
Hudson, M. K., Kennel, C. F., and Kaw, P. K.: 1973, Trans. Am. Geophys. Un. 54, p. 1147.
Hudson, M. K., and Kennel, C. F.: 1975, J. Geophys. Res. 80, p. 4581.
Kelley, M. C., Haerendel, G., Kappler, H., Valenzuela, A., Balsley, B. B., Carter, D. A., Ecklund, W. L., Carlson, C. W., Hausler, B., and Torbert, R.: 1976, Geophys. Res. Letts. 3, p. 448.
Kelley, M. C., and Ott, E.: 1978, J. Geophys. Res. 83, p. 4369.
Martyn, D. F.: 1959, J. Geophys. Res. 64, p. 2178.
McClure, J. P., Hanson, W. B., and Hoffman, J. H.: 1977, J. Geophys. Res. 82, p. 2650.
Morse, F. A., Edgar, B. C., Koons, H. C., Rice, C. J., Heikkila, W. J., Hoffman, J. H., Tinsley, B. A., Winningham, J. D., Christensen, A. B., Woodman, R. F., Pomalaza, J., and Teixeira, N. R.: 1977, J. Geophys. Res. 82, p. 578.
Ossakow, S. L., and Chaturvedi, P. K.: 1978, J. Geophys. Res. 83, p. 2085.
Ossakow, S. L., Zalesak, S. T., McDonald, B. E., and Chaturvedi, P. K.: 1978, J. Geophys. Res. (in press).
Ott, E.: 1978, J. Geophys. Res. 83, p. 2066.
Scannapieco, A. J., and Ossakow, S. L.: 1976, Geophys. Res. Letts. 3, p. 451.
Sudan, R. N., Akinrimisi, J., and Farley, D. T.: 1973, J. Geophys. Res. 78, p. 240.
Woodman, R. F., and La Hoz, C.: 1976, J. Geophys. Res. 81, p. 5447.
Zalesak, S. T., Ossakow, S. L., McDonald, B. E., and Chaturvedi, P. K.: 1978, Trans. Am. Geophys. Un. 59, p. 345.

EQUATORIAL SPREAD F: A REVIEW OF RECENT EXPERIMENTAL RESULTS

Michael C. Kelley
School of Electrical Engineering
Cornell University
Ithaca NY 14853

The near space region of the Earth is an excellent laboratory in which to study large scale plasma instabilities. This is particularly true in the equatorial ionosphere where the background conditions are fairly well understood, as opposed to the auroral zone ionosphere for example, where an extremely complex interaction occurs between the ionosphere and the hot plasma at high altitudes. In this paper we review an intense research effort aimed at understanding the large scale disruption of the equatorial F layer which often commences just after sunset, and lasts for most of the night. A very attractive explanation for the phenomena, although one not universally accepted, is that the F layer is unstable to the classic Rayleigh-Taylor condition in which a heavy fluid, the plasma, is supported against gravity by a light "fluid", the Earth's magnetic field. We conclude that a reasonable case has been made for this explanation provided that the concept is extended to include nonlinear Rayleigh-Taylor like bouyancy effects above the F peak where the linear process is stable. Internal gravity waves in neutral atmosphere seem to play an important role in seeding the Rayleigh-Taylor process with large scale finite amplitude perturbations. One of the remarkable features of this phenomena is the nearly simultaneous generation of structure with scale sizes spanning five orders of magnitude. These results may have applications in astrophysical processes where the Rayleigh-Taylor instability is thought to play a role.

I. INTRODUCTION

The purpose of this review is neither historical nor all encompassing. Rather, the salient features of recent experimental work will be highlighted and examined in relation to theory. The modern era of equatorial spread F (ESF) began with the paper by Farley et al. (1970) in which the first generation of Jicamarca Radar Observatory (JRO) studies of ESF were summarized. They concluded that no one theory presented to date could successfully explain their observations.

The basic problem was that 3m irregularities could be observed below, at, or above the F peak and during times when the F layer was moving upwards, downwards, or not at all. This seemed to rule out all theories requiring particular directional relationships between the forces acting on the plasma.

In subsequent years, trans-equatorial radio propagation experiments were conducted, radar observing techniques were greatly improved, two rockets were fired into ESF conditions, several satellites have probed the disturbed equatorial region, and scintillation and airglow observations have been made. These data, coupled with numerous theoretical studies and computer simulations have greatly increased our understanding of this complex physical phenomenon. An active experiment involving the striation of barium clouds has also shed light on the problem. Thus almost all the tools available for experimental and theoretical physics have been applied in this study.

The problem seems to naturally subdivide into three areas, bottomside ESF, topside ESF, and the production of short wavelength waves. It is generally agreed that the latter are byproducts of some primary process operating at long wavelengths. The secondary nature of waves produced at 3m, which are detected by 50MHz radar such as that at JRO, must be kept in mind when interpreting data such as reproduced in Figure 1. The gray scale in this figure indicates in, db, the reflected signal strength above that received by the vertically directed antenna due to thermal fluctuations. This radar map of ESF at 3m shows all three features mentioned above and is reproduced from a paper by Woodman and La Hoz (1976). The data shows that just after sunset a narrow band of scattering elements formed below the F peak, that is, on the bottomside of the peak in electron density. Within an hour, horizontally limited regions of structure were observed at altitudes well above the F peak (topside). In early evening events such as this, the high altitude echoing regions always connect back to the lower altitude layer, although data at later local times often show

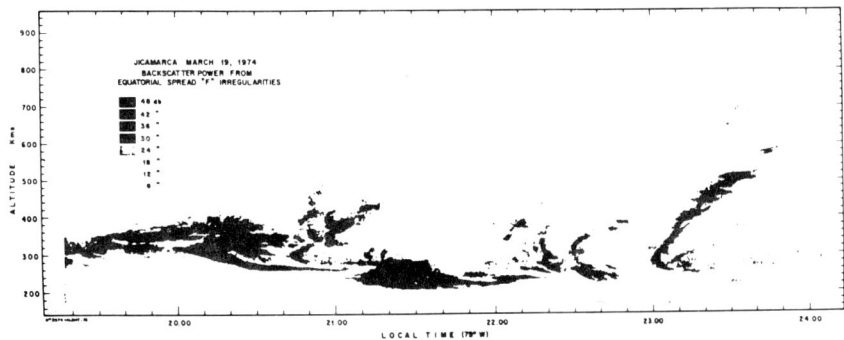

Figure 1. A radar backscatter map obtained at the Jicamarca Radar Observatory (reproduction granted by American Geophysical Union).

patches of high altitude irregularities. These structures were called plumes by Woodman and La Hoz and usually slant to the "right" on such a display. In the next sections we treat the first two of the three aspects of ESF in detail. We mention several mechanisms for the short wavelength waves, but consider their origin one of the outstanding open questions in this field and leave their detailed discussion for the future.

II. BOTTOMSIDE EQUATORIAL SPREAD F

Ionosonde studies, of course, gave first evidence for bottomside spread F (Booker and Wells, 1938). Both frequency and range spreading are observed as are large scale tilts in the normal to the F layer. When data are taken in conjunction with low power 3m backscatter data, the ionosonde responds earliest which indicates that long wavelength structures appear first in the medium (Balsley et al., 1972). Profiles of electron density obtained during two rocket flights onto bottomside ESF are reproduced in Figure 2. The upper two plots (Natal, Brazil - Kelley et al., 1976) show a smooth profile near and above the F peak (except for two pulses during the upleg when barium ion clouds were injected near the rocket). The lower data set was obtained during the EQUION rocket flight from Peru (Morse et al., 1977). All three profiles show intense km scale irregularities on the bottomside of the steep upward directed density gradient. Going clockwise from top left in the Figure the local times of observation were 1910, 1938, and 2100.

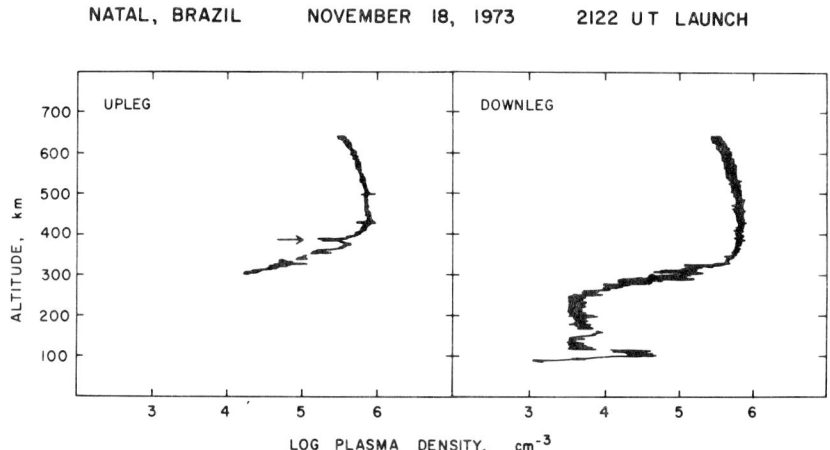

Figure 2a. Up and downleg plasma density profiles obtained during a Javelin rocket flight from Natal, Brazil.

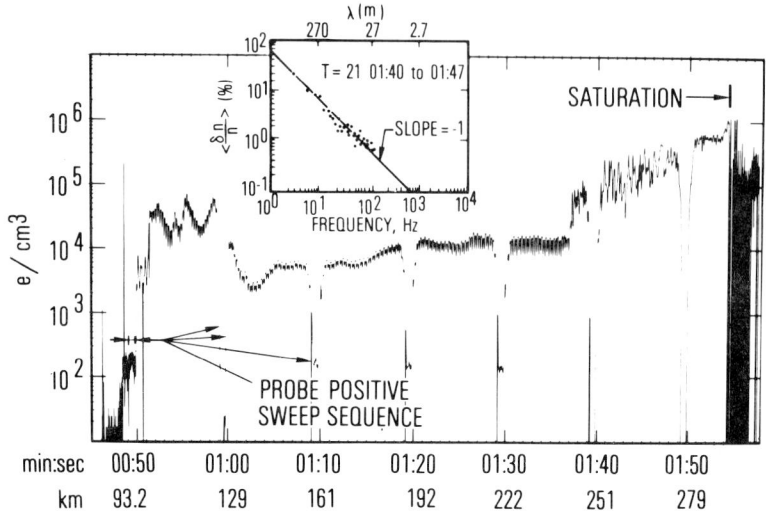

Figure 2b. Upleg plasma density profile obtained during the EQUION rocket flight from Chilca, Peru (reproductions granted by American Geophysical Union).

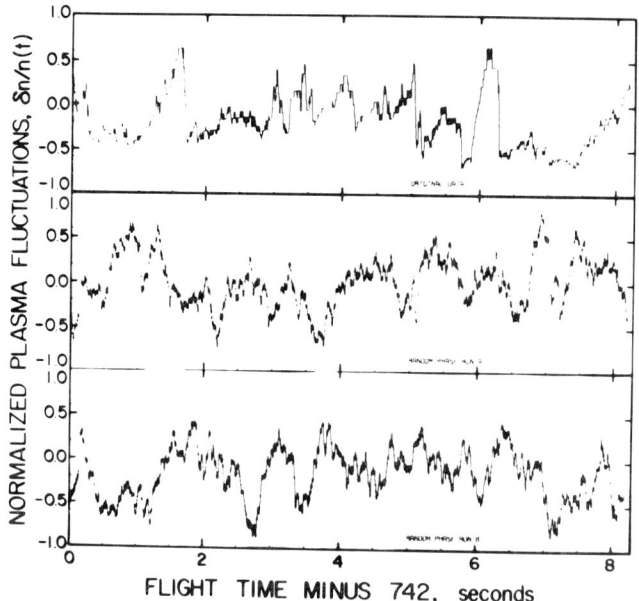

Figure 3. Top panel: detrended data from the downleg of Natal rocket flight. Lower two panels: random phase reconstruction of data in top panel (reproduction granted by American Geophysical Union).

No echoes at 3m were detected by a low power radar during the time when the earliest data set was obtained. Weak topside echoes were seen at the launch site during the downleg of the rocket experiment (top right plot), although the vehicle was well to the east at the time and did not penetrate any topside structure. Strong 3m echoes were detected on the topside by JRO throughout the Peruvian rocket flight, but unfortuantely the rocket sensor was saturated above the F peak.

An expanded plot of the Natal data is reproduced in the top portion of Figure 3 which shows the intensity of the fluctuations (variations of nearly an order of magnitude) and an assymmetry on one edge relative to the other. The power spectrum of this data is plotted as the crosses in Figure 4. The spectrum of the Peruvian data is plotted as the insert in the lower portion of Figure 2. Both display decreasing power with increasing wavelength and a power law with index between -2 and -2.2. Spectra with a similar power law index were reported by Dyson et al. (1974) as the most common signature of satellite irregularity measurements in the disturbed equatorial zone, an example of which is shown in Figure 5.

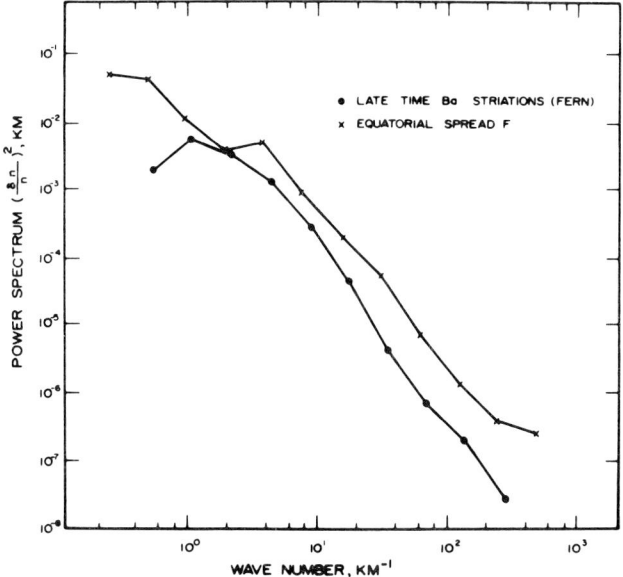

Figure 4. Power spectral analysis of plasma density fluctuations in bottomside equatorial spread F and late time barium clouds (reproduction granted by American Geophysical Union).

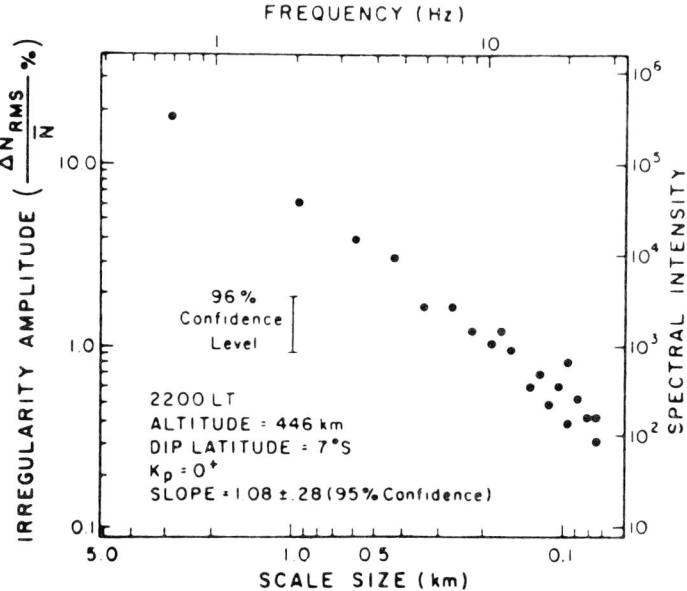

Figure 5. Power spectral analysis of plasma density fluctuations obtained in the equatorial regions by a satellite sensor (reproduction granted by American Geophysical Union).

Power law spectra such as these can be due to turbulent cascade mechanisms such as observed in fluids and some plasma processes. Woodman (personal communication, 1975), however, suggested that steep gradients in the medium could also cause the same effect. Costa and Kelley (1978a) tested this explanation as follows. The downleg Natal data was detrended (top trace in Figure 3) then Fourier analyzed (crosses in Figure 4). To each complex number in the FFT an arbitrary phase angle was added from a random number table and the data reassembled in the time domain. Two runs are shown in the other traces of Figure 3. All three sets have *identical* power spectra. Costa and Kelley argued that the dominant feature of sharp edges in the top trace is not present in the lower two and that steep edges indeed are an important factor in the power spectrum. This seems to indicate that nonlinear steepening is the saturation mechanism for bottomside spread F. Chaturvedi and Ossakow (1977) have shown that such a process should occur in the nonlinear state of the Rayleigh-Taylor gravitational instability.

The linear instability process, first suggested by Dungey (1956)

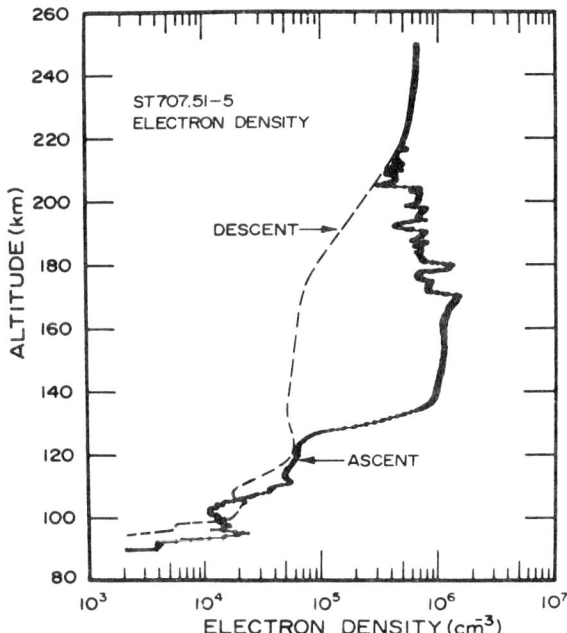

Figure 6. Plasma density profile obtained in a large barium release 42 minutes after injection into the mid-latitude ionosphere (reproduction granted by American Geophysical Union).

and expanded by Haerendel (1974) and Hudson and Kennel (1975), was rejected by Farley et al. (1970) since $\bar{g}$ must be anti-parallel to $\bar{\nabla}n$ which only holds on the bottomside. The $\bar{E}x\bar{B}$ instability (Martyn, 1959; Linson and Workman, 1970) was also rejected since the ionosphere must be going upward (downward) on the bottomside (topside) for instability. However, for a pure bottomside process, which, we now know can be separately explained, (see next section) both of these processes are tenable and can be re-examined.

An active experiment has been performed which sheds light on this matter. First, it should be noted that the $\bar{E}x\bar{B}$ instability and the Rayleigh-Taylor process are identical in a collisional plasma if $\bar{E}$ is replaced by $\nu_{in}^{-1}(\bar{g}x\hat{B})$ in the equations, where $\hat{B}$ is the unit vector in the direction of the magnetic field. Probe data from a rocket fired through a large striated barium cloud is reproduced in Figure 6 (Kelley et al., 1979) with the power spectrum plotted as the solid circle data points in Figure 4. The spectra are in remarkable agreement to the equatorial data also plotted. Furthermore, using the same analysis technique as shown in Figure 2, Kelley et al. (1979) argued for steepened structures in late time barium cloud striations. The cloud power spectra also agree with a two level numerical simulations of the $\bar{E}x\bar{B}$ process (Scannapieco et al., 1976). Now $\bar{g}$ is always anti-parallel to

$\bar{V}n$ on the bottomside and the growth rates are sufficiently large, when compared to $\bar{E}x\bar{B}$ process in barium clouds, that the collisionless Rayleigh-Taylor instability seems a viable explanation for bottomside ESF. Since ESF is often observed to commense during the rapid sunset uplift of plasma at the equator (Farley et al., 1970) and to start again during anomalous drift reversals from down to up during magnetically active times (Fejer et al., 1976), Kelley et al. (1979) suggested a complimentary role for the $\bar{E}x\bar{B}$ instability in ESF.

As discussed in detail in another URSI review paper, traveling ionospheric disturbances in the form of gravity waves have been suggested as the prime cause for ESF (Booker, 1978). Detailed calculations by Klostermeyer (1978) show that the ionosphere must be traveling downward if these waves are to come into spatial resonance with the plasma. Since even at 3m, ESF is often observed to begin during plasma uplift (Farley et al., 1970), it seems unlikely that an important aspect of topside spread F is due to these large scale neutral atmosphere oscillations, however, and this aspect is discussed in the next section.

Scintillations of signals from satellites is also used to study ESF (Aarons, 1977). Most of the important new results from this method, which is sensitive to km scale irregularities, relate to the topside. Costa and Kelley (1976) did show that the bottomside spectrum in Figure 2 were sufficiently intense to cause a moderate amplitude scintillation but could not cause the observed saturated VHF and gigahertz scintillations.

III. TOPSIDE EQUATORIAL SPREAD F

At the Gordon Conference in 1975, Woodman presented the first 3m irregularity maps (see Figure 1) and suggested that depleted regions of plasma density rise into the region above the F peak and cause the plume-like structure. He also presented the schematic shown in Figure 7 (published along with numerous radar maps by Woodman and La Hoz, 1976) of a "bubble" rising above the F peak via bouyancy forces and leaving a wake of short wavelength irregularities behind. At the same conference, Kelley showed simultaneous radar, rocket, and barium cloud data from Natal (see Kelley et al., 1976) which indicated that the "hole", indicated by the arrow in Figure 2, rose into the topside less than two minutes after the rocket traversal. The irregularity patch moved at a velocity in excess of that deduced from simultaneous barium cloud measurements.

Definitive experimental results from the satellite AE-C concerning topside bubbles and their motion was presented by McClure et al. (1977). An example is plotted in Figure 8. In the lower trace, the plasma density is plotted and in the top, two components of the ion drift velocity perpendicular to $\bar{B}$. The relative drift is quite accurate and

SPREAD F: EXPERIMENTAL RESULTS

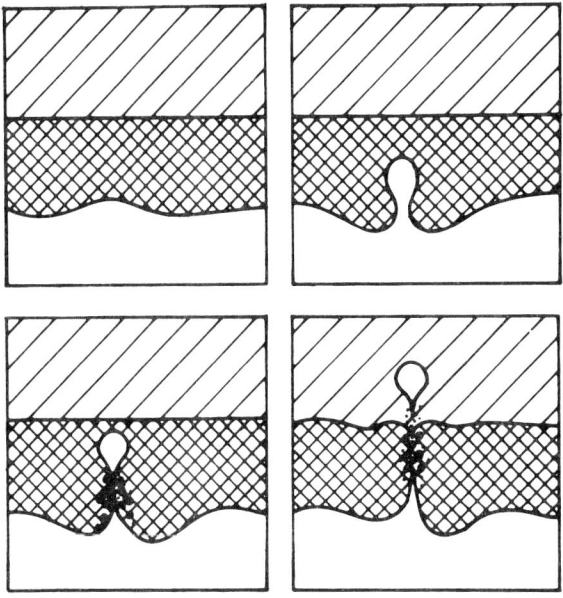

Figure 7. Schematic bubble diagram proposed by Woodman (reproduction granted by American Geophysical Union).

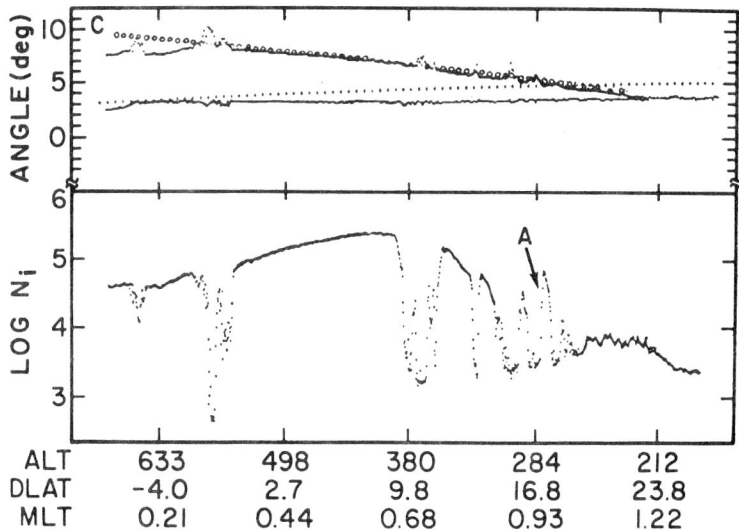

Figure 8. Ion density and drift velocity data obtained on Atmospheric Explorer C (reproduction granted by American Geophysical Union).

should be judged with respect to a trend line in the data. Plasma depletions evident in the bottom curve are associated with large deviations of the ion velocity data from the trend line. McClure et al. showed that these perturbations are usually upward and westward in the plasma frame. This is the direction predicted by Woodman based on plume structures which usually slant to the right in the Jicamarca plot format. Since the radar looked only vertically in the early measurements of Woodman and La Hoz, the plume-like feature must be due either to a wake of irregularities behind the upward moving feature or a slanted structure which itself is greatly extended in altitude. This point is discussed further below.

A series of numerical simulations of ESF have been conducted by the NRL group in the past several years. The first published by Scannapieco and Ossakow (1976) started with a finite amplitude ($\delta n/n = 4\%$) perturbation with a 3km east-west wavelength. The analysis was two dimensional and so tacitly assumed no E region conductivity to short out perturbation electric fields. The electric field and neutral wind were set equal to zero so that only the gravitational effects were studied. The region of plasma depletion slowly moved upward, increased in intensity and reached a height above the F peak. Later simulations, one of which is reproduced in Figure 9, which used more

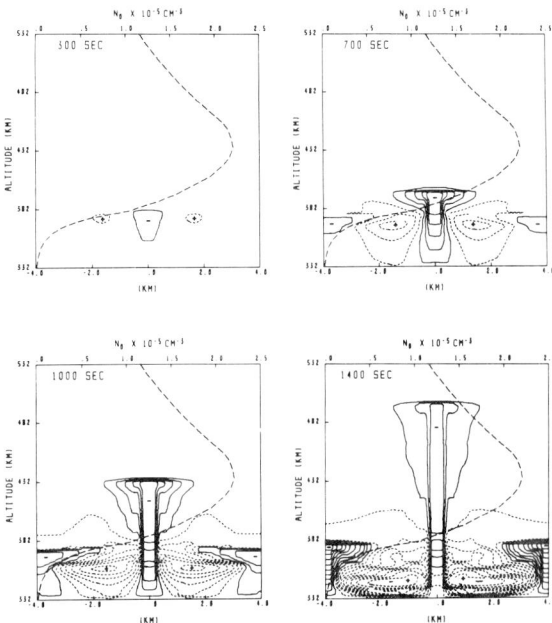

Figure 9. Computer simulation of the development of a finite amplitude density depletion subject to the forces at the F region equator (reproduction granted by American Geophysical Union).

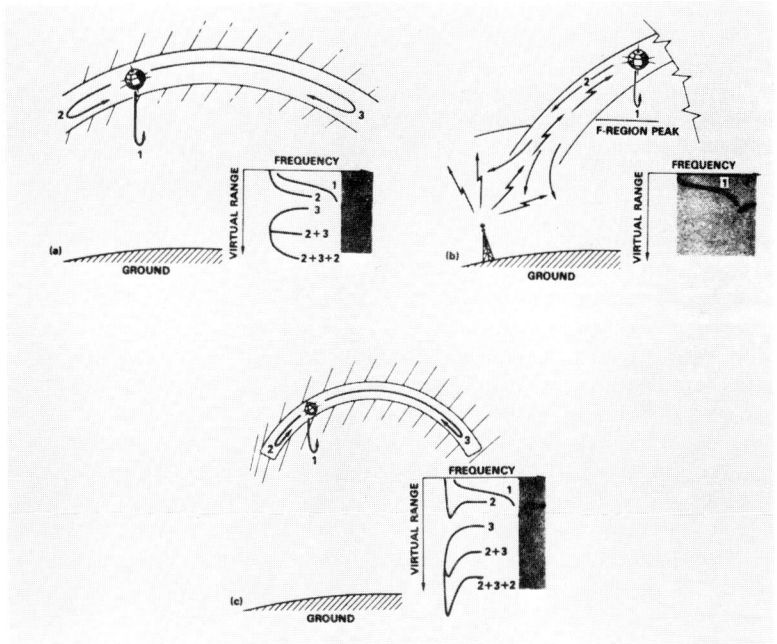

Figure 10. Schematic diagram of HF propagation paths for various density depletion geometries (reproduction granted by American Geophysical Union).

realistic conditions of bottomside scale length and a higher F peak altitude, showed similar developments but much larger perturbation electric fields (Ossakow et al., 1978). Upward velocities of 150m/s were found, in good agreement with the experimental results (Kelley et al., 1976; McClure et al., 1977). The review by Ossakow in this volume discusses these results and other theoretical studies in detail.

Some of the simulations show the bubbles pinching off at the bottom while others indicate wedge shapes. In situ and radar data are ambiguous on this point since with one vertical or horizontal cut it is impossible to distinguish other dimensional characteristics. Dyson and Benson (1978) have shown that the two-dimensionality assumed by theorists is valid using topside ionosonde data from ISIS I. They summarize their results in Figure 10 which in the center portion shows how a field aligned depleted structure extending to the base of the ionosphere guide the ISIS HF waves to a ground site, as observed. McClure et al. (1978) presented total electron content (TEC) data which suggests a wedge shape depletion (also presumably field aligned, of course) throughout a 100km horizontal scale and 600km vertical scale in which the electron content was less than 10% of background.

There may very well be a class of bubbles which pinch off and others which do not. One feature which may complicate the interpretation of the plumes is that the rising bubbles may inject velocity (electrostatic) turbulence into the background plasma which mixes the vertical density gradient and creates small scale structure (Kelley and Ott, 1978). This may create a trailing wake but also make the upwelling region appear larger due to the tendency for two-dimensional turbulence to expand in real space as well as velocity space (dual-cascade).

It is important to note that for finite amplitude depleted regions the eastward gravitationally driven electric current will continue to deposit positive charges on the west side of bubbles (or wedges) and negative charge on the east side even on the topside. Thus even though the linear Rayleigh-Taylor instability is damped above the F peak, the nonlinear process can continue. Furthermore, even though we have shown that the bottomside process seems to be due to a linear $\bar{E}x\bar{B}$ or Rayleigh-Taylor instability, the topside structure is not necessarily entirely due to this initiation process. In fact based on the correlation between ESF and TID's discussed in detail by Booker (1978), and the recent work by Rottger (1973a,b, 1976), Beer (1974), and Klostermeyer (1978), it seems likely that the largest scale topside depletions (100km scale size) are seeded by the spatial resonance mechanism for TID's, (Whitehead, 1971). An example of the Rottger's data on the transequatorial propagation delays in HF radio wave propagation between Lindau and Tsumba are shown in Figure 11, relative to a great circle path. Quite regular patterns are indicated which are interpreted in terms of organization of plasma by a gravity wave. TEC measurements from the wideband satellite also show regular modulations with horizontal scale comparable to that of TID's (Livingston and Baron, private communication, 1978).

The plasma must be moving downward to resonate with gravity waves which have upward group velocities. Thus when the post sunset drift reversal occurs, this density layering should occur. Calculations by Klostermeyer show this effect. The separation distance of large plumes is also similar to the horizontal wavelength of the TID's which argues for their importance. On the topside however, the plumes usually tilt in the opposite direction from that expected from eastward and downward tilted gravity waves phase fronts (Woodman, private communication, 1978). Thus once the depleted region is initiated by a gravity wave, it seems that the bouyancy effect takes over to launch the depleted region upward. Note that the westward tilt is explicible via the gravitational mechanism, at least for bubbles of finite vertical extent, when neutral winds are taken into account (Woodman and La Hoz, 1976).

Simultaneous scintillation and Jicamarca measurements shown in Figure 12 relate km scale structure responsible for VHF scintillations with the 3m radar maps for a common interaction volume. Early in the evening the two regions are equivalent but later intense scintillation is observed with no corresponding 3m backscatter (Basu et al., 1978a).

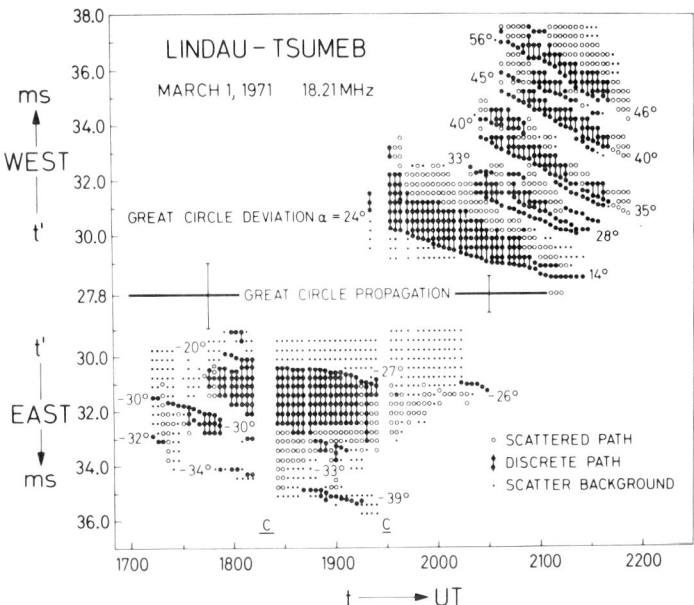

Figure 11. Delay of transequatorial HF transmissions relative to a great circle path (reproduction granted by J. Atmos. Terr. Phys.).

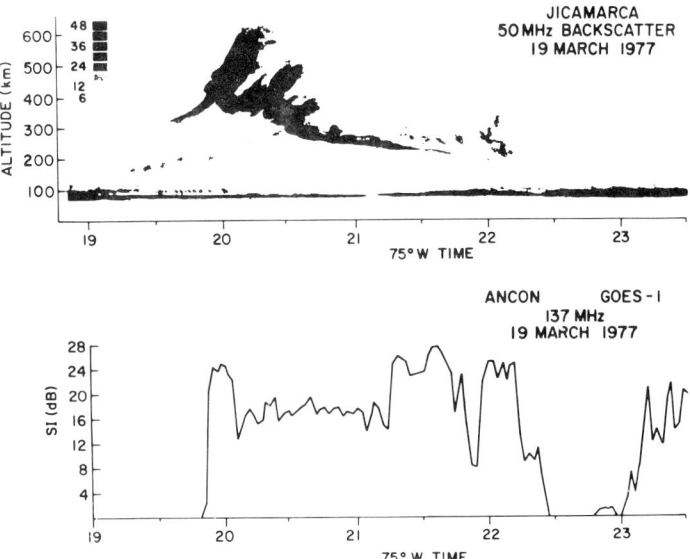

Figure 12. Simultaneous observations of 3m backscatter and VHF scintillation (reproduction granted by American Geophysical Union).

Also, the power spectrum of 3m amplitude fluctuations shows a cutoff at later local times not present in the early evening (Basu et al., 1978b). This shows that in "mature" structures km scales are still quite intense even though short wavelength waves are not. Sharp density gradients result in drift and other short wavelength waves (Costa and Kelley, 1978b; Huba et al., 1978) which tend to destroy the gradient responsible for their growth (Kadomstev, 1965, Burke et al., 1978). Diffusion will then damp the existing short waves and smoother density gradients result albeit still irregular on the scale of kilometers. Decaying turbulence in the background will also result in a relative decrease in the intensity ratio of short to long wavelength waves (Kelley and Ott, 1978).

IV. CONCLUSIONS

Considerable progress has occurred in the understanding of equatorial spread F through application of almost every technique available to study plasmas in the near space region of the Earth. Free energy stored in the equatorial plasma, which is supported against gravity by the Earth's magnetic field, is released on a time scale of an hour and results in large amplitude depletions in plasma density at altitudes up to several thousand km. Turbulent cascade of energy to larger and smaller wavelengths as well as secondary plasma instabilities operating on the long wavelength waves act to spread the irregular structure over length scales from 100km to fractions of a meter.

Although plasma processes are very important, there is persuasive evidence that internal gravity waves in the neutral atmosphere act to seed the largest depletions and hence may control some of the curious seasonal and longitude dependences observed.

Outstanding questions involve the spectra and source of short wavelength waves and the actual two-dimensional structure of upwelling bubbles. These two questions are coupled since a "pinched-off" bubble must leave behind a wake of irregularities if the radar plumes are to be explained. A wedge shaped structure may continuously generate short wavelength waves until the gradient scale lengths on the edges become to large. Simultaneous radar and in situ measurements are necessary to sort out these questions adequately.

REFERENCES

Aarons, J., 1977, IEEE Trans. Ant. and Propagat., AP25, p.729...
Balsley, B.B., G. Haerendel, and R.A. Greenwald, 1972, J. Geophys. Res., 77, p.5625.
Basu, S., S. Basu, J. Aarons, J.P. McClure, and M.D. Cousins, 1978, J. Geophys. Res., p.83.

Basu, S., S. Basu, J. Aarons, J. Brickman, J.P. McClure, and
M.D. Cousins, 1978b, URSI Meeting, Helsinki.
Beer, T., 1974, Aust. J. Phys., 27, p.391.
Booker, H.G., and H.W. Wells, 1938, Terrest. Mag. and Atmos. Elect.
(now J. Geophys. Res.) 43, p. 249.
Booker, H.G., URSI, Helsinki, August 1978.
Burke, W.J., D.E. Donatelli, R.C. Sagalyn, and M.C. Kelley, 1978,
Planet Sp. Sci., in press.
Chaturvedi, P.K., and S.L. Ossakow, 1977, Geophys. Res., L., 4,
p. 558.
Costa, E., and M.C. Kelley, 1976, Geophys. Res., L., 3, p. 677.
Costa, E., and M.C. Kelley, 1978a, J. Geophys. Res., 83, p. 4359.
Costa, E., and M.C. Kelley, 1978b, J. Geophys. Res., 83, p. 4365.
Dungey, J.W., 1956, J. Atmos. Terr. Phys., 9, p. 304.
Dyson, P.L., J.P. McClure, and W.B. Hanson, 1974, J. Geophys. Res., 79,
p. 1497.
Dyson, P.L., and R.F. Benson, 1978, Geophsy. Res. L., 5, p. 795.
Farley, D.T., B.B. Balsley, R.F. Woodman, and J.P. McClure, 1970,
J. Geophys. Res., p. 7199.
Fejer, B.G., D.T. Farley, B.B. Balsley, and R.F. Woodman, 1976,
J. Geophys. Res., 81 p. 4621.
Haerendel, G., 1974, rpt. Max-Planck-Institute fur Phys. und Astrophys.,
Garching, Federal Republic of Germany.
Hudson, M.K., and C.F. Kennel, 1975a, J. Geophys. Res., 80, p. 4581.
Huba, J.D., P.K. Chaturvedi, and S.L. Ossakow, 1978, Geophys. Res. L.,
5, p. 695.
Kadomtsev, B.B., 1965, *Plasma Turbulence*, Academic Press, NY.
Kelley, M.C., G. Haerendel, H. Kappler, A. Valenzuela, B.B. Balsley,
C.A. Carter, W.L. Ecklund, C.W. Carlson, B. Hausler, and R. Torbert,
1976, Geophys. Res. L., 3, p. 448.
Kelley, M.C., and E. Ott, 1978, J. Geophys. Res., 83, p. 4369.
Kelley, M.C., K.D. Baker, and J.C. Ulwick, 1979, J. Geophys. Res.,
in press.
Klostermeyer, J., 1978, J. Geophys. Res., 83, p. 3753.
Linson, L.M., and J.B. Workman, 1970, J. Geophsy. Res., 75, p. 3211.
Martyn, D.F., 1959, J. Geophys. Res., 64, p. 2178.
McClure, J.P., W.G. Hanson, and J.H. Hoffman, 1977, J. Geophys. Res.,
83 p. 2650.
McClure, J.P., C.A. Valladares, and W.B. Hanson, 1978, URSI meeting,
Helsinki.
Morse, F.A., B.C. Edgar, H.C. Koons, C.J. Rice, W.J. Heikkila,
J.H. Hoffman, B.A. Tinsely, J.D. Winningham, A.B. Christensen,
R.F. Woodman, J. Pomalaza, and N.R. Teixeira, 1977, J. Geophys. Res.,
p. 578.
Ossakow, S.L., S.T. Zalesak, B.E. McDonald, and P.K. Chaturvedi,
1978, proceedings of 1978 Symp on the effect of the ionosphere on
space and terrestrial systems, paper 2-7, NRL.
Rottger, J., 1973a, Geophys., 39, p. 799.
Rottger, J., 1973b, J. Atmos. Terr. Phys., 35, p. 1195.
Rottger, J., 1976, J. Atmos. Terr. Phys., 38, p. 97.

Scannapieco, A.J. and S.L. Ossakow, 1976, Geophys. Res. Lett., 3, p. 451.
Scannapieco, A.J., S.L. Ossakow, S.R. Goldman, and J.M. Pierre, 1976, J. Geophys. Res., 81, p. 6037.
Whitehead, J.D., 1971, J. Geophys. Res., 76, p. 238.
Woodman, R.F., and C. La Hoz, 1976, J. Geophys. Res., 81, p. 5447.

INDEX OF SUBJECTS

Anomalous electron transport 143
Anomalous resistivity 144,245,252
Artificially stimulated emissions (ASE) 163-190, 191-203,205-216
Auroral hiss 4
Auroral kilometric radiation 4

Bernstein-Green Kruskal waves 136
Buneman instability 135
Brehmstrahlung radiation 48
Broadband electrostatic noise 4

Chorus 4,28,51,55
  equatorial chorus 56
  high latitude chorus 59
Collisionless drift wave 283
Computer simulations
  of Spread F 271-277,300-301
  of instabilities in a current carrying plasma 135-147
  of whistler interactions 185-186
Controlled VLF experiments 163-203
Current driven instability 109, 145

Dash-dot anomaly 165
Decay instability 144,250,251
Detrapping 217
Diffusion
  nonresonant 139
  resonant 145
Double layers 83-134
  arc starvation 130
  and beam breakup 133
  Bohm condition 96
  cathode double layer 129
  constriction double layer 130
  and current driven instabilities 103
  in double plasma devices 110
  formation and maintenance 101,119
  general properties 84,124
  in laboratory plasmas 109-128
  Langmuir condition 94
  in low pressure discharges 101,112-119,129
  in the magnetosphere 103
  and pondermotive force 105, 129
  potential drop 90-94
  rectification mechanism 132
  relativistic double layers 93, 95
  in space plasmas 83-84
  Space Shuttle experiments 134
  stability 98,122
  striations 130
  strong double layers 89,90,96
  time dependent double layers 100
  time independent double layers 89
  and the two stream instability 104
Drift cyclotron instability 284

Electron cyclotron model 218
Electron slot 37
Electrostatic cyclotron harmonic wave instabilities 65-82
  convective vs non convective behavior 68-78
  diffusion and heating 80
  linear growth rates 68-78
  nonlinear saturation 79
  sources of free energy 78
Electrostatic ion cyclotron instability 145
ELF noise
  associated with high latitude trough 21

intensity 3,11
ionospheric hiss 5
sferics 47
source regions 5,38
spectra 7
Embryo emission (EE) 170
Equatorial electrojet instabilities 149-159
Equatorial Spread F - see Spread F
EQION rocket 293

Field aligned current 143
Flux-tube interchange 44,48

Inhomogeneous medium, effects on wave-particle interaction 163,172-182,206-215, 217-222
Instabilities in the equatorial electrojet 149-159
Ion sound instability 135,143

Langmuir soliton collapse 233
Langmuir waves 247,253,254
Lions roar 4
Lower hybrid drift instability 284

Micropulsations 4
Modulational instabilities 233,249,252
  direct collapse 252
  dispersion relation 249-250

Nonlinear Schrodinger equation 231
Nonlinear whistler mode interaction 163-190,205-216,218
Nonthermal continuum radiation 4, 66
Nonlinear transition of wave spectrum 144

Odd half harmonic emissions 65-82
Oscillating two stream instability 233,245,249-260
  daughter waves 250,251
  dispersion relation 249-250

growth rates 251,257
pump waves 249-252
Oscillation pileup 247

Parametric decay instability 144,250,251
Phase bunching and phase trapping 174, 206
Pitch angle scattering 143
Plasmaspheric hiss 4, 41
Plumes (Spread F) 293
Pondermotive force 104,129, 231
Power line radiation 27, 37
  association with thunderstorm activity 37
  electron pitch angle diffusion by 46
  and inner radiation belt 46
  observed in Antarctica 41
  and triggered emissions 168
Pulsation phenomena 166

Quenching of VLF hiss by monochromatic waves 217
Quiet band 217

Radiation processes 225,237-240
Rayleigh-Taylor instability 265-266,268-269,296

Satellites
  Ariel 3: 13,31,37
  Ariel 4: 13,37
  Explorer 45: 10
  Geos 1: 6,10,16,66
  Imp 6: 66
  Injun 3: 10
  Injun 5: 12
  Intercosmos 3: 13,15
  Intercosmos 4: 8
  Intercosmos 5: 7,8,11
  Intercosmos 14: 9
  ISEE 66,79
  OGO 3: 28,51
  OGO 4: 31
  OGO 5: 51,65,79
  OGO 6: 21

Side band line 217

Siple station experiments 167,191-203,205-216
Solar type III radio bursts 225-261
  distribution function in solar streams 253-255
  frequency drift rates and beam velocity 260
  quasilinear effects 228-230
  radiation processes 237-240
  strong turbulence effects 230-237
Solitons 233,256,259
South Atlantic Anomaly 49
Spread F 265-306
  and barium cloud power spectra compared 295-297
  coherent mode coupling 277
  collisionless drift wave 283
  computer simulations 271-277
  drift cyclotron instability 284
  equatorial spread F bubbles 265,268-269, 272-283, 298-302
  EQION rocket 293
  and gravity waves 298,302, 304
  ionosonde studies 293
  Jicamarca radar observations 291,292,303
  linear plasma instabilities 265-268,297-298
  lower hybrid drift instability 284
  plasma fluid equations 270
  power spectrum 268-269,295-296
  Rayleigh-Taylor instability 265-266,268-269,296
  rocket measurements 292-294
  scintillation of radio signals 298,302-303
  small scale irregularities 283-287
  two-dimensional fluids 280
  velocity turbulence 302
Strong turbulence effects 230-237, 247-260

Threshold sensitivities 79
Two stream instability
  and double layer formation 104
  in the equatorial electrojet 149, 153
Trapped particles 174,207,219
Trapping force 218
Trapping length 221

Upstream waves 4

VLF noise
  associated with high latitude trough 21
  intensity 3,11
  quenching by monochromatic waves 217
  source regions 5,38
  spectra 7
  triggered emissions 163-190, 191-203,205-216

Wave induced precipitation 27,202
Wave-particle interaction 143,172,191,205
Whistler model 27,163-190,191-203, 205-216

ASTROPHYSICS AND SPACE SCIENCE LIBRARY

Edited by

J. E. Blamont, R. L. F. Boyd, L. Goldberg, C. de Jager, Z. Kopal, G. H. Ludwig, R. Lüst,
B. M. McCormac, H. E. Newell, L. I. Sedov, Z. Švestka, and W. de Graaff

1. C. de Jager (ed.), *The Solar Spectrum, Proceedings of the Symposium held at the University of Utrecht, 26–31 August, 1963.* 1965, XIV + 417 pp.
2. J. Ortner and H. Maseland (eds.), *Introduction to Solar Terrestrial Relations, Proceedings of the Summer School in Space Physics held in Alpbach, Austria, July 15–August 10, 1963 and Organized by the European Preparatory Commission for Space Research.* 1965, IX + 506 pp.
3. C. C. Chang and S. S. Huang (eds.), *Proceedings of the Plasma Space Science Symposium, held at the Catholic University of America, Washington, D.C., June 11–14, 1963.* 1965, IX + 377 pp.
4. Zdeněk Kopal, *An Introduction to the Study of the Moon.* 1966, XII + 464 pp.
5. B. M. McCormac (ed.), *Radiation Trapped in the Earth's Magnetic Field. Proceedings of the Advanced Study Institute, held at the Chr. Michelsen Institute, Bergen, Norway, August 16–September 3, 1965.* 1966, XII + 901 pp.
6. A. B. Underhill, *The Early Type Stars.* 1966, XII + 282 pp.
7. Jean Kovalevsky, *Introduction to Celestial Mechanics.* 1967, VIII + 427 pp.
8. Zdeněk Kopal and Constantine L. Goudas (eds.), *Measure of the Moon. Proceedings of the 2nd International Conference on Selenodesy and Lunar Topography, held in the University of Manchester, England, May 30–June 4, 1966.* 1967, XVIII + 479 pp.
9. J. G. Emming (ed.), *Electromagnetic Radiation in Space. Proceedings of the 3rd ESRO Summer School in Space Physics, held in Alpbach, Austria, from 19 July to 13 August, 1965.* 1968, VIII + 307 pp.
10. R. L. Carovillano, John F. McClay, and Henry R. Radoski (eds.), *Physics of the Magnetosphere, Based upon the Proceedings of the Conference held at Boston College, June 19–28, 1967.* 1968, X + 686 pp.
11. Syun-Ichi Akasofu, *Polar and Magnetospheric Substorms.* 1968, XVIII + 280 pp.
12. Peter M. Millman (ed.), *Meteorite Research. Proceedings of a Symposium on Meteorite Research, held in Vienna, Austria, 7–13 August, 1968.* 1969, XV + 941 pp.
13. Margherita Hack (ed.), *Mass Loss from Stars. Proceedings of the 2nd Trieste Colloquium on Astrophysics, 12–17 September, 1968.* 1969, XII + 345 pp.
14. N. D'Angelo (ed.), *Low-Frequency Waves and Irregularities in the Ionosphere. Proceedings of the 2nd ESRIN-ESLAB Symposium, held in Frascati, Italy, 23–27 September, 1968.* 1969, VII + 218 pp.
15. G. A. Partel (ed.), *Space Engineering. Proceedings of the 2nd International Conference on Space Engineering, held at the Fondazione Giorgio Cini, Isola di San Giorgio, Venice, Italy, May 7–10, 1969.* 1970, XI + 728 pp.
16. S. Fred Singer (ed.), *Manned Laboratories in Space. Second International Orbital Laboratory Symposium.* 1969, XIII + 133 pp.
17. B. M. McCormac (ed.), *Particles and Fields in the Magnetosphere. Symposium Organized by the Summer Advanced Study Institute, held at the University of California, Santa Barbara, Calif., August 4–15, 1969.* 1970, XI + 450 pp.
18. Jean-Claude Pecker, *Experimental Astronomy.* 1970, X + 105 pp.
19. V. Manno and D. E. Page (eds.), *Intercorrelated Satellite Observations related to Solar Events. Proceedings of the 3rd ESLAB/ESRIN Symposium held in Noordwijk, The Netherlands, September 16–19, 1969.* 1970, XVI + 627 pp.
20. L. Mansinha, D. E. Smylie, and A. E. Beck, *Earthquake Displacement Fields and the Rotation of the Earth, A NATO Advanced Study Institute Conference Organized by the Department of Geophysics, University of Western Ontario, London, Canada, June 22–28, 1969.* 1970, XI + 308 pp.
21. Jean-Claude Pecker, *Space Observatories.* 1970, XI + 120 pp.
22. L. N. Mavridis (ed.), *Structure and Evolution of the Galaxy. Proceedings of the NATO Advanced Study Institute, held in Athens, September 8–19, 1969.* 1971, VII + 312 pp.
23. A. Muller (ed.), *The Magellanic Clouds. A European Southern Observatory Presentati,n: Principal Prospects, Current Observational and Theoretical Approaches, and Prospects for Future Research, Based on the Symposium on the Magellanic Clouds, held in Santiago de Chile, March 1969, on the Occasion of the Dedication of the European Southern Observatory.* 1971, XII + 189 pp.

24. B. M. McCormac (ed.), *The Radiating Atmosphere*. Proceedings of a Symposium Organized by the Summer Advanced Study Institute, held at Queen's University, Kingston, Ontario, August 3–14, 1970. 1971, XI + 455 pp.
25. G. Fiocco (ed.), *Mesospheric Models and Related Experiments*. Proceedings of the 4th ESRIN-ESLAB Symposium, held at Frascati, Italy, July 6–10, 1970. 1971, VIII + 298 pp.
26. I. Atanasijević, *Selected Exercises in Galactic Astronomy*. 1971, XII + 144 pp.
27. C. J. Macris (ed.), *Physics of the Solar Corona*. Proceedings of the NATO Advanced Study Institute on Physics of the Solar Corona, held at Cavouri-Vouliagmeni, Athens, Greece, 6–17 September 1970. 1971, XII + 345 pp.
28. F. Delobeau, *The Environment of the Earth*. 1971, IX + 113 pp.
29. E. R. Dyer (general ed.), *Solar-Terrestrial Physics/1970*. Proceedings of the International Symposium on Solar-Terrestrial Physics, held in Leningrad, U.S.S.R., 12–19 May 1970. 1972, VIII + 938 pp.
30. V. Manno and J. Ring (eds.), *Infrared Detection Techniques for Space Research*. Proceedings of the 5th ESLAB-ESRIN Symposium, held in Noordwijk, The Netherlands, June 8–11, 1971. 1972, XII + 344 pp.
31. M. Lecar (ed.), *Gravitational N-Body Problem*. Proceedings of IAU Colloquium No. 10, held in Cambridge, England, August 12–15, 1970. 1972, XI + 441 pp.
32. B. M. McCormac (ed.), *Earth's Magnetospheric Processes*. Proceedings of a Symposium Organized by the Summer Advanced Study Institute and Ninth ESRO Summer School, held in Cortina, Italy, August 30–September 10, 1971. 1972, VIII + 417 pp.
33. Antonin Rükl, *Maps of Lunar Hemispheres*. 1972, V + 24 pp.
34. V. Kourganoff, *Introduction to the Physics of Stellar Interiors*. 1973, XI + 115 pp.
35. B. M. McCormac (ed.), *Physics and Chemistry of Upper Atmospheres*. Proceedings of a Symposium Organized by the Summer Advanced Study Institute, held at the University of Orléans, France, July 31–August 11, 1972. 1973, VIII + 389 pp.
36. J. D. Fernie (ed.), *Variable Stars in Globular Clusters and in Related Systems*. Proceedings of the IAU Colloquium No. 21, held at the University of Toronto, Toronto, Canada, August 29–31, 1972. 1973, IX + 234 pp.
37. R. J. L. Grard (ed.), *Photon and Particle Interaction with Surfaces in Space*. Proceedings of the 6th ESLAB Symposium, held at Noordwijk, The Netherlands, 26–29 September, 1972. 1973, XV + 577 pp.
38. Werner Israel (ed.), *Relativity, Astrophysics and Cosmology*. Proceedings of the Summer School, held 14–26 August, 1972, at the BANFF Centre, BANFF, Alberta, Canada. 1973, IX + 323 pp.
39. B. D. Tapley and V. Szebehely (eds.), *Recent Advances in Dynamical Astronomy*. Proceedings of the NATO Advanced Study Institute in Dynamical Astronomy, held in Cortina d'Ampezzo, Italy, August 9–12, 1972. 1973, XIII + 468 pp.
40. A. G. W. Cameron (ed.), *Cosmochemistry*. Proceedings of the Symposium on Cosmochemistry, held at the Smithsonian Astrophysical Observatory, Cambridge, Mass., August 14–16, 1972. 1973, X + 173 pp.
41. M. Golay, *Introduction to Astronomical Photometry*. 1974, IX + 364 pp.
42. D. E. Page (ed.), *Correlated Interplanetary and Magnetospheric Observations*. Proceedings of the 7th ESLAB Symposium, held at Saulgau, W. Germany, 22–25 May, 1973. 1974, XIV + 662 pp.
43. Riccardo Giacconi and Herbert Gursky (eds.), *X-Ray Astronomy*. 1974, X + 450 pp.
44. B. M. McCormac (ed.), *Magnetospheric Physics*. Proceedings of the Advanced Summer Institute, held in Sheffield, U.K., August 1973. 1974, VII + 399 pp.
45. C. B. Cosmovici (ed.), *Supernovae and Supernova Remnants*. Proceedings of the International Conference on Supernovae, held in Lecce, Italy, May 7–11, 1973. 1974, XVII + 387 pp.
46. A. P. Mitra, *Ionospheric Effects of Solar Flares*. 1974, XI + 294 pp.
47. S.-I. Akasofu, *Physics of Magnetospheric Substorms*. 1977, XVIII + 599 pp.
48. H. Gursky and R. Ruffini (eds.), *Neutron Stars, Black Holes and Binary X-Ray Sources*. 1975, XII + 441 pp.
49. Z. Švestka and P. Simon (eds.), *Catalog of Solar Particle Events 1955–1969. Prepared under the Auspices of Working Group 2 of the Inter-Union Commission on Solar-Terrestrial Physics*. 1975, IX + 428 pp.
50. Zdeněk Kopal and Robert W. Carder, *Mapping of the Moon*. 1974, VIII + 237 pp.
51. B. M. McCormac (ed.), *Atmospheres of Earth and the Planets*. Proceedings of the Summer Advanced Study Institute, held at the University of Liège, Belgium, July 29–August 8, 1974. 1975, VII + 454 pp.
52. V. Formısano (ed.), *The Magnetospheres of the Earth and Jupiter*. Proceedings of the Neil Brice Memorial Symposium, held in Frascati, May 28–June 1, 1974. 1975, XI + 485 pp.

53. R. Grant Athay, *The Solar Chromosphere and Corona: Quiet Sun.* 1976, XI + 504 pp.
54. C. de Jager and H. Nieuwenhuijzen (eds.), *Image Processing Techniques in Astronomy.* Proceedings of a Conference, held in Utrecht on March 25–27, 1975, XI + 418 pp.
55. N. C. Wickramasinghe and D. J. Morgan (eds.), *Solid State Astrophysics.* Proceedings of a Symposium, held at the University College, Cardiff, Wales, 9–12 July 1974. 1976, XII + 314 pp.
56. John Meaburn, *Detection and Spectrometry of Faint Light.* 1976, IX + 270 pp.
57. K. Knott and B. Battrick (eds.), *The Scientific Satellite Programme during the International Magnetospheric Study.* Proceedings of the 10th ESLAB Symposium, held at Vienna, Austria, 10–13 June 1975. 1976, XV + 464 pp.
58. B. M. McCormac (ed.), *Magnetospheric Particles and Fields.* Proceedings of the Summer Advanced Study School, held in Graz, Austria, August 4–15, 1975. 1976, VII + 331 pp.
59. B. S. P. Shen and M. Merker (eds.), *Spallation Nuclear Reactions and Their Applications.* 1976, VIII + 235 pp.
60. Walter S. Fitch (ed.), *Multiple Periodic Variable Stars.* Proceedings of the International Astronomical Union Colloquium No. 29, Held at Budapest, Hungary, 1–5 September 1975. 1976, XIV + 348 pp.
61. J. J. Burger, A. Pedersen, and B. Battrick (eds.), *Atmospheric Physics from Spacelab.* Proceedings of the 11th ESLAB Symposium, Organized by the Space Science Department of the European Space Agency, held at Frascati, Italy, 11–14 May 1976. 1976, XX + 409 pp.
62. J. Derral Mulholland (ed.), *Scientific Applications of Lunar Laser Ranging.* Proceedings of a Symposium held in Austin, Tex., U.S.A., 8–10 June, 1976. 1977, XVII + 302 pp.
63. Giovanni G. Fazio (ed.), *Infrared and Submillimeter Astronomy.* Proceedings of a Symposium held in Philadelphia, Penn., U.S.A., 8–10 June, 1976. 1977, X + 226 pp.
64. C. Jaschek and G. A. Wilkins (eds.), *Compilation, Critical Evaluation and Distribution of Stellar Data.* Proceedings of the International Astronomical Union Colloquium No. 35, held at Strasbourg, France, 19-21 August, 1976. 1977, XIV + 316 pp.
65. M. Friedjung (ed.), *Novae and Related Stars.* Proceedings of an International Conference held by the Institut d'Astrophysique, Paris, France, 7-9 September, 1976. 1977, XIV + 228 pp.
66. David N. Schramm (ed.), *Supernovae.* Proceedings of a Special IAU Session on Supernovae held in Grenoble, France, 1 September, 1976. 1977, X + 192 pp.
67. Jean Audouze (ed.), *CNO Isotopes in Astrophysics.* Proceedings of a Special IAU Session held in Grenoble, France, 30 August, 1976. 1977, XIII + 195 pp.
68. Z. Kopal, *Dynamics of Close Binary Systems,* forthcoming.
69. A. Bruzek and C. J. Durrant (eds.), *Illustrated Glossary for Solar and Solar-Terrestrial Physics.* 1977, approx. 216 pp.
70. H. van Woerden (ed.), *Topics in Interstellar Matter.* 1977, VIII + 295 pp.
71. M. A. Shea, D. F. Smart, and T. S. Wu (eds.), *Study of Travelling Interplanetary Phenomena.* 1977, XII + 439 pp.
72. V. Szebehely (ed.), *Dynamics of Planets and Satellites and Theories of Their Motion.* Proceedings of IAU Colloquium No. 41, held in Cambridge, England, 17-19 August 1976. 1978, xii + 375 pp.
73. James R. Wertz (ed.), *Spacecraft Attitude Determination and Control.* 1978, xvi + 858 pp.